Information Science and Statistics

Series Editors:
M. Jordan
J. Kleinberg
B. Schölkopf

Information Science and Statistics

Akaike and Kitagawa: The Practice of Time Series Analysis.

Bishop: Pattern Recognition and Machine Learning.

Cowell, Dawid, Lauritzen, and Spiegelhalter: Probabilistic Networks and Expert Systems.

Doucet, de Freitas, and Gordon: Sequential Monte Carlo Methods in Practice.

Fine: Feedforward Neural Network Methodology.

Hawkins and Olwell: Cumulative Sum Charts and Charting for Quality Improvement.

Jensen and Nielsen: Bayesian Networks and Decision Graphs, Second Edition.

Lee and Verleysen: Nonlinear Dimensionality Reduction.

Marchette: Computer Intrusion Detection and Network Monitoring: A Statistical Viewpoint.

Rissanen: Information and Complexity in Statistical Modeling.

Rubinstein and Kroese: The Cross-Entropy Method: A Unified Approach to Combinatorial Optimization, Monte Carlo Simulation, and Machine Learning.

Studený: Probabilistic Conditional Independence Structures.

Vapnik: The Nature of Statistical Learning Theory, Second Edition.

Wallace: Statistical and Inductive Inference by Minimum Massage Length.

Finn V. Jensen and Thomas D. Nielsen

Bayesian Networks and Decision Graphs

February 8, 2007

Springer

Berlin Heidelberg NewYork
HongKong London
Milan Paris Tokyo

Finn V. Jensen
Department of Computer Science
Aalborg University
Fredrik Bajers Vej 7, building E
DK-9220 Aalborg EAST,
Denmark
fvj@cs.aau.dk

Thomas D. Nielsen
Department of Computer Science
Aalborg University
Fredrik Bajers Vej 7, building E
DK-9220 Aalborg EAST,
Denmark
tdn@cs.aau.dk

Series Editors:

Michael Jordan
Division of Computer
  Science and Department
  of Statistics
University of California,
  Berkeley
Berkeley, CA 94720
USA

Jon Kleinberg
Department of Computer
  Science
Cornell University
Ithaca, NY 14853
USA

Bernhard Schölkopf
Max Planck Institute for
  Biological Cybernetics
Spemannstrasse 38
72076 Tübingen
Germany

ISBN-13: 978-1-4419-2394-3   eISBN-13: 978-0-387-68282-2

Printed on acid-free paper.

9 8 7 6 5 4 3 2 1

springer.com

Preface

Ever since the first machines were constructed, artists and scientists have shared a vision of a human-like machine: an autonomous self-moving machine that acts and reasons like a human being. Much effort has been put into this dream, but we are still very far from having androids with even the tiniest similarity to humans.

This does not mean that all of these efforts have been wasted. As a spin-off, we have seen a long series of inventions that can take over very specialized sections of human work. These inventions fall into two categories: machines that can make physical changes in the world and thereby substitute human labor, and machines that can perform activities usually thought of as requiring intellectual skills.

In contemporary science and engineering, we still have this split into two categories. The activity of the first category is mainly concentrated on the construction of robots. The aim is to construct autonomous machines performing "sophisticated" actions such as searching for a cup, finding a way from the office to a lavatory, driving a vehicle in a deserted landscape, or walking on two legs. Construction of such robots requires computers to perform certain kinds of artificial intelligence. Basically, it is the kind of intelligence that humans share with most mammals. It involves skills such as visual recognition of items, sound recognition, learning to abstract crucial items from a scene, or control of balance and position in 3-D space. Although they are very challenging research tasks, and they certainly require enormous computing power and very sophisticated algorithms, you would not say that these skills are intellectual, and the basis for the activity is the physical appearance of a device that moves. To put it another way: the success criterion is how the algorithms work when controlling a physical machine in real time.

The activity in the second category is basically concerned with reasoning and human activities that we presumably do not share with other animals. The activity is separated from matter. When performed, no changes in the physical world need to take place. The first real success was the automated calculator: a machine that can perform very large and complicated arithmetic calculations. Automated calculation skill is nowadays hardly considered artificial intelligence, and we are now acquainted with computers performing tasks that decades ago were considered highly intellectual (e.g. taking derivatives of functions or performing reduction of mathematical expressions). When an activity has been so well understood that it can be formalized, it will soon be performed by computers, and gradually we acknowledge that this activity does not really require intelligence.

A branch of research in the second category has to do with reasoning. The first successes were in logical reasoning. Propositional logic is fully formalized, and although some tasks are NP-complete and therefore in some situations intractable for a computer, we have for propositional logic completed the transition from "intellectual task" to "we have computers to do this for us."

Unfortunately, logical reasoning is very limited in scope. It deals with how to infer from propositions that you know are true. Very often you do not know the truth of a proposition for certain, but you still need to perform inference from your incomplete and uncertain knowledge. Actually, this is the most common situation for human reasoning. Reasoning under uncertainty is not yet so well understood that it can be formalized entirely for computers. There are several approaches to reasoning under uncertainty. The approach taken in this book is (subjective) probability theory. When the reasoning ends up in a conclusion on a decision, we use *utilities*, and we assume that the decision taken is the one that maximizes the expected utility. In other words, the approach prescribes a certain behavior. We may not always expect this behavior from human beings, and therefore the approach is also termed *normative*. There are alternative approaches to reasoning under uncertainty. Most prominent is *possibility theory*, which in certain contexts is called *fuzzy logic*. The interested reader may consult the wide literature on these approaches.

The aim of normative systems can in short be termed human *wisdom*: to take decisions on the basis of accumulated and processed experience. The tasks are of the following types:

- using observations to interpret a situation;
- focusing a search for more information;
- choosing an appropriate intervening action;
- adapting to changing environments;
- learning from experience.

A damping factor for properly exploiting the advances in artificial intelligence has for a long time been the lack of successes in robotics. An autonomous agent that moves, observes, and changes the world must carry a not easily controllable body.

Therefore, the advances have been exploited mainly in *decision support systems*, computer systems that provide advice for humans on highly specialized tasks. With the Internet, the scope of artificial intelligence has widened considerably. The Internet is an ideal nonphysical world for intelligent agents, which are pure spirits without bodies. In the years to come, we will experience a flood of intelligent agents on the Internet, and companies as well as private persons will be able to launch their own agents to explore and collect information on the Internet. Also, we will experience the dark sides of human endeavor. Some agents will destroy, intrude, tell lies and so on, and we will have to defend ourselves against them. Agents will meet agents, and they will have to decide how to treat each other, they will have to learn from past experience, and they will have to adapt to changing environments.

During the 1990s, Bayesian networks and decision graphs attracted a great deal of attention as a framework for building normative systems, not only in research institutions but also in industry. Contrary to most other frameworks for handling uncertainty, a good deal of theoretical insight as well as practical experience is required in order to exploit the opportunities provided by Bayesian networks and decision graphs.

On the other hand, many scientists and engineers wish to exploit the possibilities of normative systems without being experts in the field. This book should meet that demand. It is intended for both classroom use and self-study, and it addresses persons who are interested in exploiting the approach for the construction of decision support systems or bodyless agents.

The theoretical exposition in the book is self-contained, and the mathematical prerequisite is some prior exposure to calculus and elementary graph theory. Throughout the book we alternate between theoretical exposition and practical examples for gaining experience with the use of Bayesian networks and decision graphs, and we have assumed that the reader has access to a computer system for handling Bayesian networks and influence diagrams (the exercises marked with an ^{E} require such a system). There are many systems, academic as well as commercial. A comprehensive list of systems can be found at www.cs.berkeley.edu/~murphyk/Bayes/bnsoft.html. Several of the commercial systems have an academic version, which can be downloaded free of charge. In several chapters the presentation is based on examples, and for overview purposes there is a summary section at the end of each chapter.

A hands-on course could cover Sections 1.1–1.3, Chapter 2, Chapter 3, Sections 6.1–6.2, 7.1, 8.1–8.3, 9.1–9.4, and Sections 11.1–11.2. A first-year graduate course could cover Chapters 1–3, Sections 4.1–4.6, 5.2–5.3, 5.5, 5.7, 6.1–6.3, 7.1–7.3, Chapters 8–9, Sections 10.1–10.2, and Chapter 11.

The book is an introduction to Bayesian networks and decision graphs. Many results are not mentioned or just treated superficially. The following textbooks and monographs can be used for further study:

- Judea Pearl, *Probabilistic Reasoning in Intelligent Systems*, Morgan Kaufmann Publishers, 1988.
- Russell Almond, *Graphical Belief Modelling*, Chapman & Hall, 1995.
- Steffen L. Lauritzen, *Graphical Models*, Oxford University Press, 1996.
- Enrique Castillo, José M. Gutiérrez, and Ali S. Hadi, *Expert Systems and Probabilistic Network Models*, Springer-Verlag, 1997.

- Robert G. Cowell, A. Philip Dawid, and Steffen L. Lauritzen, *Probabilistic Networks and Expert Systems*, Springer-Verlag, 1999.
- Kevin B. Korb and Ann E. Nicholson, *Bayesian Artificial Intelligence*, Chapman & Hall 2004.
- Richard E. Neapolitan, *Learning Bayesian Networks*, Pearson Prentice Hall, 2004.

The annual *Conference on Uncertainty in Artificial Intelligence* (www.auai .org) is the main forum for researchers working with Bayesian networks and decision graphs, so the best general references for further reading are the proceedings from these conferences.

Another relevant conference is the biannual *European Conference on Symbolic and Quantitative Approaches to Reasoning with Uncertainty* (EC-SQARU). The conference deals with various approaches to uncertainty calculus, and the proceedings are published in the Springer-Verlag series *Lecture Notes in Artificial Intelligence*.

The book is supported by a web site, bndg.cs.aau.dk, which provides readers with solutions and models for selected exercises, a list of errata, special exercises, and other links relevant to the issues in the book.

Changes from the First Edition

In the second edition, we have added several subjects. Primarily, we have included chapters presenting commonly used methods for learning graphical models, and we have extended the treatment of graphical languages for modeling decision problems. We have also reorganized the material such that Part I is devoted to Bayesian networks and Part II deals with decision graphs.

The mathematical treatment is intended to be at the same level as in the first edition. However, many of the new issues in the book are mathematically rather demanding, particularly learning. Some of the sections are marked with an asterisk to indicate that they are not required for reading any of the unmarked sections.

Acknowledgments

We wish to express our gratitude to several people for ideas and comments during the preparation of the book. First, we thank present and previous colleagues in the Machine Intelligence group, Olav Bangsø, Søren L. Dittmer, Uffe Kjærulff, Tomáš Kočka, Anders L. Madsen, Dennis Nilsson, Kristian G. Olesen, Jose Peña, Jiří Vomlel, and Marta Vomlelová. We also thank the many academic colleagues around the world with whom we have had the pleasure of exchanging ideas, in particular Poul S. Eriksen, Linda van der Gaag, Helge Langseth, Steffen L. Lauritzen, Serafín Moral, Prakash Shenoy,

Antonio Salmerón, Claus Skaanning, Marco Valtorta, Yang Xiang, and Nevin Zhang. Special thanks to Søren Holbech Nielsen for assistance with figures, bibliography, and exercises.

We also thank several years' worth of undergraduate students who have had to cope with unfinished drafts of notes for parts of their course on decision support systems.

Aalborg, February 2007   Finn V. Jensen and Thomas D. Nielsen

Table of Contents

Preface v

1 Prerequisites on Probability Theory 1
   1.1 Two Perspectives on Probability Theory................... 1
   1.2 Fundamentals of Probability Theory 2
      1.2.1 Conditional Probabilities 4
      1.2.2 Probability Calculus 5
      1.2.3 Conditional Independence 6
   1.3 Probability Calculus for Variables 7
      1.3.1 Calculations with Probability Tables: An Example..... 11
   1.4 An Algebra of Potentials 13
   1.5 Random Variables 15
      1.5.1 Continuous Distributions 15
   1.6 Exercises 16

Part I Probabilistic Graphical Models

2 Causal and Bayesian Networks 23
   2.1 Reasoning Under Uncertainty 23
      2.1.1 Car Start Problem................................ 23
      2.1.2 A Causal Perspective on the Car Start Problem 24
   2.2 Causal Networks and d-Separation....................... 26
      2.2.1 d-separation 30
   2.3 Bayesian Networks 32
      2.3.1 Definition of Bayesian Networks 32
      2.3.2 The Chain Rule for Bayesian Networks.............. 35
      2.3.3 Inserting Evidence................................ 39
      2.3.4 Calculating Probabilities in Practice 41
   2.4 Graphical Models – Formal Languages for Model Specification 42
   2.5 Summary.. 44

2.6   Bibliographical Notes 45
2.7   Exercises 45

3   Building Models 51
3.1   Catching the Structure 51
      3.1.1   Milk Test 52
      3.1.2   Cold or Angina? 54
      3.1.3   Insemination 55
      3.1.4   A Simplified Poker Game 57
      3.1.5   Naive Bayes Models 58
      3.1.6   Causality 60
3.2   Determining the Conditional Probabilities 60
      3.2.1   Milk Test 60
      3.2.2   Stud Farm 62
      3.2.3   Poker Game 66
      3.2.4   Transmission of Symbol Strings 68
      3.2.5   Cold or Angina? 71
      3.2.6   Why Causal Networks? 72
3.3   Modeling Methods 73
      3.3.1   Undirected Relations 73
      3.3.2   Noisy-Or 75
      3.3.3   Divorcing 78
      3.3.4   Noisy Functional Dependence 80
      3.3.5   Expert Disagreements 81
      3.3.6   Object-Oriented Bayesian Networks 84
      3.3.7   Dynamic Bayesian Networks 91
      3.3.8   How to Deal with Continuous Variables 93
      3.3.9   Interventions 96
3.4   Special Features 97
      3.4.1   Joint Probability Tables 98
      3.4.2   Most-Probable Explanation 98
      3.4.3   Data Conflict 98
      3.4.4   Sensitivity Analysis 99
3.5   Summary 100
3.6   Bibliographical Notes 101
3.7   Exercises 102

4   Belief Updating in Bayesian Networks 109
4.1   Introductory Examples 109
      4.1.1   A Single Marginal 110
      4.1.2   Different Evidence Scenarios 111
      4.1.3   All Marginals 114
4.2   Graph-Theoretic Representation 115
      4.2.1   Task and Notation 115
      4.2.2   Domain Graphs 116

4.3 Triangulated Graphs and Join Trees 119
    4.3.1 Join Trees 122
4.4 Propagation in Junction Trees 124
    4.4.1 Lazy Propagation in Junction Trees 127
4.5 Exploiting the Information Scenario 130
    4.5.1 Barren Nodes 130
    4.5.2 d-Separation 131
4.6 Nontriangulated Domain Graphs 132
    4.6.1 Triangulation of Graphs 134
    4.6.2 Triangulation of Dynamic Bayesian Networks 137
4.7 Exact Propagation with Bounded Space 140
    4.7.1 Recursive Conditioning 140
4.8 Stochastic Simulation in Bayesian Networks 145
    4.8.1 Probabilistic Logic Sampling 146
    4.8.2 Likelihood Weighting 148
    4.8.3 Gibbs Sampling 150
4.9 Loopy Belief Propagation 152
4.10 Summary 154
4.11 Bibliographical Notes 156
4.12 Exercises 157

5 Analysis Tools for Bayesian Networks 167
5.1 IEJ Trees 168
5.2 Joint Probabilities and A-Saturated Junction Trees 169
    5.2.1 A-Saturated Junction Trees 169
5.3 Configuration of Maximal Probability 171
5.4 Axioms for Propagation in Junction Trees 173
5.5 Data Conflict 174
    5.5.1 Insemination 175
    5.5.2 The Conflict Measure conf 175
    5.5.3 Conflict or Rare Case 176
    5.5.4 Tracing of Conflicts 177
    5.5.5 Other Approaches to Conflict Detection 179
5.6 SE Analysis 179
    5.6.1 Example and Definitions 179
    5.6.2 h-Saturated Junction Trees and SE Analysis 182
5.7 Sensitivity to Parameters 184
    5.7.1 One-Way Sensitivity Analysis 187
    5.7.2 Two-Way Sensitivity Analysis 188
5.8 Summary 188
5.9 Bibliographical Notes 190
5.10 Exercises 191

6   Parameter Estimation 195
    6.1   Complete Data 195
        6.1.1   Maximum Likelihood Estimation 196
        6.1.2   Bayesian Estimation 197
    6.2   Incomplete Data 200
        6.2.1   Approximate Parameter Estimation: The EM Algorithm 201
        6.2.2   *Why We Cannot Perform Exact Parameter Estimation 207
    6.3   Adaptation 207
        6.3.1   Fractional Updating 210
        6.3.2   Fading 211
        6.3.3   *Specification of an Initial Sample Size 212
        6.3.4   Example: Strings of Symbols 213
        6.3.5   Adaptation to Structure 214
        6.3.6   *Fractional Updating as an Approximation 215
    6.4   Tuning 218
        6.4.1   Example 220
        6.4.2   Determining **grad** dist(x, y) as a Function of t 222
    6.5   Summary 223
    6.6   Bibliographical Notes 225
    6.7   Exercises 226

7   Learning the Structure of Bayesian Networks 229
    7.1   Constraint-Based Learning Methods 230
        7.1.1   From Skeleton to DAG 231
        7.1.2   From Independence Tests to Skeleton 234
        7.1.3   Example 235
        7.1.4   Constraint-Based Learning on Data Sets 237
    7.2   Ockham's Razor 240
    7.3   Score-Based Learning 241
        7.3.1   Score Functions 242
        7.3.2   Search Procedures 245
        7.3.3   Chow–Liu Trees 250
        7.3.4   *Bayesian Score Functions 253
    7.4   Summary 258
    7.5   Bibliographical Notes 260
    7.6   Exercises 261

8   Bayesian Networks as Classifiers 265
    8.1   Naive Bayes Classifiers 266
    8.2   Evaluation of Classifiers 268
    8.3   Extensions of Naive Bayes Classifiers 270
    8.4   Classification Trees 272
    8.5   Summary 274
    8.6   Bibliographical Notes 275
    8.7   Exercises 276

Part II Decision Graphs

9   Graphical Languages for Specification of Decision Problems 279
   9.1   One-Shot Decision Problems 280
         9.1.1   Fold or Call? 281
         9.1.2   Mildew.. 282
         9.1.3   One Decision in General........................... 283
   9.2   Utilities 284
         9.2.1   Instrumental Rationality 287
   9.3   Decision Trees 290
         9.3.1   A Couple of Examples 293
         9.3.2   Coalesced Decision Trees 295
         9.3.3   Solving Decision Trees 296
   9.4   Influence Diagrams 302
         9.4.1   Extended Poker Model 302
         9.4.2   Definition of Influence Diagrams 305
         9.4.3   Repetitive Decision Problems 308
   9.5   Asymmetric Decision Problems........................... 310
         9.5.1   Different Sources of Asymmetry 314
         9.5.2   Unconstrained Influence Diagrams 316
         9.5.3   Sequential Influence Diagrams 322
   9.6   Decision Problems with Unbounded Time Horizons 324
         9.6.1   Markov Decision Processes 324
         9.6.2   Partially Observable Markov Decision Processes ... 330
   9.7   Summary.. 332
   9.8   Bibliographical Notes 337
   9.9   Exercises 337

10 Solution Methods for Decision Graphs 343
   10.1 Solutions to Influence Diagrams 343
         10.1.1 The Chain Rule for Influence Diagrams 345
         10.1.2 Strategies and Expected Utilities 346
         10.1.3 An Example 352
   10.2 Variable Elimination 353
         10.2.1 Strong Junction Trees............................ 355
         10.2.2 Required Past 358
         10.2.3 Policy Networks 360
   10.3 Node Removal and Arc Reversal.......................... 362
         10.3.1 Node Removal 362
         10.3.2 Arc Reversal..................................... 363
         10.3.3 An Example 365
   10.4 Solutions to Unconstrained Influence Diagrams 367
         10.4.1 Minimizing the S-DAG............................. 367
         10.4.2 Determining Policies and Step Functions 371

10.5 Decision Problems Without a Temporal Ordering:
      Troubleshooting 373
      10.5.1 Action Sequences 373
      10.5.2 A Greedy Approach 375
      10.5.3 Call Service.......................................378
      10.5.4 Questions 378
10.6 Solutions to Decision Problems with Unbounded Time Horizon 380
      10.6.1 A Basic Solution 380
      10.6.2 Value Iteration 381
      10.6.3 Policy Iteration 385
      10.6.4 Solving Partially Observable Markov Decision Processes*388
10.7 Limited Memory Influence Diagrams 392
10.8 Summary...395
10.9 Bibliographical Notes 400
10.10Exercises 401

11 **Methods for Analyzing Decision Problems** 407
11.1 Value of Information 407
      11.1.1 Test for Infected Milk? 407
      11.1.2 Myopic Hypothesis-Driven Data Request 409
      11.1.3 Non-Utility-Based Value Functions 411
11.2 Finding the Relevant Past and Future of a Decision Problem . . 413
      11.2.1 Identifying the Required Past 415
11.3 Sensitivity Analysis 420
      11.3.1 Example 421
      11.3.2 One-Way Sensitivity Analysis in General 423
11.4 Summary...426
11.5 Bibliographical Notes 427
11.6 Exercises 427

List of Notation 429

References...431

Index 441

1

Prerequisites on Probability Theory

In this chapter we review some standard results and definitions from probability theory. The reader is assumed to have had some contact with probability theory before, and the purpose of this section is simply to brush up on some of the basic concepts and to introduce some of the notation used in the later chapters. Sections 1.1–1.3 are prerequisites for Section 2.3 and thereafter, Section 1.4 is a prerequisite for Chapter 4, and Section 1.5 is a prerequisite for Chapter 6 and Chapter 7.

1.1 Two Perspectives on Probability Theory

In many domains, the probability of seeing a certain outcome of an experiment can be interpreted as the *relative frequency* of seeing this particular outcome in all of the experiments performed. For instance, if you throw a six-sided die, then you would say that the probability of obtaining a three is $1/6$, because if we throw this die a large number of times we would expect to see a three in approximately $1/6$ of the throws. Along the same line of reasoning, we would also say that if we randomly draw a card from a deck consisting of 52 cards, then the probability that it will be a spade is $13/52$. This interpretation of probability rests on the assumption that there is some stochastic process that can be repeated several times and from which the relative frequencies can be counted. On the other hand, we often talk about the probability of seeing a certain event although we cannot specify a frequency for it. For example, I may estimate that the probability that the Danish soccer team will win the World Cup in 2010 is p. This probability is my own personal judgment of how likely it is that the Danish team will actually win, and it is based on my belief, experience, and current state of information. However, another person may specify another probability for the same event, and it has no meaning to look for ways of determining which of us is right, if either. These probabilities are referred to as *subjective probabilities*. One way to interpret my subjective probability of Denmark winning the world cup in 2010 is to imagine the following two wagers:

1. If the Danish soccer team wins the world cup in 2010, I will receive $100.
2. I will draw a ball from an urn containing 100 balls out of which n are white and $100 - n$ are black. If the ball drawn is white then I will receive $100 in 2010.

If all the balls are white then I will prefer the second wager, and if all the balls are black then I will prefer the first. However, for a certain n between 0 and 100 I will be indifferent about the two wagers, and for this n, $n/100$ will be my subjective probability that the Danish soccer team will win the World Cup.

1.2 Fundamentals of Probability Theory

For both views on probability described above, we will refer to the set of possible outcomes of an experiment as the *sample space* of the experiment. Here we use the somewhat abstract term "experiment" to refer to any type of process for which the outcome is uncertain, e.g., the throw of a die and the winner of the World Cup. We shall also assume that the sample space of an experiment contains all possible outcomes of the experiment, and that each pair of outcomes are mutually exclusive. These assumptions ensure that the experiment is guaranteed to end up in exactly one of the specified outcomes in the sample space. For instance, for the die example above, the sample space would be $\mathcal{S} = \{1, 2, 3, 4, 5, 6\}$, and for the soccer example the sample space would be $\mathcal{S} = \{\text{yes}, \text{no}\}$, assuming that I am interested only in whether the Danish team will win; both of the sample spaces satisfy the assumptions above. A subset of a sample space is called an *event*. For example, the event that we will get a value of three or higher with a six-sided die corresponds to the subset $\{3, 4, 5, 6\} \subseteq \{1, 2, 3, 4, 5, 6\}$, and the event will occur if the outcome of the throw is an element in the set. In general, we say that an event $\mathcal{A}$ is *true* for an experiment if the outcome of the experiment is an element of $\mathcal{A}$. When an event contains only one element, we will also refer to the event as an outcome.

To measure our degree of uncertainty about an experiment we assign a probability $P(\mathcal{A})$ to each event $\mathcal{A} \subseteq S$. These probabilities must obey the following three axioms:

The event $\mathcal{S}$ that we will get an outcome in the sample space is certain to occur and is therefore assigned the probability 1.

Axiom 1 $P(\mathcal{S}) = 1$.

Any event $\mathcal{A}$ must have a nonnegative probability.

Axiom 2 *For all $\mathcal{A} \subseteq \mathcal{S}$ it holds that $P(\mathcal{A}) \geq 0$.*

If two events $\mathcal{A}$ and $\mathcal{B}$ are disjoint (see Figure 1.1(a)), then the probability of the combined event is the sum of the probabilities for the two individual events:

Axiom 3 *If $\mathcal{A} \subseteq \mathcal{S}$, $\mathcal{B} \subseteq \mathcal{S}$ and $\mathcal{A} \cap \mathcal{B} = \emptyset$, then $P(\mathcal{A} \cup \mathcal{B}) = P(\mathcal{A}) + P(\mathcal{B})$.*

For example, the event that a die will turn up 3, $\mathcal{B} = \{3\}$, and the event that the die will have an even number, $\mathcal{A} = \{2, 4, 6\}$, are two disjoint events, and the probability that one of these two events will occur is therefore

$$P(\mathcal{A} \cup \mathcal{B}) = P(\mathcal{A}) + P(\mathcal{B}) = \frac{1}{6} + \frac{3}{6} = \frac{4}{6}.$$

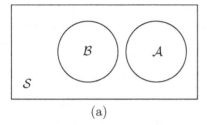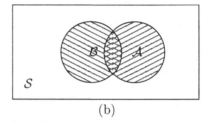

(a)   (b)

Fig. 1.1. In figure (a) the two events $\mathcal{A}$ and $\mathcal{B}$ are disjoint, whereas in figure (b), $\mathcal{A} \cap \mathcal{B} \neq \emptyset$.

On the other hand, if $\mathcal{A}$ and $\mathcal{B}$ are not disjoint (see Figure 1.1(b)), then it can easily be shown that

$$P(\mathcal{A} \cup \mathcal{B}) = P(\mathcal{A}) + P(\mathcal{B}) - P(\mathcal{A} \cap \mathcal{B}),$$

where $\mathcal{A} \cap \mathcal{B}$ is the intersection between $\mathcal{A}$ and $\mathcal{B}$ and it represents the event that *both* $\mathcal{A}$ and $\mathcal{B}$ will occur. Consider again a deck with 52 cards. The event $\mathcal{A}$ that I will draw a spade and the event $\mathcal{B}$ that I will draw a king are clearly not disjoint events; their intersection specifies the event that I will draw the king of spades, $\mathcal{A} \cap \mathcal{B} = \{\text{king of spades}\}$. Thus, the probability that I will draw either a king or a spade is

$$P(\mathcal{A} \cup \mathcal{B}) = P(\mathcal{A}) + P(\mathcal{B}) - P(\mathcal{A} \cap \mathcal{B}) = \frac{13}{52} + \frac{4}{52} - \frac{1}{52} = \frac{16}{52}.$$

Notation: Sometimes we will emphasize that a probability is based on a frequency (rather than being a subjective probability), in which case we will use the notation $P^{\#}$. If the event $\mathcal{A}$ contains only one outcome a, we write $P(a)$ rather than $P(\{a\})$.

1.2.1 Conditional Probabilities

Whenever a statement about the probability $P(\mathcal{A})$ of an event $\mathcal{A}$ is given, then it is implicitly given conditioned on other known factors. For example, a statement such as "the probability of the die turning up 6 is $\frac{1}{6}$" usually has the unsaid prerequisite that it is a fair die, or rather, as long as I know nothing further, I assume it to be a fair die. This means that the statement should be "given that it is a fair die, the probability ...." In this way, any statement on probabilities is a statement conditioned on what else is known. These types of probabilities are called *conditional probabilities* and are generally statements of the following kind:

"Given the event $\mathcal{B}$, the probability of the event $\mathcal{A}$ is p."

The notation for the preceding statement is $P(\mathcal{A}|\mathcal{B}) = p$. It should be stressed that $P(\mathcal{A}|\mathcal{B}) = p$ does not mean that whenever $\mathcal{B}$ is true, then the probability of $\mathcal{A}$ is p. It means that if $\mathcal{B}$ is true, and *everything else is irrelevant for* $\mathcal{A}$, then the probability of $\mathcal{A}$ is p.

Assume that we have assigned probabilities to all subsets of the sample space $\mathcal{S}$, and let $\mathcal{A}$ and $\mathcal{B}$ be subsets of $\mathcal{S}$ (Figure 1.1(b)). The question is whether the probability assignment for $\mathcal{S}$ can be used to calculate $P(\mathcal{A}|\mathcal{B})$. If we know the event $\mathcal{B}$, then all possible outcomes are elements of $\mathcal{B}$, and the outcomes for which $\mathcal{A}$ can be true are $\mathcal{A} \cap \mathcal{B}$. So, we look for the probability assignment for $\mathcal{A} \cap \mathcal{B}$ given that we know $\mathcal{B}$. Knowing $\mathcal{B}$ does not change the proportion between the probabilities of $\mathcal{A} \cap \mathcal{B}$ and another set $\mathcal{C} \cap \mathcal{B}$ (if, for example, I will bet twice as much on $\mathcal{A} \cap \mathcal{B}$ as on $\mathcal{C} \cap \mathcal{B}$, then after knowing $\mathcal{B}$, I will still bet twice as much on $\mathcal{A} \cap \mathcal{B}$ as on $\mathcal{C} \cap \mathcal{B}$). We can conclude that the proportions $P(\mathcal{A} \cap \mathcal{B})/P(\mathcal{C} \cap \mathcal{B})$ and $P(\mathcal{A}|\mathcal{B})/P(\mathcal{C}|\mathcal{B})$ must be the same. Setting $\mathcal{C} = \mathcal{B}$, and since we know from Axiom 1 that $P(\mathcal{B}|\mathcal{B}) = 1$, we have justified the following property, which should be considered an axiom.

Property 1.1 (Conditional probability). For two events $\mathcal{A}$ and $\mathcal{B}$, with $P(\mathcal{B}) > 0$, the conditional probability for $\mathcal{A}$ given $\mathcal{B}$ is

$$P(\mathcal{A}\,|\,\mathcal{B}) = \frac{P(\mathcal{A} \cap \mathcal{B})}{P(\mathcal{B})}.$$

For example, the conditional probability that a die will come up 4 given that we get an even number is $P(\mathcal{A} = \{4\}\,|\,\mathcal{B} = \{2, 4, 6\}) = P(\{4\})/P(\{2, 4, 6\})$, and by assuming that the die is fair we get $\frac{1/6}{3/6} = \frac{1}{3}$.

Obviously, when working with conditional probabilities we can also condition on more than one event, in which case the definition of a conditional probability generalizes as

$$P(\mathcal{A}\,|\,\mathcal{B} \cap \mathcal{C}) = \frac{P(\mathcal{A} \cap \mathcal{B} \cap \mathcal{C})}{P(\mathcal{B} \cap \mathcal{C})}.$$

1.2.2 Probability Calculus

The expression in Property 1.1 can be rewritten so that we obtain the so-called *fundamental rule* for probability calculus:

Theorem 1.1 (The fundamental rule).

$$P(\mathcal{A} \mid \mathcal{B})P(\mathcal{B}) = P(\mathcal{A} \cap \mathcal{B}). \qquad (1.1)$$

That is, the fundamental rule tells us how to calculate the probability of seeing both $\mathcal{A}$ and $\mathcal{B}$ when we know the probability of $\mathcal{A}$ given $\mathcal{B}$ and the probability of $\mathcal{B}$.

By conditioning on another event $\mathcal{C}$, the fundamental rule can also be written as

$$P(\mathcal{A} \mid \mathcal{B} \cap \mathcal{C})P(\mathcal{B} \mid \mathcal{C}) = P(\mathcal{A} \cap \mathcal{B} \mid \mathcal{C}).$$

Since $P(\mathcal{A} \cap \mathcal{B}) = P(\mathcal{B} \cap \mathcal{A})$ (and also $P(\mathcal{A} \cap \mathcal{B} \mid \mathcal{C}) = P(\mathcal{B} \cap \mathcal{A} \mid \mathcal{C})$), we get that $P(\mathcal{A} \mid \mathcal{B})P(\mathcal{B}) = P(\mathcal{A} \cap \mathcal{B}) = P(\mathcal{B} \mid \mathcal{A})P(\mathcal{A})$ from the fundamental rule. This yields the well-known *Bayes' rule*:

Theorem 1.2 (Bayes' rule).

$$P(\mathcal{A} \mid \mathcal{B}) = \frac{P(\mathcal{B} \mid \mathcal{A})P(\mathcal{A})}{P(\mathcal{B})}.$$

Bayes' rule provides us with a method for updating our beliefs about an event $\mathcal{A}$ given that we get information about another event $\mathcal{B}$. For this reason $P(\mathcal{A})$ is usually called the *prior* probability of $\mathcal{A}$, whereas $P(\mathcal{A} \mid \mathcal{B})$ is called the *posterior* probability of $\mathcal{A}$ given $\mathcal{B}$; the probability $P(\mathcal{B} \mid \mathcal{A})$ is called the *likelihood* of $\mathcal{A}$ given $\mathcal{B}$. For an explanation of this strange use of the term, see Example 1.1.

Finally, as for the fundamental rule, we can also state Bayes' rule in a context $\mathcal{C}$:

$$P(\mathcal{A} \mid \mathcal{B}, \mathcal{C}) = \frac{P(\mathcal{B} \mid \mathcal{A}, \mathcal{C})P(\mathcal{A} \mid \mathcal{C})}{P(\mathcal{B} \mid \mathcal{C})}.$$

Example 1.1. We have two diseases a_1 and a_2, both of which can cause the symptom b. Let $P(b \mid a_1) = 0.9$ and $P(b \mid a_2) = 0.3$. Assume that the prior probabilities for a_1 and a_2 are the same ($P(a_1) = P(a_2)$). Now, if b occurs, Bayes' rule gives

$$P(a_1 \mid b) = \frac{P(b \mid a_1)P(a_1)}{P(b)} = 0.9 \cdot \frac{P(a_1)}{P(b)};$$

$$P(a_2 \mid b) = \frac{P(b \mid a_2)P(a_2)}{P(b)} = 0.3 \cdot \frac{P(a_2)}{P(b)}.$$

Even though we cannot calculate the posterior probabilities, we can conclude that a_1 is three times as likely as a_2 given the symptom b.

If we furthermore know that a_1 and a_2 are the only possible causes of b, we can go even further (assuming that the probability of having both diseases is 0). Then $P(a_1 \mid b) + P(a_2 \mid b) = 1$, and we get

$$\frac{P(a_1)}{P(b)} = \frac{P(a_2)}{P(b)} = \frac{1}{0.9 + 0.3} = \frac{1}{1.2},$$

$P(a_1 \mid b) = 0.9/1.2 = 0.75$, and $P(a_2 \mid b) = 0.3/1.2 = 0.25$.

1.2.3 Conditional Independence

Sometimes information on one event $\mathcal{B}$ does not change our belief about the occurrence of another event $\mathcal{A}$, and in this case we say that $\mathcal{A}$ and $\mathcal{B}$ are *independent*.

Definition 1.1 (Independence). *The events $\mathcal{A}$ and $\mathcal{B}$ are independent if*

$$P(\mathcal{A} \mid \mathcal{B}) = P(\mathcal{A}).$$

For example, if we throw two fair dice, then seeing that the first die turns up 2 will not change our beliefs about the outcome of the second die.

This notion of independence is symmetric, so that if $\mathcal{A}$ is independent of $\mathcal{B}$, then $\mathcal{B}$ is independent of $\mathcal{A}$:

$$P(\mathcal{B} \mid \mathcal{A}) = \frac{P(\mathcal{A} \cap \mathcal{B})}{P(\mathcal{A})} = \frac{P(\mathcal{A} \mid \mathcal{B})P(\mathcal{B})}{P(\mathcal{A})} = \frac{P(\mathcal{A})P(\mathcal{B})}{P(\mathcal{A})} = P(\mathcal{B}).$$

The proof requires that $P(\mathcal{A}) > 0$, so if $P(\mathcal{A}) = 0$, the calculations are not valid. However, for our considerations it does not matter; if $\mathcal{A}$ is impossible why bother considering it?

When two events are independent, then the fundamental rule can be rewritten as

$$P(\mathcal{A} \cap \mathcal{B}) = P(\mathcal{A} \mid \mathcal{B})P(\mathcal{B}) = P(\mathcal{A}) \cdot P(\mathcal{B}).$$

That is, we can calculate the probability that both events will occur by multiplying the probabilities for the individual events.

The concept of independence also appears when we are conditioning on several events. Specifically, if information about the event $\mathcal{B}$ does not change our belief about the event $\mathcal{A}$ when we already know the event $\mathcal{C}$, then we say that $\mathcal{A}$ and $\mathcal{B}$ are *conditionally independent* given $\mathcal{C}$.

Definition 1.2 (Conditional independence). *The events $\mathcal{A}$ and $\mathcal{B}$ are conditionally independent given the event $\mathcal{C}$ if*

$$P(\mathcal{A} \mid \mathcal{B} \cap \mathcal{C}) = P(\mathcal{A} \mid \mathcal{C}).$$

Similar to the situation above, the conditional independence statement is symmetric. If $\mathcal{A}$ is conditionally independent of $\mathcal{B}$ given $\mathcal{C}$, then $\mathcal{B}$ is conditionally independent of $\mathcal{A}$ given $\mathcal{C}$:

$$P(\mathcal{B}\mid\mathcal{A}\cap\mathcal{C}) = \frac{P(\mathcal{A}\cap\mathcal{B}\mid\mathcal{C})P(\mathcal{C})}{P(\mathcal{A}\mid\mathcal{C})P(\mathcal{C})} = \frac{P(\mathcal{A}\mid\mathcal{B}\cap\mathcal{C})P(\mathcal{B}\mid\mathcal{C})}{P(\mathcal{A}\mid\mathcal{C})} = \frac{P(\mathcal{A}\mid\mathcal{C})P(\mathcal{B}\mid\mathcal{C})}{P(\mathcal{A}\mid\mathcal{C})}$$
$$= P(\mathcal{B}\mid\mathcal{C}).$$

Furthermore, when two events are conditionally independent, then we can use a multiplication rule similar to the one above when calculating the probability that both of the events will occur:

$$P(\mathcal{A}\cap\mathcal{B}\mid\mathcal{C}) = P(\mathcal{A}\mid\mathcal{C})\cdot P(\mathcal{B}\mid\mathcal{C}).$$

Note that when two events are independent it is actually a special case of conditional independence but with $\mathcal{C} = \emptyset$.

1.3 Probability Calculus for Variables

So far we have talked about probabilities of simple events and outcomes with respect to a certain sample space. In this book, however, we will be working with a collection of sample spaces, also called *variables*, and we will now extend the concepts above to probabilities over variables. A variable can be considered an experiment, and for each outcome of the experiment the variable has a corresponding *state*. The set of states associated with a variable A is denoted by $\mathrm{sp}(A) = (a_1, a_2, \ldots, a_n)$, and similar to the sample space these states should be *mutually exclusive* and *exhaustive*. The last assumption ensures that the variable is in one of its states (although we may not know which one), and the first assumption ensures that the variable is in only one state. For example, if we let D be a variable representing the outcome of rolling a die, then its state space would be $\mathrm{sp}(D) = (1, 2, 3, 4, 5, 6)$. We will use uppercase letters for variables and lowercase letters for states, and unless otherwise stated, a variable has a finite number of states.

For a variable A with states $a_1, \ldots, a_n$, we express our uncertainty about its state through a probability distribution $P(A)$ over these states:

$$P(A) = (x_1, \ldots, x_n); \qquad x_i \geq 0; \qquad \sum_{i=1}^{n} x_i = x_1 + \cdots + x_n = 1,$$

where x_i is the probability of A being in state a_i. A distribution is called *uniform* (or *even*) if all probabilities are equal.

Notation: In general, the probability of A being in state a_i is denoted by $P(A = a_i)$, and denoted by $P(a_i)$ if the variable is obvious from the context.

As we talked about conditional probabilities for events, we can also talk about *conditional probabilities* for variables: If the variable B has states $b_1, \ldots, b_m$, then $P(A \mid B)$ contains $n \cdot m$ conditional probabilities $P(a_i \mid b_j)$ that specify the probability of seeing a_i given b_j. That is, the conditional probability for a variable given another variable is a set of probabilities (usually organized in an $n \times m$ table) with one probability for each configuration of the states of the variables involved (see Table 1.1 for an example). Moreover, since $P(A \mid B)$ specifies a probability distribution for each event $B = b_j$, we know from Axiom 1 that the probabilities over A should sum to 1 for each state of B:

$$\sum_{i=1}^{n} P(A = a_i \mid B = b_j) = 1 \text{ for each } b_j.$$

	b_1	b_2	b_3
a_1	0.4	0.3	0.6
a_2	0.6	0.7	0.4

Table 1.1. An example of a conditional probability table $P(A \mid B)$ for the binary variable A given the ternary variable B. Note that for each state of B the probabilities of A sum up to 1.

The probability of seeing joint outcomes for different experiments can be expressed by the *joint probability* for two or more variables: For each configuration (a_i, b_j) of the variables A and B, $P(A, B)$ specifies the probability of seeing both $A = a_i$ *and* $B = b_j$. Hence, $P(A, B)$ consists of $n \cdot m$ numbers, and, similar to $P(A \mid B)$, $P(A, B)$ is usually represented in an $n \times m$ table (see Table 1.2 for an example). Note that since the state spaces of both A and B are mutually exclusive and exhaustive, it follows that all combinations of their states (the Cartesian product) are also mutually exclusive and exhaustive, and they can therefore be considered a sample space. Hence, by Axiom 1,

$$P(A, B) = \sum_{i=1}^{n} \sum_{j=1}^{m} P(A = a_i, B = b_j) = 1.$$

	b_1	b_2	b_3
a_1	0.16	0.12	0.12
a_2	0.24	0.28	0.08

Table 1.2. An example of a joint probability table $P(A, B)$ for the binary variable A and the ternary variable B. Note that the sum of all entries is 1.

When the fundamental rule (equation (1.1)) is used on variables A and B, the procedure is to apply the rule to each of the $n \cdot m$ configurations (a_i, b_j) of the two variables:

$$P(a_i \mid b_j)P(b_j) = P(a_i, b_j).$$

This means that in the table $P(A \mid B)$, each probability in $P(A \mid b_j)$ is multiplied by $P(b_j)$ to obtain the table $P(A, b_j)$, and by doing this for each b_j we get $P(A, B)$. If $P(B) = (0.4, 0.4, 0.2)$, then Table 1.2 is the result of using the fundamental rule on Table 1.1 (see also Table 1.3).

$$P(A,B) = \begin{array}{c|ccc} & b_1 & b_2 & b_3 \\ \hline a_1 & 0.4 \cdot 0.4 & 0.3 \cdot 0.4 & 0.6 \cdot 0.2 \\ a_2 & 0.6 \cdot 0.4 & 0.7 \cdot 0.4 & 0.4 \cdot 0.2 \end{array} = \begin{array}{c|ccc} & b_1 & b_2 & b_3 \\ \hline a_1 & 0.16 & 0.12 & 0.12 \\ a_2 & 0.24 & 0.28 & 0.08 \end{array}$$

Table 1.3. The joint probability table $P(A, B)$ in Table 1.2 can be found by multiplying $P(B) = (0.4, 0.4, 0.2)$ by $P(A \mid B)$ in Table 1.1.

When applied to variables, the fundamental rule is expressed as follows:

Theorem 1.3 (The fundamental rule for variables).

$$P(A, B) = P(A \mid B)P(B),$$

and conditioned on another variable C we have

$$P(A, B \mid C) = P(A \mid B, C)P(B \mid C).$$

From a joint probability table $P(A, B)$, the probability distribution $P(A)$ can be calculated by considering the outcomes of B that can occur together with each state a_i of A. There are exactly m different outcomes for which A is in state a_i, namely the mutually exclusive outcomes $(a_i, b_1), \ldots, (a_i, b_m)$. Therefore, by Axiom 3,

$$P(a_i) = \sum_{j=1}^{m} P(a_i, b_j).$$

This calculation is called *marginalization*, and we say that the variable B is marginalized out of $P(A, B)$ (resulting in $P(A)$). The notation is

$$P(A) = \sum_{B} P(A, B).$$

By marginalizing B out of Table 1.2, we get

$$P(A) = (0.16 + 0.12 + 0.12, 0.24 + 0.28 + 0.08) = (0.4, 0.6),$$

and by marginalizing out A we get

$$P(B) = (0.16 + 0.24, 0.12 + 0.28, 0.12 + 0.08) = (0.4, 0.4, 0.2).$$

That is, the marginalization operation allows us to remove variables from a joint probability distribution.

Bayes' rule for events (Theorem 1.2) can also be extended to variables, by treating the division in the same way as we treated multiplication above.

Theorem 1.4 (Bayes' rule for variables).

$$P(B \mid A) = \frac{P(A \mid B)P(B)}{P(A)} = \frac{P(A, B)}{\sum_B P(A, B)},$$

and conditioned on another variable C we have

$$P(B \mid A, C) = \frac{P(A \mid B, C)P(B \mid C)}{P(A \mid C)} = \frac{P(A, B \mid C)}{\sum_B P(A, B \mid C)}.$$

Note that the two equalities in the equations follow from (1) the fundamental rule and (2) the marginalization operator described above.

By applying Bayes' rule using $P(A)$, $P(B)$, and $P(A \mid B)$ as specified above, we get $P(B \mid A)$ shown in Table 1.4.

$$P(B \mid A) = \frac{P(A \mid B)P(B)}{P(A)} =$$

	a_1	a_2
b_1	$\frac{0.4 \cdot 0.4}{0.4}$	$\frac{0.6 \cdot 0.4}{0.6}$
b_2	$\frac{0.3 \cdot 0.4}{0.4}$	$\frac{0.7 \cdot 0.4}{0.6}$
b_3	$\frac{0.6 \cdot 0.2}{0.4}$	$\frac{0.4 \cdot 0.2}{0.6}$

$=$

	a_1	a_2
b_1	0.4	0.4
b_2	0.3	0.47
b_3	0.3	0.13

Table 1.4. The conditional probability $P(B \mid A)$ obtained by applying Bayes' rule to $P(A \mid B)$ in Table 1.1, $P(A) = (0.4, 0.6)$, and $P(B) = (0.4, 0.4, 0.2)$. Note that the probabilities over B sum to 1 for each state of A.

The concept of (conditional) independence is also defined for variables.

Definition 1.3 (Conditional independence for variables). *Two variables A and C are said to be* conditionally independent *given the variable B if*

$$P(a_i \mid c_k, b_j) = P(a_i \mid b_j)$$

for each $a_i \in \text{sp}(A)$, $b_j \in \text{sp}(B)$, and $c_k \in \text{sp}(C)$.

As a shorthand notation we will sometimes write $P(A \mid C, B) = P(A \mid B)$.

This means that when the state of B is known, then no knowledge of C will alter the probability of A. Observe that we require the independence statement to hold for each state of B; if the conditioning set is empty then we say that A and C are *marginally independent* or just independent (written as $P(A\,|\,C) = P(A)$).

When two variables A and C are conditionally independent given B, then the fundamental rule (Theorem 1.3) can be simplified:

$$P(A, C\,|\,B) = P(A\,|\,B, C)P(C\,|\,B) = P(A\,|\,B)P(C\,|\,B).$$

In the expression above, we multiply two conditional probability tables over different domains. Fortunately, the method for doing this multiplication is a straightforward extension of what we have done so far:

$$P(a_i, c_k\,|\,b_j) = P(a_i\,|\,b_j)P(c_k\,|\,b_j).$$

For example, by multiplying $P(A\,|\,B)$ and $P(C\,|\,B)$ (specified in Table 1.1 and Table 1.5, respectively) we get the joint probability $P(A, C\,|\,B)$ in Table 1.6.

	b_1	b_2	b_3
c_1	0.2	0.9	0.3
c_2	0.05	0.05	0.2
c_3	0.75	0.05	0.5

Table 1.5. The conditional probability table $P(C\,|\,B)$ for the ternary variable C given the ternary variable B.

$P(A, C\,|\,B) = P(A\,|\,B)P(C\,|\,B)$

		b_1	b_2	b_3
=	c_1	$(0.2 \cdot 0.4, 0.2 \cdot 0.6)$	$(0.9 \cdot 0.3, 0.9 \cdot 0.7)$	$(0.3 \cdot 0.6, 0.3 \cdot 0.4)$
	c_2	$(0.05 \cdot 0.4, 0.05 \cdot 0.6)$	$(0.05 \cdot 0.3, 0.05 \cdot 0.7)$	$(0.2 \cdot 0.6, 0.2 \cdot 0.4)$
	c_3	$(0.75 \cdot 0.4, 0.75 \cdot 0.6)$	$(0.05 \cdot 0.3, 0.05 \cdot 0.7)$	$(0.5 \cdot 0.6, 0.5 \cdot 0.4)$

		b_1	b_2	b_3
=	c_1	$(0.08, 0.12)$	$(0.27, 0.63)$	$(0.18, 0.12)$
	c_2	$(0.02, 0.03)$	$(0.015, 0.035)$	$(0.12, 0.08)$
	c_3	$(0.3, 0.45)$	$(0.015, 0.035)$	$(0.3, 0.2)$

Table 1.6. If A and C are conditionally independent given B, then $P(A, C\,|\,B)$ can be found by multiplying $P(A\,|\,B)$ and $P(C\,|\,B)$ as specified in Table 1.1 and Table 1.5, respectively.

1.3.1 Calculations with Probability Tables: An Example

To illustrate the theorems above, assume that we have three variables, A, B, and C, with the probabilities as in Table 1.7. We receive evidence $A = a_2$ and

$C = c_1$ and we would now like to calculate the conditional probability table $P(B \mid a_2, c_1)$.

	b_1	b_2	b_3
a_1	(0, 0.05, 0.05)	(0.05, 0.05, 0)	(0.05, 0.05, 0.05)
a_2	(0.1, 0.1, 0)	(0.1, 0, 0.1)	(0.2, 0, 0.05)

Table 1.7. A joint probability table for the variables A, B, and C. The three numbers in each entry correspond to the states c_1, c_2, and c_3.

First, we focus on the part of the table corresponding to $A = a_2$ and $C = c_1$, and we get

$$P(a_2, B, c_1) = (0.1, 0.1, 0.2). \tag{1.2}$$

To calculate $P(B \mid a_2, c_1)$, we can use Theorem 1.4:

$$P(B \mid a_2, c_1) = \frac{P(a_2, B, c_1)}{P(a_2, c_1)} = \frac{P(a_2, B, c_1)}{\sum_B P(a_2, B, c_1)}. \tag{1.3}$$

By marginalizing B out of equation (1.2) we get

$$P(a_2, c_1) = 0.1 + 0.1 + 0.2 = 0.4.$$

Finally, by performing the division in equation (1.3) we get

$$P(B \mid a_2, c_1) = \left(\frac{0.1}{0.4}, \frac{0.1}{0.4}, \frac{0.2}{0.4} \right) = (0.25, 0.25, 0.5).$$

Another way of doing the same is to say that we wish to transform $P(a_2, B, c_1)$ into a probability distribution. Because the numbers do not add up to one, we *normalize* the distribution by dividing each number by the sum of all the numbers.

Suppose now that we were given only the evidence $A = a_2$, and we want to calculate $P(B \mid a_2, C)$. The calculation of this probability table follows the same steps as above, except that we now work with tables during the calculations. As before, we start by focusing on the part of $P(A, B, C)$ corresponding to $A = a_2$ and we get the result in Table 1.8.

To calculate $P(B \mid a_2, C)$ we use

$$P(B \mid a_2, C) = \frac{P(a_2, B, C)}{P(a_2, C)} = \frac{P(a_2, B, C)}{\sum_B P(a_2, B, C)}. \tag{1.4}$$

The probability $P(a_2, C)$ is found by marginalizing B out of Table 1.8:

$$P(a_2, C) = (0.1 + 0.1 + 0.2, 0.1 + 0 + 0, 0 + 0.1 + 0.05) = (0.4, 0.1, 0.15), \tag{1.5}$$

and by inserting this in equation (1.4) we get the result shown in Table 1.2.

	b_1	b_2	b_3
c_1	0.1	0.1	0.2
c_2	0.1	0	0
c_2	0	0.1	0.05

Table 1.8. The probability table $P(a_2, B, C)$ that corresponds to the part of the probability table in Table 1.8 restricted to $A = a_2$.

$$P(B\,|\,a_2, C) = \begin{array}{c|ccc} & b_1 & b_2 & b_3 \\ \hline c_1 & \frac{0.1}{0.4} & \frac{0.1}{0.4} & \frac{0.2}{0.4} \\ c_2 & \frac{0.1}{0.1} & \frac{0}{0.1} & \frac{0}{0.1} \\ c_2 & \frac{0}{0.15} & \frac{0.1}{0.15} & \frac{0.05}{0.15} \end{array} = \begin{array}{c|ccc} & b_1 & b_2 & b_3 \\ \hline c_1 & 0.25 & 0.25 & 0.5 \\ c_2 & 1 & 0 & 0 \\ c_2 & 0 & 2/3 & 1/3 \end{array}$$

Table 1.9. The calculation of $P(B\,|\,a_2, C)$ using $P(a_2, B, C)$ (Table 1.1) and $P(a_2, C)$ (equation (1.5)).

1.4 An Algebra of Potentials

Below we list some properties of the algebra of multiplication and marginalization of tables. The tables need not be (conditional) probabilities, and they are generally called *potentials*.

A potential ϕ is a real-valued function over a *domain* of finite variables $\mathcal{X}$:

$$\phi : \mathrm{sp}(\mathcal{X}) \to \mathbb{R}$$

The domain of a potential is denoted by $\mathrm{dom}\,(\phi)$. For example, the domain of the potential $P(A, B\,|\,C)$ is $\mathrm{dom}\,(P(A, B\,|\,C)) = \{A, B, C\}$.

Two potentials can be *multiplied*, denoted by an (often suppressed) dot. Multiplication has the following properties:

1. $\mathrm{dom}\,(\phi_1 \phi_2) = \mathrm{dom}\,(\phi_1) \cup \mathrm{dom}\,(\phi_2)$.
2. **The commutative law:** $\phi_1 \phi_2 = \phi_2 \phi_1$.
3. **The associative law:** $(\phi_1 \phi_2)\phi_3 = \phi_1(\phi_2 \phi_3)$.
4. **Existence of unit:** The unit potential $\mathbf{1}$ is a potential that contains only 1's and is defined over any domain such that $\mathbf{1} \cdot \phi = \phi$, for all potentials ϕ.

The marginalization operator defined in Section 1.3 can be generalized to potentials so that $\sum_A \phi$ is a potential over $\mathrm{dom}(\phi) \backslash \{A\}$. Furthermore, marginalization is *commutative*:

$$\sum_A \sum_B \phi = \sum_B \sum_A \phi.$$

For potentials of the form $P(A\,|\,\mathcal{V})$, where $\mathcal{V}$ is a set of variables, we have

5. **The unit potential property:** $\sum_A P(A\,|\,\mathcal{V}) = 1$.

For marginalization of a product, the following holds

6. **The distributive law**: If $A \notin \mathrm{dom}(\phi_1)$, then $\sum_A \phi_1 \phi_2 = \phi_1 \sum_A \phi_2$.

The distributive law is usually known as $ab + ac = a(b + c)$, and the preceding formula is actually the same law applied to tables. To verify it, consider the calculations in Tables 1.10–1.14. Here we see that Table 1.12 and Table 1.14 are equal and correspond to the left-hand and right-hand sides of the distributive law.

$B \backslash A$	a_1 a_2
b_1	x_1 x_2
b_2	x_3 x_4

$B \backslash C$	c_1 c_2
b_1	y_1 y_2
b_2	y_3 y_4

Table 1.10. $\phi_1(A, B)$ and $\phi_2(C, B)$.

$B \backslash A$	a_1	a_2
b_1	(x_1y_1, x_1y_2)	(x_2y_1, x_2y_2)
b_2	(x_3y_3, x_3y_4)	(x_4y_3, x_4y_4)

Table 1.11. $\phi_1(A, B) \cdot \phi_2(C, B)$. The two numbers in each entry correspond to the states c_1 and c_2.

$B \backslash A$	a_1	a_2
b_1	$x_1y_1 + x_1y_2$	$x_2y_1 + x_2y_2$
b_2	$x_3y_3 + x_3y_4$	$x_4y_3 + x_4y_4$

Table 1.12. $\sum_C \phi_1(A, B) \cdot \phi_2(C, B)$.

B	
b_1	$y_1 + y_2$
b_2	$y_3 + y_4$

Table 1.13. $\sum_C \phi_2(C, B)$.

We also use the term *projection* for marginalization. For example, if A and B are marginalized out of $\phi(A, B, C)$, we may say that ϕ is *projected* down to C, and we use the notation $\phi^{\downarrow C}$. With this notation, the properties of marginalization look as follows ($\mathcal{V}$ and $\mathcal{W}$ denote sets of variables):

$B \setminus A$	a_1	a_2
b_1	$x_1(y_1 + y_2)$	$x_2(y_1 + y_2)$
b_2	$x_3(y_3 + y_4)$	$x_4(y_3 + y_4)$

Table 1.14. $\phi_1(A, B) \sum_C \phi_2(C, B)$.

7. **The commutative law:** $(\phi^{\downarrow V})^{\downarrow W} = (\phi^{\downarrow W})^{\downarrow V}$.
8. **The distributive law:** If $\text{dom}(\phi_1) \subseteq V$, then $(\phi_1 \phi_2)^{\downarrow V} = \phi_1(\phi_2^{\downarrow V})$.

1.5 Random Variables

Let $\mathcal{S}$ be a sample space. A *random variable* is a real-valued function on $\mathcal{S}$; $V : \mathcal{S} \to \mathbb{R}$. If, for example, you throw a die, and you win \$1 if you get 4 or above, and you lose \$1 if you get 3 or below, then the corresponding random variable is a function with value -1 on $\{1, 2, 3\}$ and 1 on $\{4, 5, 6\}$.

The *mean value* of a random variable V on $\mathcal{S}$ is defined as

$$\mu(V) = \sum_{s \in \mathcal{S}} V(s)P(s). \tag{1.6}$$

For the example above, the mean value is $-1\frac{1}{6} + -1\frac{1}{6} + -1\frac{1}{6} + \frac{1}{6} + \frac{1}{6} + \frac{1}{6} = 0$ (provided that the die is fair). The mean value is also called the *expected value*.

A measure of how much a random variable varies between its values is the *variance*, σ^2. It is defined as the mean of the square of the difference between value and mean:

$$\sigma^2(V) = \sum_{s \in \mathcal{S}} (V(s) - \mu(V))^2 P(s). \tag{1.7}$$

For the example above we have

$$\sigma^2 = 3(-1 - 0)^2 \frac{1}{6} + 3(1 - 0)^2 \frac{1}{6} - 1.$$

1.5.1 Continuous Distributions

Consider an experiment, where an arrow is thrown at the $[0, 1] \times [0, 1]$ square. The possible outcomes are the points (x, y) in the unit square. Since the probability is zero for any particular outcome, the probability distribution is assigned to subsets of the unit square. We may think of this assignment as a process of distributing a probability mass of 1 over the sample space. We may, for example, assign a probability for landing in the small square $[x, x+\epsilon] \times [y, y+\epsilon]$. To be more systematic, let n be a natural number, then the unit square can be partitioned into small squares of the type $[\frac{i}{n}, \frac{i+1}{n}] \times [\frac{j}{n}, \frac{j+1}{n}]$, and we can assign probabilities $P([\frac{i}{n}, \frac{i+1}{n}] \times [\frac{j}{n}, \frac{j+1}{n}])$ to these squares with area

$\frac{1}{n^2}$. Now, if $P([\frac{i}{n}, \frac{i+1}{n}] \times [\frac{j}{n}, \frac{j+1}{n}]) = x$, then you can say that the probability mass x is distributed over the small square with an average density of $n^2 x$, and we define the *density function* (also called the *frequency function*) $f(x, y)$ as

$$f(x, y) = \lim_{n \to \infty} n^2 P\left(\left[x, x + \frac{1}{n}\right] \times \left[y, y + \frac{1}{n}\right]\right).$$

In general, if S is a continuous sample space, the density function is a nonnegative real-valued function f on S, for which it holds that for any subset A of S,

$$\int_A f(s)ds = P(A).$$

In particular,

$$\int_S f(s)ds = 1.$$

When S is an interval $[a, b]$ (possibly infinite), the outcomes are real numbers (such as height or weight), and you may be interested in the mean (height or weight). It is defined as

$$\mu = \int_a^b x f(x)dx,$$

and the variance is given by

$$\sigma^2 = \int_a^b (\mu - x)^2 f(x)dx.$$

Mathematically, the mean and variance are the mean and variance of the identity function $I(x) = x$, but we use the term "mean and variance of the *distribution.*"

1.6 Exercises

Exercise 1.1. Given Axioms 1 to 3, prove that

$$P(A \cup B) = P(A) + P(B) - P(A \cap B).$$

Exercise 1.2. Consider the experiment of rolling a red and a blue fair six-sided die. Give an example of a sample space for the experiment along with probabilities for each outcome. Suppose then that we are interested only in the sum of the dice (that is, the experiment consists in rolling the dice and adding up the numbers). Give another example of a sample space for this experiment and probabilities for the outcomes.

Exercise 1.3. Consider the experiment of flipping a fair coin, and if it lands heads, rolling a fair four-sided die, and if it lands tails, rolling a fair six-sided die. Suppose that we are interested only in the number rolled by the die, and a sample space $\mathcal{S}_A$ for the experiment could thus be the numbers from 1 to 6. Another sample space could be $\mathcal{S}_B = \{t1, \ldots, t6, h1, \ldots, h4\}$, with for example $t2$ meaning "tails and a roll of 2" and $h4$ meaning "heads and a roll of 4." Choose either $\mathcal{S}_A$ or $\mathcal{S}_B$ and associate probabilities with it. According to your sample space and probability distribution, what is the probability of rolling either 3 or 5.

Exercise 1.4. Draw a Venn diagram (like that in Figure 1.1) over $\mathcal{S}_B$ defined in Exercise 1.3. The diagram should show the events corresponding to "rolling a 3," "flipping tails," and "flipping tails and rolling a 3."

Exercise 1.5. Let $\mathcal{S}_B$ be defined as in Exercise 1.3, but with a loaded coin and loaded dice. A probability distribution is given in Table 1.15. What is the probability that the loaded coin lands "tails"? What is the conditional probability of rolling a 4, given that the coin lands tails? Which of the loaded dice has the highest chance of rolling 4 or more?

$t1$	$\frac{5}{18}$	$t6$	$\frac{1}{18}$
$t2$	$\frac{1}{9}$	$h1$	$\frac{1}{24}$
$t3$	$\frac{1}{9}$	$h2$	$\frac{1}{24}$
$t4$	$\frac{1}{18}$	$h3$	$\frac{1}{8}$
$t5$	$\frac{1}{18}$	$h4$	$\frac{1}{8}$

Table 1.15. Probabilities for $\mathcal{S}_B$ in Exercise 1.5.

Exercise 1.6. Prove that

$$P(\mathcal{A} \mid \mathcal{B} \cup \mathcal{C})P(\mathcal{B} \mid \mathcal{C}) = P(\mathcal{A} \cap \mathcal{B} \mid \mathcal{C}).$$

Exercise 1.7. A farmer has a cow, which he suspects is pregnant. He administers a test to the urine of the cow to determine whether it is pregnant. There are four outcomes in this experiment:

1. The cow is pregnant and the test is positive.
2. The cow is pregnant, but the test is negative.
3. The cow is not pregnant, but the test is positive.
4. The cow is not pregnant, and the test is negative.

The prior probability of the event that the cow is pregnant is 0.05, the probability of the event that the test is positive, when the cow indeed is pregnant, is 0.98 and the probability that the test is negative, when the cow is not pregnant, is 0.999. The test turns out to be positive. What is the posterior probability of the cow being pregnant?

Exercise 1.8. Consider the following two experiments: One consists in throwing a red six-sided die, and one consists in throwing a blue six-sided die. We let R be a variable representing the roll of the red die, having a set of states $\{r1, r2, r3, r4, r5, r6\}$, and B be a variable representing the roll of the blue die (states $\{b1, b2, b3, b4, b5, b6\}$). Assume that the red die is fair so that $P(R = r1) = \cdots = P(R = r6) = \frac{1}{6}$, and that the variable for the blue die has probabilities $P(B = b1) = P(B = b2) = P(B = b3) = \frac{1}{12}$ and $P(B = b4) = P(B = b5) = P(B = b6) = \frac{1}{4}$. Give an example of a sample space for an experiment consisting of throwing both the red and the blue die. Using $P(R)$ and $P(B)$, what is the probability distribution for your sample space?

Exercise 1.9. Consider the sample space S_B from Exercise 1.3, with probability distribution as defined in Table 1.15. Recast the sample space as variables. What is the probability distribution for each variable?

Exercise 1.10. Prove the fundamental rule for variables:

$$P(A, B) = P(A \mid B)P(B) .$$

Exercise 1.11. Calculate $P(A)$, $P(B)$, $P(A \mid B)$, and $P(B \mid A)$ from the table for $P(A, B)$ (Table 1.16).

	b_1	b_2	b_3
a_1	0.05	0.10	0.05
a_2	0.15	0.00	0.25
a_3	0.10	0.20	0.10

Table 1.16. $P(A, B)$ for Exercise 1.11.

Exercise 1.12. Table 1.17 describes a test T for an event A. The number 0.01 is the frequency of *false negatives*, and the number 0.001 is the frequency of *false positives*.

(*i*) The police can order a blood test on drivers under the suspicion of having consumed too much alcohol. The test has the above characteristics. Experience says that 20% of the drivers under suspicion do in fact drive with too much alcohol in their blood. A suspicious driver has a positive blood test. What is the probability that the driver is guilty of driving under the influence of alcohol?

(*ii*) The police block a road, take blood samples of all drivers, and use the same test. It is estimated that one out of 1,000 drivers have too much alcohol in their blood. A driver has a positive test result. What is the probability that the driver is guilty of driving under the influence of alcohol?

	$A = yes$	$A = no$
$T = yes$	0.99	0.001
$T = no$	0.01	0.999

Table 1.17. Table for Exercise 1.12. Conditional probabilities $P(T \mid A)$ characterizing test T for A.

Exercise 1.13. In Table 1.18, a joint probability table for the binary variables A, B, and C is given.

- Calculate $P(B, C)$ and $P(B)$.
- Are A and C independent given B?

	b_1	b_2
a_1	$(0.006, 0.054)$	$(0.048, 0.432)$
a_2	$(0.014, 0.126)$	$(0.032, 0.288)$

Table 1.18. $P(A, B, C)$ for Exercise 1.13.

Exercise 1.14. Write a short algorithm that given an $n \times m$ potential $\phi(A, B)$ calculates $\sum_A \phi$. Use your algorithm on the joint probability table $P(A, B)$ in Table 1.2 and on the conditional probability table $P(A|B)$ in Table 1.1.

Exercise 1.15. Prove that the associative, commutative, and distributive laws hold for potentials.

Exercise 1.16. Let $\phi(x) = ax$ be a distribution on $[0, 1]$. Determine a. What are the mean and the variance of ϕ?

Exercise 1.17. Let $\phi(x) - a\sin(x)$ be a distribution on $[0, \pi]$. Determine a and the mean of ϕ.

Part I

Probabilistic Graphical Models

2

Causal and Bayesian Networks

In this chapter we introduce causal networks, which are the basic graphical feature for (almost) everything in this book. We give rules for reasoning about relevance in causal networks; is knowledge of A relevant for my belief about B? These sections deal with reasoning under uncertainty in general. Next, Bayesian networks are defined as causal networks with the strength of the causal links represented as conditional probabilities. Finally, the chain rule for Bayesian networks is presented. The chain rule is the property that makes Bayesian networks a very powerful tool for representing domains with inherent uncertainty. The sections on Bayesian networks assume knowledge of probability calculus as laid out in Sections 1.1–1.4.

2.1 Reasoning Under Uncertainty

2.1.1 Car Start Problem

The following is an example of the type of reasoning that humans do daily.

"In the morning, my car will not start. I can hear the starter turn, but nothing happens. There may be several reasons for my problem. I can hear the starter roll, so there must be power from the battery. Therefore, the most-probable causes are that the fuel has been stolen overnight or that the spark plugs are dirty. It may also be due to dirt in the carburetor, a loose connection in the ignition system, or something more serious. To find out, I first look at the fuel meter. It shows half full, so I decide to clean the spark plugs."

To have a computer do the same kind of reasoning, we need answers to questions such as, "What made me conclude that among the probable causes "stolen fuel", and "dirty spark plugs" are the two most-probable causes?" or "What made me decide to look at the fuel meter, and how can an observation concerning fuel make me conclude on the seemingly unrelated spark plugs?" To be more precise, we need ways of representing the problem and ways of performing inference in this representation such that a computer can simulate this kind of reasoning and perhaps do it better and faster than humans.

For propositional logic, Boolean logic is the representation framework, and various derived structures, such as truth tables and binary decision diagrams, have been invented together with efficient algorithms for inference.

In logical reasoning, we use four kinds of logical connectives: conjunction, disjunction, implication, and negation. In other words, simple logical statements are of the kind, "if it rains, then the lawn is wet," "both John and Mary have caught the flu," "either they stay at home or they go to the cinema," or "the lawn is not wet." From a set of logical statements, we can deduce new statements. From the two statements "if it rains, then the lawn is wet" and "the lawn is not wet," we can infer that it is not raining.

When we are dealing with uncertain events, it would be nice if we could use similar connectives with certainties rather than truth values attached, so we may extend the truth values of propositional logic to "certainties," which are numbers between 0 and 1. A certainty 0 means "certainly not true," and the higher the number, the higher the certainty. Certainty 1 means "certainly true."

We could then work with statements such as, "if I take a cup of coffee while on break, I will with certainty 0.5 stay awake during the next lecture" or "if I take a short walk during the break, I will with certainty 0.8 stay awake during the next lecture." Now, suppose I take a walk as well as have a cup of coffee. How certain can I be to stay awake? To answer this, I need a rule for how to *combine* certainties. In other words, I need a function that takes the two certainties 0.5 and 0.8 and returns a number, which should be the certainty resulting from combining the certainty from the two statements.

The same is needed for chaining: "if a then b with certainty x," and "if b then c with certainty y." I know a, so what is the certainty of c?

It has turned out that any function for combination and chaining will in some situations lead to wrong conclusions.

Another problem, which is also a problem for logical reasoning, is abduction: I have the rule "a woman has long hair with certainty 0.7." I see a long-haired person. What can I infer about the person's sex?

2.1.2 A Causal Perspective on the Car Start Problem

A way of structuring a situation for reasoning under uncertainty is to construct a graph representing causal relations between events.

Example 2.1 (A reduced Car Start Problem).
    To simplify the situation, assume that we have the events $\{yes, no\}$ for *Fuel?*, $\{yes, no\}$ for *Clean Spark Plugs?*, $\{full, \frac{1}{2}, empty\}$ for *Fuel Meter*, and $\{yes, no\}$ for *Start?*. In other words, the events are clustered around variables, each with a set of outcomes, also called *states*. We know that the state of *Fuel?* and the state of *Clean Spark Plugs?* have a causal impact on

the state of *Start?*. Also, the state of *Fuel?* has an impact on the state of *Fuel Meter Standing*. This is represented by the graph in Figure 2.1.

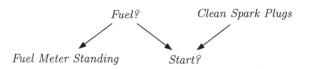

Fig. 2.1. A causal network for the reduced Car Start Problem.

If we add a direction from *no* to *yes* inside each variable (and from *empty* to *full*), we can also represent directions of the impact. For the present situation, we can say that all the impacts are positive (with the direction); that is, the more the certainty of the cause is moved in a positive direction, the more the certainty of the affected variable will also be moved in a positive direction. To indicate this, we can label the links with the sign "+" as is done in Figure 2.2.

Fig. 2.2. A causal network for the reduced Car Start Problem with a sign indicating direction of impact.

We can use the graph in Figure 2.2 to perform some reasoning. Obviously, if I know that the spark plugs are not clean, then the certainty for no start will increase. However, my situation is the opposite. I realize that I have a start problem. As my certainty on *Start?* is moved in a negative direction, I find the possible causes (*Clean Spark Plugs?* and *Fuel?*) for such a move more certain; that is, the sign "+" is valid for both directions. Now, because the certainty on for *Fuel?* = *no* has increased, I will have a higher expectation that *Fuel Meter Standing* is in state *empty*.

The movement of the certainty for *Fuel Meter Standing* tells me that by reading the fuel meter I will get information related to the start problem. I read the fuel meter, it says $\frac{1}{2}$, and reasoning backward yields that the certainty on *Fuel?* is moved in a negative direction.

So far, the reasoning has been governed by simple rules that can easily be formalized. The conclusion is harder: "Lack of fuel does not seem to be the reason for my start problem, so most probably the spark plugs are not clean." Is there a formalized rule that allows this kind of reasoning on a causal network to be computerized? We will return to this problem in Section 2.2.

Note: The reasoning has focused on changes of certainty. In certainty calculus, if the actual certainty of a specific event must be calculated, then knowledge of certainties prior to any information is also needed. In particular, prior certainties are required for the events that are not effects of causes in the network. If, for example, my car cannot start, the actual certainty that the fuel has been stolen depends on my neighborhood.

2.2 Causal Networks and d-Separation

A causal network consists of a set of *variables* and a set of *directed links* (also called *arcs*) between variables. Mathematically, the structure is called a *directed graph*. When talking about the relations in a directed graph, we use the wording of family relations: if there is a link from A to B, we say that B is a *child* of A, and A is a *parent* of B.

The variables represent propositions (or sample spaces), see also Section 1.3. A variable can have any number of states (or outcomes). A variable may, for example, be the color of a car (states *blue, green, red, brown*), the number of children in a specific family (states 0, 1, 2, 3, 4, 5, 6, > 6), or a disease (states *bronchitis, tuberculosis, lung cancer*). Variables may have a countable or a continuous state set, but we consider only variables with a finite number of states (we shall return to the issue of continuous state spaces in Section 3.3.8).

In a causal network, a variable represents a set of possible states of affairs. A variable is in exactly one of its states; which one may be unknown to us.

As illustrated in Section 2.1.2, causal networks can be used to follow how a change of certainty in one variable may change the certainty for other variables. We present in this section a set of rules for that kind of reasoning. The rules are independent of the particular calculus for uncertainty.

Serial Connections

Consider the situation in Figure 2.3. Here A has an influence on B, which in turn has an influence on C. Obviously, evidence about A will influence the certainty of B, which then influences the certainty of C. Similarly, evidence about C will influence the certainty of A through B. On the other hand, if the state of B is known, then the channel is blocked, and A and C become independent; we say that A and C are d-separated given B. When the state of a variable is known, we say that the variable is *instantiated*.

We conclude that evidence may be transmitted through a serial connection unless the state of the variable in the connection is known.

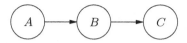

Fig. 2.3. Serial connection. When B is instantiated, it blocks communication between A and C.

Example 2.2. Figure 2.4 shows a causal model for the relations between *Rainfall* (*no, light, medium, heavy*), *Water level* (*low, medium, high*), and *Flooding* (*yes, no*). If I have not observed the water level, then knowing that there has been a flooding will increase my belief that the water level is high, which in turn will tell me something about the rainfall. The same line of reasoning holds in the other direction. On the other hand, if I already know the water level, then knowing that there has been flooding will not tell me anything new about rainfall.

Fig. 2.4. A causal model for *Rainfall, Water level,* and *Flooding.*

Diverging Connections

The situation in Figure 2.5 is called a *diverging* connection. Influence can pass between all the children of A unless the state of A is known. That is, $B, C, \ldots, E$ are d-separated given A.

Evidence may be transmitted through a diverging connection unless it is instantiated.

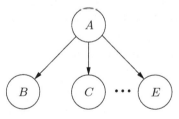

Fig. 2.5. Diverging connection. If A is instantiated, it blocks communication between its children.

Example 2.3. Figure 2.6 shows the causal relations between *Sex* (*male, female*), *length of hair* (*long, short*), and *stature* (<168 cm, ≥168 cm).

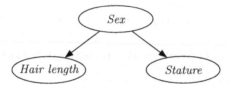

Fig. 2.6. Sex has an impact on length of hair as well as stature.

If we do not know the sex of a person, seeing the length of his/her hair will tell us more about the sex, and this in turn will focus our belief on his/her stature. On the other hand, if we know that the person is a man, then the length of his hair gives us no extra clue on his stature.

Converging Connections

A description of the situation in Figure 2.7 requires a little more care. If nothing is known about A except what may be inferred from knowledge of its parents $B, \ldots, E$, then the parents are independent: evidence about one of them cannot influence the certainties of the others through A. Knowledge of one possible cause of an event does not tell us anything about the other possible causes. However, if anything is known about the consequences, then information on one possible cause may tell us something about the other causes. This is the *explaining away* effect illustrated in the car start problem: the car cannot start, and the potential causes include dirty spark plugs and an empty fuel tank. If we now get the information that there is fuel in the tank, then our certainty in the spark plugs being dirty will increase (since this will explain why the car cannot start). Conversely, if we get the information that there is no fuel on the car, then our certainty in the spark plugs being dirty will decrease (since the lack of fuel explains why the car cannot start). In Figure 2.8, two examples are shown. Observe that in the second example we observe only A indirectly through information about F; knowing the state of F tells us something about the state of E, which in turn tells us something about A.

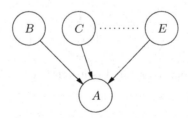

Fig. 2.7. Converging connection. If A changes certainty, it opens communication between its parents.

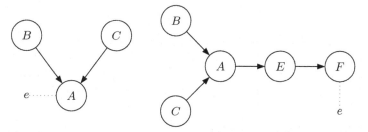

Fig. 2.8. Examples in which the parents of A are dependent. The dotted lines indicate insertion of evidence.

The conclusion is that evidence may be transmitted through a converging connection only if either the variable in the connection or one of its descendants has received evidence.

Remark: Evidence about a variable is a statement of the certainties of its states. If the variable is instantiated, we call it *hard* evidence; otherwise, it is called *soft*. In the example above, we can say that hard evidence about the variable F provides soft evidence about the variable A. Blocking in the case of serial and diverging connections requires hard evidence, whereas opening in the case of converging connections holds for all kinds of evidence.

Example 2.4. Figure 2.9 shows the causal relations among *Salmonella* infection, *flu*, *nausea*, and *pallor*.

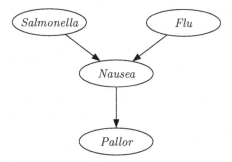

Fig. 2.9. Salmonella and flu may cause nausea, which in turn causes pallor.

If we know nothing of nausea or pallor, then the information on whether the person has a *Salmonella* infection will not tell us anything about flu. However, if we have noticed that the person is pale, then the information that he/she does not have a *Salmonella* infection will make us more ready to believe that he/she has the flu.

2.2.1 d-separation

The three preceding cases cover all ways in which evidence may be transmitted through a variable, and following the rules it is possible to decide for any pair of variables in a causal network whether they are independent given the evidence entered into the network. The rules are formulated in the following definition.

Definition 2.1 (d-separation). *Two distinct variables A and B in a causal network are d-separated ("d" for "directed graph") if for all paths between A and B, there is an intermediate variable V (distinct from A and B) such that either*

- *the connection is serial or diverging and V is instantiated*
  or
- *the connection is converging, and neither V nor any of V's descendants have received evidence.*

If A and B are not d-separated, we call them d-connected.

Figure 2.10 gives an example of a larger network. The evidence entered at B and M represents instantiations. If evidence is entered at A, it may be transmitted to D. The variable B is blocked, so the evidence cannot pass through B to E. However, it may be passed to H and K. Since the child M of K has received evidence, evidence from H may pass to I and further to E, C, F, J, and L, so the path $A - D - H - K - I - E - C - F - J - L$ is a d-connecting path. Figure 2.11 gives two other examples.

Note that although A and B are d-connected, changes in the belief in A will not necessarily change the belief in B. To stress this difference, we will sometimes say that A and B are *structurally independent* if they are d-separated (see also Exercise 2.23).

In connection to d-separation, a special set of nodes for a node A is the so-called *Markov blanket* for A:

Definition 2.2. *The* Markov blanket *of a variable A is the set consisting of the parents of A, the children of A, and the variables sharing a child with A.*

The Markov blanket has the property that when instantiated, A is d-separated from the rest of the network (see Figure 2.12).

You may wonder why we have introduced d-separation as a definition rather than as a theorem. A theorem should be as follows.

Claim: If A and B are d-separated, then changes in the certainty of A have no impact on the certainty of B.

However, the claim cannot be established as a theorem without a more-precise description of the concept of "certainty." You can take d-separation as a property of human reasoning and require that any certainty calculus should comply with the claim.

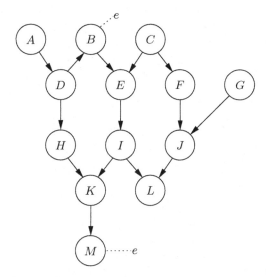

Fig. 2.10. A causal network with M and B instantiated. The node A is d-separated from G only.

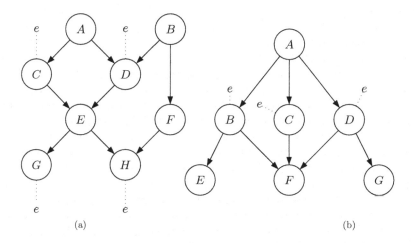

Fig. 2.11. Causal networks with hard evidence entered (the variables are instantiated). (a) Although all neighbors of E are instantiated, it is d-connected to F, B, and A. (b) F is d-separated from the remaining uninstantiated variables.

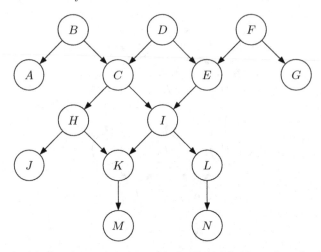

Fig. 2.12. The Markov blanket for I is $\{C, E, H, K, L\}$. Note that if only I's neighbors are instantiated, then J is not d-separated from I.

From the definition of d-separation we see that in order to test whether two variables, say A and B, are d-separated given hard evidence on a set of variables C you would have to check whether all paths connecting A and B are d-separating paths. An easier way of performing this test, without having to consider the various types of connections, is as follows: First you construct the so-called *ancestral graph* consisting of A, B, and C together with all nodes from which there is a directed path to either A, B, or C (see Figure 2.13(a)). Next, you insert an undirected link between each pair of nodes with a common child and then you make all links undirected. The resulting graph (see Figure 2.13(b)) is known as the *moral graph* for Figure 2.13(a). The moral graph can now be used to check whether A and B are d-separated given C: if all paths connecting A and B intersect C, then A and B are d-separated given C.

The above procedure generalizes straightforwardly to the case in which we work with sets of variables rather than single variables: you just construct the ancestral graph using these sets of variables and perform the same steps as above: $\mathcal{A}$ and $\mathcal{B}$ are then d-separated given $\mathcal{C}$ if all paths connecting a variable in $\mathcal{A}$ with a variable in $\mathcal{B}$ intersect a variable in $\mathcal{C}$.

2.3 Bayesian Networks

2.3.1 Definition of Bayesian Networks

Causal relations also have a quantitative side, namely their *strength*. This can be expressed by attaching numbers to the links.

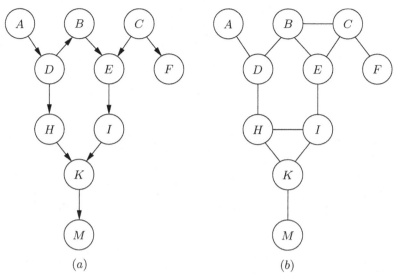

(a)   (b)

Fig. 2.13. To test whether A is d-separated from F given evidence on B and M in Figure 2.10, we first construct the ancestral graph for $\{A, B, F, M\}$ (figure (a)). Next we add an undirected link between pairs of nodes with a common child and then the direction is dropped on all links (figure (b)). In the resulting graph we have that the path $A - D - H - K - I - E - C - F$ does not intersect B and M, hence A and F are d-connected given B and M.

Let A be a parent of B. Using probability calculus, it would be natural to let $P(B \mid A)$ be the strength of the link. However, if C is also a parent of B, then the two conditional probabilities $P(B \mid A)$ and $P(B \mid C)$ alone do not give any clue about how the impacts from A and C interact. They may cooperate or counteract in various ways, so we need a specification of $P(B \mid A, C)$.

It may happen that the domain to be modeled contains causal feedback cycles (see Figure 2.14).

Feedback cycles are difficult to model quantitatively. For causal networks, no calculus has been developed that can cope with feedback cycles, but certain noncausal models have been proposed to deal with this issue. For Bayesian networks we require that the network does not contain cycles.

Definition 2.3. *A Bayesian network consists of the following:*

- *A set of variables[1] and a set of directed edges between variables.*
- *Each variable has a finite set of mutually exclusive states.*
- *The variables together with the directed edges form an* acyclic directed graph *(traditionally abbreviated DAG); a directed graph is* acyclic *if there is no directed path $A_1 \rightarrow \cdots \rightarrow A_n$ so that $A_1 = A_n$.*

---

[1] When we wish to emphasize that this kind of variable represents a sample space we call it a *chance variable*.

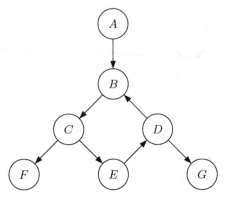

Fig. 2.14. A directed graph with a feedback cycle. This is not allowed in Bayesian networks.

— *To each variable A with parents $B_1, \ldots, B_n$, a conditional probability table $P(A \mid B_1, \ldots, B_n)$ is attached.*

Note that if A has no parents, then the table reduces to the unconditional probability table $P(A)$. For the DAG in Figure 2.15, the prior probabilities $P(A)$ and $P(B)$ must be specified. It has been claimed that prior probabilities are an unwanted introduction of bias to the model, and calculi have been invented in order to avoid it. However, as discussed in Section 2.1.2, prior probabilities are necessary not for mathematical reasons but because prior certainty assessments are an integral part of human reasoning about certainty (see also Exercise 1.12).

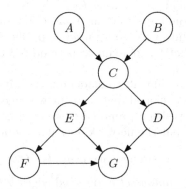

Fig. 2.15. A directed acyclic graph (DAG). The probabilities to specify are $P(A)$, $P(B)$, $P(C \mid A, B)$, $P(E \mid C)$, $P(D \mid C)$, $P(F \mid E)$, and $P(G \mid D, E, F)$.

The definition of Bayesian networks does not refer to causality, and there is no requirement that the links represent causal impact. That is, when building the structure of a Bayesian network model, we need not insist on having the links go in a causal direction. However, we then need to check the model's d-separation properties and ensure that they correspond to our perception of the world's conditional independence properties. The model should not include conditional independences that do not hold in the real world.

This also means that if A and B are d-separated given evidence e, then the probability calculus used for Bayesian networks must yield $P(A \mid e) = P(A \mid B, e)$ (see Section 2.3.2).

Example 2.5 (A Bayesian network for the Car Start Problem).
The Bayesian network for the reduced Car Start Problem is the one in Figure 2.16.

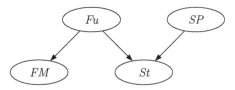

Fig. 2.16. The causal network for the reduced car start problem. We have used the abbreviations *Fu (Fuel?)*, *SP (Clean Spark Plugs?)*, *St (Start?)*, and *FM (Fuel Meter Standing)*.

For the quantitative modeling, we need the probability assessments $P(Fu)$, $P(SP), P(St \mid Fu, SP), P(FM \mid Fu)$. To avoid having to deal with numbers that are too small, let $P(Fu) = (0.98, 0.02)$ and $P(SP) = (0.96, 0.04)$. The remaining tables are given in Table 2.1. Note that the table for $P(FM \mid Fu)$ reflects the fact that the fuel meter may be malfunctioning, and the table for $P(St \mid Fu, SP)$ leaves room for causes other than no fuel and dirty spark plugs by assigning $P(St = no \mid Fu = yes, SP = yes) > 0$.

2.3.2 The Chain Rule for Bayesian Networks

Let $\mathcal{U} = \{A_1, \ldots, A_n\}$ be a universe of variables. If we have access to the joint probability table $P(\mathcal{U}) = P(A_1, \ldots, A_n)$, then we can also calculate $P(A_i)$ as well as $P(A_i \mid e)$, where e is evidence about some of the variables in the Bayesian network (see, e.g., Section 1.3.1). However, $P(\mathcal{U})$ grows exponentially with the number of variables, and $\mathcal{U}$ need not be very large before the table becomes intractably large. Therefore, we look for a more compact *representation* of $P(\mathcal{U})$, i.e., a way of storing information from which $P(\mathcal{U})$ can be calculated if needed.

Let BN be a Bayesian network over $\mathcal{U}$, and let $P(\mathcal{U})$ be a probability distribution reflecting the properties specified by BN: (i) the conditional probabilities for a variable given its parents in $P(\mathcal{U})$ must be as specified in BN, and (ii) if the variables A and B are d-separated in BN given the set $\mathcal{C}$, then A and B are independent given $\mathcal{C}$ in $P(\mathcal{U})$.

	$Fu = yes$	$Fu = no$
$FM = full$	0.39	0.001
$FM = \frac{1}{2}$	0.60	0.001
$FM = empty$	0.01	0.998

$$P(FM \mid Fu)$$

	$Fu = yes$	$Fu = no$
$Sp = yes$	(0.99, 0.01)	(0,1)
$Sp = no$	(0.01, 0.99)	(0,1)

$$P(St \mid Fu, Sp)$$

Table 2.1. Conditional probabilities for the model in Figure 2.16. The numbers (x, y) in the lower table represent $(St = yes, St = no)$.

Based on these two properties, what other properties can be deduced about $P(\mathcal{U})$? If the universe consists of only one variable A, then BN specifies $P(A)$, and $P(\mathcal{U})$ is uniquely determined. We shall show that this holds in general.

For probability distributions over sets of variables, we have an equation called *the chain rule*. For Bayesian networks this equation has a special form. First we state the general chain rule:

Proposition 2.1 (The general chain rule). *Let $\mathcal{U} = \{A_1, \ldots, A_n\}$ be a set of variables. Then for any probability distribution $P(\mathcal{U})$ we have*

$$P(\mathcal{U}) = P(A_n \mid A_1, \ldots, A_{n-1})P(A_{n-1} \mid A_1, \ldots, A_{n-2}) \ldots P(A_2 \mid A_1)P(A_1).$$

Proof. Iterative use of the fundamental rule:

$$P(\mathcal{U}) = P(A_n \mid A_1, \ldots, A_{n-1})P(A_1, \ldots, A_{n-1}),$$
$$P(A_1, \ldots, A_{n-1}) = P(A_{n-1} \mid A_1, \ldots, A_{n-2})P(A_1, \ldots, A_{n-2}),$$
$$\vdots$$
$$P(A_1, A_2) = P(A_2 \mid A_1)P(A_1).$$

$\square$

Theorem 2.1 (The chain rule for Bayesian networks). *Let BN be a Bayesian network over $\mathcal{U} = \{A_1, \ldots, A_n\}$. Then BN specifies a unique joint probability distribution $P(\mathcal{U})$ given by the product of all conditional probability tables specified in BN:*

$$P(\mathcal{U}) = \prod_{i=1}^{n} P(A_i \mid \mathrm{pa}(A_i)),$$

where pa(A_i) *are the parents of* A_i *in BN, and* $P(\mathcal{U})$ *reflects the properties of BN.*

Proof.   First we should show that $P(\mathcal{U})$ is indeed a probability distribution. That is, we need to show that Axioms 1–3 hold. This is left as an exercise (see Exercise 2.15).

Next we prove that the specification of BN is consistent, so that $P(\mathcal{U})$ reflects the properties of BN. It is not hard to prove that the probability distribution specified by the product in the chain rule reflects the conditional probabilities from BN (see Exercise 2.16). We also need to prove that the product reflects the d-separation properties. This is done through induction in the number of variables in BN.

When BN has one variable, it is obvious that the d-separation properties specified by BN hold for the product of all specified conditional probabilities.

Assume that for any Bayesian network with $n-1$ variables and a distribution $P(\mathcal{U})$ specified as the product of all conditional probabilities, it holds that if A and B are d-separated given $\mathcal{C}$, then $P(A \mid B, \mathcal{C}) = P(A \mid \mathcal{C})$. Let BN be a Bayesian network with n variables $\{A_1, \ldots, A_n\}$. Assume that A_n has no children and let BN' be the result of removing A_n from BN. Clearly BN' is a Bayesian network with the same conditional probability distributions as BN (except for A_n) and with the same d-separation properties over $\{A_1, \ldots, A_{n-1}\}$ as BN. Moreover,

$$P(\mathcal{U} \setminus \{A_n\}) = \sum_{A_n} P(\mathcal{U}) = \sum_{A_n} \prod_{i=1}^{n} P(A_i \mid \mathrm{pa}(A_i))$$

$$= \prod_{i=1}^{n-1} P(A_i \mid \mathrm{pa}(A_i)) \sum_{A_n} P(A_n \mid \mathrm{pa}(A_n))$$

$$= \prod_{i=1}^{n-1} P(A_i \mid \mathrm{pa}(A_i)) \mathbf{1} = \prod_{i=1}^{n-1} P(A_i \mid \mathrm{pa}(A_i)),$$

and by the induction hypothesis $P(\mathcal{U} \setminus \{A_n\})$ reflects the properties of BN'. Now, if A and B are d-separated given $\mathcal{C}$ in BN, then they are also d-separated in BN', and therefore $P(A \mid B, \mathcal{C}) = P(A \mid \mathcal{C})$. To prove that it also holds for d-separation properties involving A_n, we consider the case in which $A_n \in \mathcal{C}$ and the case in which $A = A_n$. For the first case we have that since A_n participates only in a converging connection, it holds that if A and B are d-separated given $\mathcal{C}$, then they are also d-separated given $\mathcal{C} \setminus \{A_n\}$ and we get the situation above. For the second case, we first note that

$$P(A_n \mid B, \mathcal{C}) = \sum_{\mathrm{pa}(A_n)} P(A_n \mid B, \mathcal{C}, \mathrm{pa}(A_n)) P(\mathrm{pa}(A_n) \mid B, \mathcal{C}).$$

Now, if A_n and B are d-separated given $\mathcal{C}$, then pa(A_n) and B are also d-separated given $\mathcal{C}$, and since A_n is not involved, we have $P(\mathrm{pa}(A_n) \mid B, \mathcal{C}) =$

$P(\mathrm{pa}(A_n)\,|\,\mathcal{C})$. So we need to prove only that $P(A_n\,|\,B,\mathcal{C},\mathrm{pa}(A_n)) = P(A_n\,|\,\mathrm{pa}(A_n))$. Using the fundamental rule and the chain rule, we get

$$
P(A_n\,|\,B,\mathcal{C},\mathrm{pa}(A_n)) = \frac{P(A_n,B,\mathcal{C},\mathrm{pa}(A_n))}{P(B,\mathcal{C},\mathrm{pa}(A_n))} = \frac{\sum_{\mathcal{U}\backslash\{A_n,B,\mathcal{C},\mathrm{pa}(A_n)\}} P(\mathcal{U})}{\sum_{\mathcal{U}\backslash\{B,\mathcal{C},\mathrm{pa}(A_n)\}} P(\mathcal{U})}
$$

$$
= \frac{\sum_{\mathcal{U}\backslash\{A_n,B,\mathcal{C},\mathrm{pa}(A_n)\}} \prod_{i=1}^{n} P(A_i\,|\,\mathrm{pa}(A_i))}{\sum_{\mathcal{U}\backslash\{B,\mathcal{C},\mathrm{pa}(A_n)\}} \prod_{i=1}^{n} P(A_i\,|\,\mathrm{pa}(A_i))}
$$

$$
= \frac{P(A_n\,|\,\mathrm{pa}(A_n)) \sum_{\mathcal{U}\backslash\{A_n,B,\mathcal{C},\mathrm{pa}(A_n)\}} \prod_{i=1}^{n-1} P(A_i\,|\,\mathrm{pa}(A_i))}{\sum_{\mathcal{U}\backslash\{A_n,B,\mathcal{C},\mathrm{pa}(A_n)\}} \prod_{i=1}^{n-1} P(A_i\,|\,\mathrm{pa}(A_i)) \sum_{A_n} P(A_n\,|\,\mathrm{pa}(A_n))}
$$

$$
= \frac{P(A_n\,|\,\mathrm{pa}(A_n)) \sum_{\mathcal{U}\backslash\{A_n,B,\mathcal{C},\mathrm{pa}(A_n)\}} \prod_{i=1}^{n-1} P(A_i\,|\,\mathrm{pa}(A_i))}{\sum_{\mathcal{U}\backslash\{A_n,B,\mathcal{C},\mathrm{pa}(A_n)\}} \prod_{i=1}^{n-1} P(A_i\,|\,\mathrm{pa}(A_i))\mathbf{1}}
$$

$$
= P(A_n\,|\,\mathrm{pa}(A_n)).
$$

To prove uniqueness, let $\{A_1,\ldots,A_n\}$ be a topological ordering of the variables. Then, for each variable A_i with parents $\mathrm{pa}(A_i)$ we have that A_i is d-separated from $\{A_1,\ldots,A_{i-1}\} \backslash \mathrm{pa}(A_i)$ given $\mathrm{pa}(A_i)$ (see Exercise 2.11). This means that for any distribution P reflecting the specifications by BN we must have $P(A_i\,|\,A_1,\ldots,A_{i-1}) = P(A_i\,|\,\mathrm{pa}(A_i))$. Substituting this in the general chain rule yields that any distribution reflecting the specifications by BN must be the product of the conditional probabilities specified in BN. $\square$

The chain rule yields that a Bayesian network is a compact representation of a joint probability distribution. The following example illustrates how to exploit that for reasoning under uncertainty.

Example 2.6 (The Car Start Problem revisited).

In this example, we apply the rules of probability calculus to the Car Start Problem. This is done to illustrate that probability calculus can be used to perform the reasoning in the example, in particular, explaining away. In Chapter 4, we give general algorithms for probability updating in Bayesian networks. We will use the Bayesian network from Example 2.5 to perform the reasoning in Section 2.1.1.

We will use the joint probability table for the reasoning. The joint probability table is calculated from the chain rule for Bayesian networks,

$$
P(\mathit{Fu}, \mathit{FM}, \mathit{SP}, \mathit{St}) = P(\mathit{Fu})P(\mathit{SP})P(\mathit{FM}\,|\,\mathit{Fu})P(\mathit{St}\,|\,\mathit{Fu}, \mathit{SP}).
$$

The result is given in Tables 2.2 and 2.3.

The evidence $\mathit{St} = \mathit{no}$ tells us that we are in the context of Table 2.3. By marginalizing FM and Fu out of Table 2.3 (summing each row), we get

$$
P(\mathit{SP}, \mathit{St} = \mathit{no}) = (0.02864, 0.03965).
$$

	$FM = full$	$FM = \frac{1}{2}$	$FM = empty$
$Sp = yes$	$(0.363, 0)$	$(0.559, 0)$	$(0.0093, 0)$
$Sp = no$	$(0.00015, 0)$	$(0.00024, 0)$	$(3.9 \cdot 10^{-6}, 0)$

Table 2.2. The joint probability table for $P(Fu, FM, SP, St = yes)$.

	$FM = full$	$FM = \frac{1}{2}$	$FM = empty$
$Sp = yes$	$(0.00367, 1.9 \cdot 10^{-5})$	$(0.00564, 1.9 \cdot 10^{-5})$	$(9.4 \cdot 10^{-5}, 0.0192)$
$Sp = no$	$(0.01514, 8 \cdot 10^{-7})$	$(0.0233, 8 \cdot 10^{-7})$	$(0.000388, 0.000798)$

Table 2.3. The joint probability table for $P(Fu, FM, SP, St = no)$. The numbers (x, y) in the table represent $(Fu = yes, Fu = no)$.

We get the conditional probability $P(SP \mid St = no)$ by dividing by $P(St = no)$. This is easy, since $P(St = no) = P(SP = yes, St = no) + P(SP = no, St = no) = 0.02864 + 0.03965 = 0.06829$, and we get

$$P(SP \mid St = no) = \left(\frac{0.02864}{0.06829}, \frac{0.03965}{0.06829} \right) = (0.42, 0.58).$$

Another way of saying this is that the distribution we end up with will be a set of numbers that sum to 1. If they do not, normalize by dividing by the sum.

In the same way, we get $P(Fu \mid St = no) = (0.71, 0.29)$.

Next, we get the information that $FM = \frac{1}{2}$, and the context for calculation is limited to the part with $FM = \frac{1}{2}$ and $St = no$. The numbers are given in Table 2.4.

	$Fu = yes$	$Fu = no$
$Sp = yes$	0.00564	$1.9 \cdot 10^{-5}$
$Sp = no$	0.0233	$8 \cdot 10^{-7}$

Table 2.4. $P(Fu, SP, St = no, FM = \frac{1}{2})$.

By marginalizing Sp out and normalizing, we get $P(Fu \mid St = no, FM = \frac{1}{2}) = (0.999, 0.001)$, and by marginalizing Fu out and normalizing we get $P(SP \mid St = no, FM = \frac{1}{2}) = (0.196, 0.804)$. The probability of $SP = yes$ increased by observing $FM = \frac{1}{2}$, so the calculus did catch the explaining away effect.

2.3.3 Inserting Evidence

Bayesian networks are used for calculating new probabilities when you get new information. The information so far has been of the type "$A = a$," where A is a variable and a is a state of A. Let A have n states with $P(A) = (x_1, \ldots, x_n)$, and assume that we get the information e that A can be only in state i or j. This statement expresses that all states except i and j are impossible, and we have the probability distribution $P(A, e) = (0, \ldots, 0, x_i, 0, \ldots, 0, x_j, 0, \ldots, 0)$. Note that $P(e)$, the prior probability of e, is obtained by marginalizing A out of $P(A, e)$. Note also that $P(A, e)$ is the result of multiplying $P(A)$ by $(0, \ldots, 0, 1, 0, \ldots, 0, 1, 0, \ldots, 0)$, where the 1's are at the i'th and j'th places.

Definition 2.4. *Let A be a variable with n states. A finding on A is an n-dimensional table of zeros and ones.*

To distinguish between the statement e, "A is in either state i or j," and the corresponding 0/1-finding vector, we sometimes use the boldface notation $\mathbf{e}$ for the finding. Semantically, a finding is a statement that certain states of A are impossible.

Now, assume that you have a joint probability table, $P(\mathcal{U})$, and let $\mathbf{e}$ be the preceding finding. The joint probability table $P(\mathcal{U}, e)$ is the table obtained from $P(\mathcal{U})$ by replacing all entries with A not in state i or j by the value zero and leaving the other entries unchanged. This is the same as multiplying $P(\mathcal{U})$ by $\mathbf{e}$,

$$P(\mathcal{U}, e) = P(\mathcal{U}) \cdot \mathbf{e}.$$

Note that $P(e) = \sum_U P(\mathcal{U}, e) = \sum_U (P(\mathcal{U}) \cdot \mathbf{e})$. Using the chain rule for Bayesian networks, we have the following theorem.

Theorem 2.2. *Let BN be a Bayesian network over the universe $\mathcal{U}$, and let $\mathbf{e}_1, \ldots, \mathbf{e}_m$ be findings. Then*

$$P(\mathcal{U}, e) = \prod_{A \in \mathcal{U}} P(A \mid pa(A)) \cdot \prod_{i=1}^{m} \mathbf{e}_i,$$

and for $A \in \mathcal{U}$ we have

$$P(A \mid e) = \frac{\sum_{\mathcal{U} \setminus \{A\}} P(\mathcal{U}, e)}{P(e)}.$$

Some types of evidence cannot be represented as findings. You may, for example, receive a statement from someone that the chance of A being in state a_1 is twice as high as for a_2. This type of evidence is called *likelihood evidence*. It is possible to treat this kind of evidence in Bayesian networks. The preceding statement is then represented by the distribution $(0.67, 0.33)$, and Theorem 2.2 still holds. However, because it is unclear what it means that a likelihood statement is true, $P(e)$ cannot be interpreted as the probability of the evidence, and $P(\mathcal{U}, e)$ therefore has an unclear semantics. We will not deal further with likelihood evidence.

2.3.4 Calculating Probabilities in Practice

As described in Section 2.3.3 and illustrated in Example 2.6, probability up-dating in Bayesian networks can be performed using the chain rule to calculate $P(\mathcal{U})$, the joint probability table of the universe. However, $\mathcal{U}$ need not be large before $P(\mathcal{U})$ becomes intractably large. In this section, we illustrate how the calculations can be performed without having to deal with the full joint ta-ble. In Chapter 4, we give a detailed treatment of algorithms for probability updating.

Consider the Bayesian network in Figure 2.17, and assume that all vari-ables have ten states. Assume that we have the evidence $e = \{D = d, F = f\}$, and we wish to calculate $P(A \mid e)$.

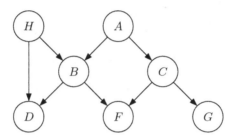

Fig. 2.17. A Bayesian network.

From the chain rule we have

$$P(\mathcal{U}, e) = P(A, B, C, d, f, G, H)$$
$$= P(A)P(H)P(B \mid A, H)P(C \mid A)P(d \mid B, H)P(f \mid B, C)P(G \mid C),$$

where for example $P(d \mid B, H)$ denotes the table over B and H resulting from fixing the D-entry to the state d. We say that the conditional probability table has been *instantiated* to $D = d$. Notice that we need not calculate the full table $P(\mathcal{U})$ with 10^7 entries. If we wait until evidence is entered, we will in this case need to work with a table with only 10^5 entries. Later, we see that we need not work with tables larger than 1000 entries.

To calculate $P(A, e)$, we marginalize the variables B, C, G, and H out of $P(A, B, C, d, f, G, H)$. The order in which we marginalize does not affect the result (Section 1.4), so let us start with G; that is, we wish to calculate

$$\sum_G P(A, B, C, d, f, G, H)$$
$$= \sum_G P(A)P(H)P(B \mid A, H)P(C \mid A)P(d \mid B, H)P(f \mid B, C)P(G \mid C).$$

In the right-hand product, only the last table contains G in its domain, and due to the distributive law (Section 1.4) we have

$$\sum_G P(A, B, C, d, f, G, H)$$
$$= P(A)P(H)P(B \mid A, H)P(C \mid A)P(d \mid B, H)P(f \mid B, C) \sum_G P(G \mid C),$$

and we need only calculate $\sum_G P(G \mid C)$. Actually, for each state c of C, we have $\sum_G P(G \mid c) = 1$; hence no calculations are necessary. We therefore get

$$P(A, B, C, d, f, H) = \sum_G P(A, B, C, d, f, G, H)$$
$$= P(A)P(H)P(B \mid A, H)P(C \mid A)P(d \mid B, H)P(f \mid B, C).$$

Next, we marginalize H out. Using the distributive law again, we get

$$\sum_H P(A, B, C, d, f, H)$$
$$= P(A)P(C \mid A)P(f \mid B, C) \sum_H P(H)P(B \mid A, H)P(d \mid B, H).$$

We multiply the three tables $P(H), P(B \mid A, H)$, and $P(d \mid B, H)$, and we marginalize H out of the product. The result is a table $T(d, B, A)$, and we have

$$P(A, B, C, d, f) = P(A)P(C \mid A)P(f \mid B, C)T(d, B, A).$$

Finally, we calculate this product and marginalize B and C out of it.

Notice that we never work with a table of more than three variables (the table produced by multiplying $P(H), P(B \mid A, H)$, and $P(d \mid B, H)$) compared to the five variables in $P(A, B, C, d, f, G, H)$.

The method we just used is called *variable elimination* and can be described in the following way: we start with a set $\mathcal{T}$ of tables, and whenever we wish to marginalize a variable X, we take from $\mathcal{T}$ all tables with X in their domains, calculate the product of them, marginalize X out of it, and place the resulting table in $\mathcal{T}$.

2.4 Graphical Models – Formal Languages for Model Specification

From a mathematical point of view, the basic property of Bayesian networks is the chain rule: a Bayesian network is a compact representation of the joint probability table over its universe. In this respect, a Bayesian network is one type of compact representation among many others. However, there is more to it than this: From a knowledge engineering point of view, a Bayesian network is a type of *graphical model*. The structure of the network is formulated in a graphical communication language for which the language features have a very simple semantics, namely causality. This does not mean that "causality" is an easy concept. It may be very difficult to experience causality, and philosophically the concept is not fully understood. However, most often humans can communicate sensibly about causal relations in a knowledge domain. Furthermore, the graphical specification also specifies the requirements for the quantitative part of the model (the conditional probabilities). In Chapter 3, we extend the modeling language, and in Part II we present other types of graphical models.

As mentioned, graphical models are communication languages. They consist of a qualitative part, where features from graph theory are used, and a quantitative part consisting of *potentials*, which are real-valued functions over sets of nodes from the graph; in Bayesian networks the potentials are conditional probability tables. The graphical part specifies the kind of potentials and their domains.

Graphical models can be used for interpersonal communication: The graphical specification is easy for humans to read, and it helps focus attention, for example in a group working jointly on building a model. For interpersonal communication, the semantics of the various graph-theoretic features must be rather welldefined if misunderstandings are to be avoided.

The next step in the use of graphical models has to do with communication to a computer. You wish to communicate a graphical model to a computer, and the computer should be able to process the model and give answers to various queries. In order to achieve this, the specification language must be formally defined with a well-defined syntax and semantics.

The first concern in constructing a graphical modeling language is to ensure that it is sufficiently welldefined so that it can be communicated to a computer. This covers the graphical part as well as the specification of potentials. The next concern is the scope of the language: what is the range of domains and tasks that you will be able to model with this language? The final concern is tractability: do you have algorithms such that in reasonable time the computer can process a model and query to provide answers?

The Bayesian network is a sufficiently welldefined language, and behind the graphical specification in the user interface, the computer systems for processing Bayesian networks have an alphanumeric specification language, which for some systems is open to the user. Actually, the language for Bayesian networks is a context-free language with a single context-sensitive aspect (no directed cycles).

The scope of the Bayesian network language is hard to define, but the examples in the next chapter show that it has a very broad scope.

Tractability is not a yes or no issue. As described in Chapter 4, there are algorithms for probability updating in Bayesian networks, but basically probability updating is NP-hard. This means that some models have an updating time exponential in the number of nodes.

On the other hand, the running times of the algorithms can be easily calculated without actually running them. In Chapter 4 and Part II, we treat complexity issues for the various graphical languages presented.

2.5 Summary

d-Separation in Causal Networks

Two distinct variables A and B in a causal network are d-separated if for all paths between A and B, there is an intermediate variable V (distinct from A and B) such that either

- the connection is serial or diverging, and V is instantiated, or
- the connection is converging, and neither V nor any of V's descendants have received evidence.

Definition of Bayesian Networks

A Bayesian network consists of the following:

- There is a set of *variables* and a set of *directed edges* between variables.
- Each variable has a finite set of mutually exclusive states.
- The variables together with the directed edges form an *acyclic directed graph* (DAG).
- To each variable A with parents $B_1, \ldots, B_n$ there is attached a conditional probability table $P(A \mid B_1, \ldots, B_n)$.

The Chain Rule for Bayesian Networks

Let BN be a Bayesian network over $\mathcal{U} = \{A_1, \ldots, A_n\}$. Then BN specifies a unique joint probability distribution $P(\mathcal{U})$ given by the product of all conditional probability tables specified in BN:

$$P(\mathcal{U}) = \prod_{i=1}^{n} P(A_i \mid \mathrm{pa}(A_i)),$$

where $\mathrm{pa}(A_i)$ are the parents of A_i in BN, and $P(\mathcal{U})$ reflects the properties of BN.

Admittance of d-Separation in Bayesian Networks

If A and B are d-separated in a Bayesian network with evidence e entered, then $P(A \mid B, e) = P(A \mid e)$.

Inserting Evidence

Let $e_1, \ldots, e_m$ be findings, and then

$$P(\mathcal{U}, e) = \prod_{i=1}^{n} P(A_i \mid \mathrm{pa}(A_i)) \prod_{j=1}^{m} e_j$$

and

$$P(A \mid e) = \frac{\sum_{\mathcal{U} \setminus \{A\}} P(\mathcal{U}, e)}{P(e)}.$$

2.6 Bibliographical Notes

The connection between causation and conditional independence was studied by Spohn (1980), and later investigated with special focus on Bayesian networks in (Pearl, 2000). The concepts of causal network, d-connection, and the definition in Section 2.2.1 are due to Pearl (1986) and Verma (1987). A proof that Bayesian networks admit d-separation can be found in (Pearl, 1988) or in (Lauritzen, 1996). Geiger and Pearl (1988) proved that d-separation is the correct criterion for directed graphical models, in the sense that for any DAG, a probability distribution can be found for which the d-separation criterion is sound and complete. Meek (1995) furthermore proved that for a given DAG, the set of discrete probability distributions for which the d-separation criterion is not complete has measure zero. That is, given a random Bayesian network, there is almost no chance that it contains conditionally independent variables that cannot be read off the graph by d-separation. The method for discovering d-separation properties using ancestral graphs was first presented in (Lauritzen et al., 1990).

Bayesian networks have a long history in statistics, and can be traced back at least to the work in (Minsky, 1963). In the first half of the 1980s they were introduced to the field of expert systems through work by Pearl (1982) and Spiegelhalter and Knill-Jones (1984). Some of the first real-world applications of Bayesian networks were Munin (Andreassen et al., 1989, 1992) and Pathfinder (Heckerman et al., 1992). The basis for the inference method presented in Section 2.3.4 originates from (D'Ambrosio, 1991) and was modified to the presented variable elimination in (Dechter, 1996). The fact that inference is NP-hard was proved in (Cooper, 1987).

2.7 Exercises

Exercise 2.1. To illustrate that simple rules cannot cope with uncertainty reasoning, consider the following two cases:

(i) I have an urn with a red ball and a white ball in it. If I add a red ball and shake it, what is the certainty of drawing a red ball in one draw? If I add a white ball instead, what is the certainty of drawing a red ball? If I combine the two actions, what is the certainty of drawing a red ball?

(ii) When shooting, I am more certain to hit the target if I close the left eye. I am also more certain to hit the target if I close the right eye. What is the combined certainty if I do both?

Exercise 2.2. Construct a causal network and follow the reasoning in the following story. Mr. Holmes is working in his office when he receives a phone call from his neighbor, who tells him that Holmes' burglar alarm has gone off. Convinced that a burglar has broken into his house, Holmes rushes to his car and heads for home. On his way, he listens to the radio, and in the news it is reported that there has been a small earthquake in the area. Knowing that earthquakes have a tendency to turn on burglar alarms, he returns to work.

Exercise 2.3. Consider the Car Start Problem in Section 2.1.1 with the causal network in Figure 2.1, and the following twist on the story: "I distinctly remember visiting the pump last night, so the fuel meter should be reading *full*. Since this is not the case, either there must be a leak in the tank, someone has stolen gasoline during the night, or the fuel meter is malfunctioning. Sniffing the air I smell no gasoline, so I conclude that a thief has been visiting last night or that the fuel meter is malfunctioning." Alter the causal network in Figure 2.1 to incorporate the above twist on the story.

Exercise 2.4. In the graphs in Figures 2.18 and 2.19, determine which variables are d-separated from A.

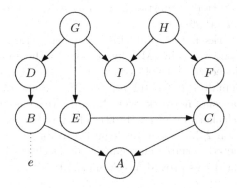

Fig. 2.18. Figure for Exercise 2.4.

Exercise 2.5. For each pair of variables in the causal network in Figure 2.1, state whether the variables can be d-separated, and if so which set(s) of variables that allow this.

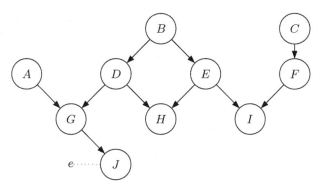

Fig. 2.19. Figure for Exercise 2.4.

Exercise 2.6. Consider the network in Figure 2.20. What are the minimal set(s) of variables required to d-separate C and E (that is, sets of variables for which no proper subset d-separates C and E)? What are the minimal set(s) of variables required to d-separate A and B? What are the maximal set(s) of variables that d-separate C and E (that is, sets of variables for which no proper superset d-separates C and E)? What are the maximal set(s) of variables that d-separate A and B?

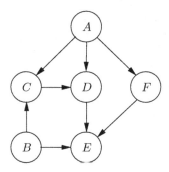

Fig. 2.20. A causal network for Exercise 2.6.

Exercise 2.7. Consider the network in Figure 2.20. What is the Markov blanket of each variable?

Exercise 2.8. Let A be a variable in a DAG. Assume that all variables in A's Markov blanket are instantiated. Show that A is d-separated from the remaining uninstantiated variables.

Exercise 2.9. Apply the procedure using the ancestral graph given in Section 2.2.1 to determine whether A is d-separated from C given B in the network in Figure 2.19.

Exercise 2.10. Let D_1 and D_2 be DAGs over the same variables. The graph D_1 is an *I-submap* of D_2 if all d-separation properties of D_1 also hold for D_2. If D_2 is also an I-submap of D_1, they are said to be *I-equivalent*. Which of the four DAGs in Figure 2.21 are I-equivalent?

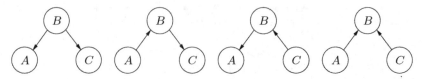

Fig. 2.21. Figure for Exercise 2.10.

Exercise 2.11. Let $\{A_1, \ldots, A_n\}$ be a topological ordering of the variables in a Bayesian network, and consider variable A_i with parents pa(A_i). Prove that A_i is d-separated from $\{A_1, \ldots, A_{i-1}\} \setminus \text{pa}(A_i)$ given $pa(A_i)$.

Exercise 2.12. Consider the network in Figure 2.20. Which conditional probability tables must be specified to turn the graph into a Bayesian network?

Exercise 2.13. In Figure 2.22 the structure of a simple Bayesian network is shown. The accompanying conditional probability tables are shown in Tables 2.5 and 2.6, and the prior probabilities for A are 0.9 and 0.1. Are A and C d-separated given B? Are A and C conditionally independent given B?

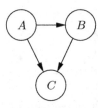

Fig. 2.22. A simple Bayesian network for Exercise 2.13.

	$A = a_1$	$A = a_2$
$B = b_1$	0.3	0.6
$B = b_2$	0.7	0.4

Table 2.5. $P(B \mid A)$.

	$A = a_1$	$A = a_2$
$B = b_1$	(0.1 ; 0.9)	(0.1 ; 0.9)
$B = b_2$	(0.2 ; 0.8)	(0.2 ; 0.8)

Table 2.6. $P(C \mid A, B)$.

Exercise 2.14. Consider the network in Figure 2.20. Using the chain rule, establish an expression for the joint distribution over the universe $\{A, B, C, D, E, F\}$. Use this expression to show that B and D are conditionally independent given A and C.

Exercise 2.15. Prove that the probability distribution $P(\mathcal{U})$ defined by the chain rule for Bayesian networks is indeed a probability distribution.

Exercise 2.16. Prove that the probability distribution $P(\mathcal{U})$ defined by the chain rule for a Bayesian network BN reflects the conditional probabilities specified in BN.

Exercise 2.17. Consider the Bayesian network from Exercise 2.13 and the finding $e = (0, 1)$ over A. What is $P(B, C, e)$?

Exercise 2.18. What steps would be taken if variable elimination were used to calculate the probability table $P(F \mid C = c_1)$ for the network in Figure 2.20? Assuming that each variable has ten states, what is the maximum size of a table during the procedure?

Exercise 2.19. Consider the DAG (a) in Exercise 2.10.

- Show that $P(B \mid A, C) = P(B \mid A)$.
- We have $P(A) = (0.1, 0.9)$ and the conditional probability tables in Table 2.7. Calculate $P(A, B, C)$.

	a_1	a_2
b_1	0.2	0.3
b_2	0.8	0.7

$P(B \mid A)$

	a_1	a_2
c_1	0.5	0.6
c_2	0.5	0.4

$P(C \mid A)$

Table 2.7. Conditional probability tables for Exercise 2.19.

Exercise 2.20. [E] Install an editor for Bayesian networks (a reference to a list of systems can be found in the preface).

Exercise 2.21. [E] Construct a Bayesian network for Exercise 1.12.

Exercise 2.22. [E] Construct a Bayesian network to follow the reasoning from Exercise 2.2. Use your own estimates of probabilities for the network.

Exercise 2.23. [E] Consider the Bayesian network in Figure 2.23 with conditional probabilities given in Table 2.8. Use your system to investigate whether A and C are independent.

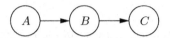

Fig. 2.23. Figure for Exercise 2.23.

	$A = yes$	$A = no$
b_1	0.6	0.2
b_2	0.1	0.5
b_3	0.2	0.1
b_4	0.1	0.2
	$P(B \mid A)$	

	b_1	b_2	b_3	b_4
$C = yes$	0.8	0.8	0.2	0.2
$C = no$	0.2	0.2	0.8	0.8
	$P(C \mid B)$			

Table 2.8. Tables for Exercise 2.23.

Exercise 2.24. [E] Use your system and Section 2.5 to perform the reasoning in Section 2.1.2.

3

Building Models

The framework of Bayesian networks is a very efficient language for building models of domains with inherent uncertainty. However, as can be seen from the calculations in Section 2.6, it is a tedious job to perform evidence transmission even for very simple Bayesian networks. Fortunately, software tools that can do the calculational job for us are available. In the rest of this book, we assume that the reader has access to such a system (some URLs are given in the preface). Therefore, we can start by concentrating on how to use Bayesian networks in model building and defer a presentation of methods for probability updating to Chapter 4.

In Section 3.1, we examine through examples the considerations you may go through when determining the structure of a Bayesian network model. Section 3.2 gives examples of estimation of conditional probabilities. The examples cover theoretically well-founded probabilities as well as probabilities taken from databases and purely subjective estimates. Section 3.3 introduces various modeling tricks to use when the quantity of numbers to acquire is overwhelming. Finally, Section 3.4 considers other types of queries that can be answered by Bayesian networks besides standard probability updating.

3.1 Catching the Structure

The first thing to have in mind when organizing a Bayesian network model is that its purpose is to give estimates of certainties for events that are not directly observable (or observable only at an unacceptable cost), and the primary task in model building is to identify these events. We call them *hypothesis events*. The hypothesis events detected are then grouped into sets of mutually exclusive and exhaustive events to form *hypothesis variable*.

The next thing to have in mind is that in order to come up with a certainty estimate, we should provide some information channels, and the task is to identify the types of achievable information that may reveal something about the hypothesis variables. These types of information are grouped into *information variablesinformation variable*, and a typical piece of information is a statement that a certain variable is in a particular state, but softer statements are also allowed.

Having identified the variables for the model, the next thing will be to establish the directed links for a causal network.

3.1.1 Milk Test

> Milk from a cow may be infected. To detect whether the milk is infected, you have a test, which may give either a *positive* or a *negative* test result. The test is not perfect. It may give a positive result on clean milk as well as a negative result on infected milk.

We have two hypothesis events: *milk infected* and *milk not infected*, and because they are mutually exclusive and exhaustive, they are grouped into the variable *Infected?* with the states *yes* and *no*. A possible information source is the test results, which can be either *positive* or *negative*. For this, we establish the variable *Test* with states *pos* and *neg*.

The causal direction between the two variables is from *Infected?* to *Test* (see Figure 3.1).

Fig. 3.1. The Bayesian network for the milk test.

Warning: Certainly, no sensible person will claim that a positive test result may infect the milk. However, our reasoning is often performed in the diagnostic direction, and in more complex situations you may therefore be tempted to wrongly direct the link from "symptom" to "disease."

From one day to another, the state of the milk can change. Cows with infected milk will heal over time, and a clean cow has a risk of having infected milk the next day. Now, imagine that the farmer performs the test each day. After a week, he has not only the current test result but also the six previous test results. For each day, we have a model like the one in Figure 3.1. These seven models should be connected such that past knowledge can be used for the current conclusion. A natural way would be to let the state of the milk yesterday have an impact on the state today. This yields the model in Figure 3.2.

The model in Figure 3.2 contains a set of hidden assumptions, which can be read from the d-separation properties.

First, the model assumes the *Markov property*: if we know the present, then the past has no influence on the future. In the language of d-separation, the assumption is that, for example, Inf_{i-1} is d-separated from Inf_{i+1} given

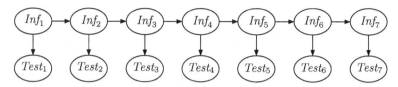

Fig. 3.2. A seven-day model for the milk test.

Inf_i. If we know that the milk on day four is infected, then this can be used to forecast the probability that the milk will be infected on day five. This forecast will not be improved by knowing that the milk was not infected on day three. For various diseases, such an assumption will not be valid. Some diseases have a natural span of time. For example, if I have the flu today but was healthy yesterday, then I will most probably have the flu the day after tomorrow. On the other hand, if I have had the flu for four days, then there is a good chance that I will be cured the day after tomorrow. If the Markov property of Figure 3.2 does not reflect reality, the model should be changed. For example, it may be argued that you also need to go an extra day back, and the model will be as in Figure 3.3.

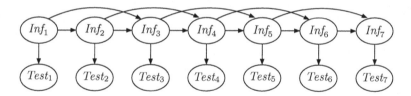

Fig. 3.3. A seven-day model with a two-day memory of infection.

Notice that although we in practice will never know the state of the infection nodes, it makes a difference whether the memory links are included. In the reasoning, we cannot exploit knowledge of the exact state of the previous infection node, but we may use a probability distribution based on a test result.

The second hidden assumption has to do with the test. Any two test nodes are d-separated given any infection node on the path. This means that the fault probability of the test is independent of whether it was previously correct. In other words, the fact that the test was wrong yesterday has no influence on whether the test will be correct today. If this does not reflect the behavior of the test, you may, for example, include its performance yesterday in the model. This is done in Figure 3.4.

A minor digression on modeling of tests: It is good to have as a rule that no test is perfect. Unless you explicitly know otherwise, a test should always

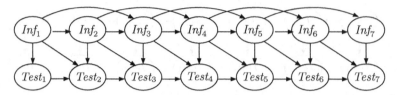

Fig. 3.4. A seven-day model with two-day memory for infection and a one-day memory of correctness of test.

be given a positive probability of false positives as well as false negatives. This is not all, though. You should also take the mechanism for false test results into account. Consider for example an HIV test with a probability of false positives of 10^{-5}, and assume that a person has received a positive test result. Now, you may have the option of repeating the test, but will this be of any help? It will depend on the mechanisms that cause the test to give a wrong result. If a test is positive because this particular person's blood is composed so that it will produce a positive test result regardless of a positive HIV infection, then a repeated test will not provide new information. If, on the other hand, the experiment is such that it now and then goes wrong, then a repeated test may be worthwhile and it will be advisable to repeat the test before the "verdict" is passed (in case the second test result is negative, a third test may be advisable). Models for these two types of failure mechanisms are shown in Figure 3.5.

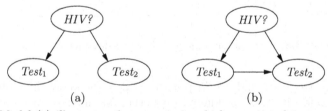

Fig. 3.5. Model (a) illustrates the scenario in which a repeated test may provide new information, and model (b) shows the situation in which repeating a test always produces the same result.

3.1.2 Cold or Angina?

I wake up in the morning with a sore throat. It may be the beginning of a cold or I may suffer from angina (inflammation of the throat). If it is severe angina, I will not go to work. To gain more insight, I can take my temperature, and I can look down my throat for yellow spots.

Here we have five hypothesis events *Cold?* {*no, yes*} and *Angina?* {*no, mild, severe*}. The hypothesis events must be organized into a set of variables with mutually exclusive and exhaustive states. We may use the variables indicated previously, but we may also use only one variable *Sick?* with states {*no, cold, mild angina, severe angina*}. In the latter case, suffering from both cold and angina is excluded as a possibility. We choose to use the two variables *Cold?* and *Angina?*.

The information variables are *Sore Throat?* {*no, yes*}, *See Spots?* {*no, yes*}, and *Fever?* {*no, low, high*}. The variable *Fever?* causes a problem because it really is continuous. In Section 3.3.8, we give methods on how to deal with continuous variables.

Now it is time to consider the causal structure between the variables. We need not worry about how information is transmitted through the network. The only thing to worry about is which variables have a direct causal impact on other variables.

In this example, we have that *Cold?* has a causal impact on *Sore Throat?* and *Fever?* while *Angina?* has an impact on all information variables. The model is given in Figure 3.6.

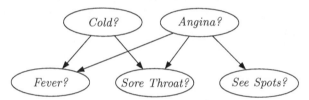

Fig. 3.6. A model for *Cold?* or *Angina?*.

The next thing to check is whether the conditional independences laid down in the model correspond to reality. For example, the model in Figure 3.6 yields that if we know the state of *Angina?*, then seeing spots will not have an impact on the expectation either for *Fever?* or for *Sore Throat?*. If we do not agree, we may introduce a link from *See Spots?* to, for example, *Fever?*. For now, we accept the conditional independences given by the model.

3.1.3 Insemination

Six weeks after insemination of a cow, you can perform two tests to determine whether the cow is pregnant: a blood test and a urine test.

Following the method from Section 3.1.1, we construct a model as in Figure 3.7. The variable *Pr* {*yes,no*} represents a possible pregnancy, and *BT* {*pos,neg*} and *UT* {*pos,neg*} represent the results of the blood test and the urine test, respectively.

Next, we will analyze the conditional independences stated by the model. We ask the expert whether it is correct that the outcomes of the two tests

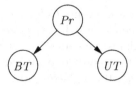

Fig. 3.7. A model for pregnancy.

are independent given *Pr*. More specifically, assume that we know the cow is pregnant. From this, we infer some expectations for the test results. Now, if we get a negative test result from the blood test, will this change our expectation for the urine test? The experts say that it will, and we must conclude that the model is not a proper reflection of reality.

There are several ways to change the model. You might, for example, introduce a link between the two test nodes, but there is no natural direction. To find out what to do, you must study the process more carefully, and it turns out that what the two tests actually do is to trace indications of hormonal changes in the cow. A more-refined model will involve a variable *Ho*, reflecting whether hormonal changes have taken place in the cow, and the model will be as in Figure 3.8.

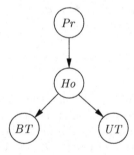

Fig. 3.8. A more correct model for pregnancy. Both the blood test (*BT*) and the urine test (*UT*) measure the hormonal state (*Ho*).

For the model in Figure 3.8, it does not hold that *BT* and *UT* are independent given *Pr*. The model states that *BT* and *UT* are independent given *Ho* (which should be checked). If the model in Figure 3.7 is used for diagnosing a possible pregnancy, a negative outcome of both the blood test and the urine test will be counted as two independent pieces of evidence and therefore over-estimate the probability for the insemination to have failed (see Exercise 3.8).

In the model in Figure 3.8, we have introduced the variable *Ho*, which is neither a hypothesis variable nor an information variable. Such variables are called *mediating variables*. Mediating variables are often introduced when two variables are not (conditionally) independent as opposed to the situation in the current model. Some standard situations are illustrated in Figure 3.9.

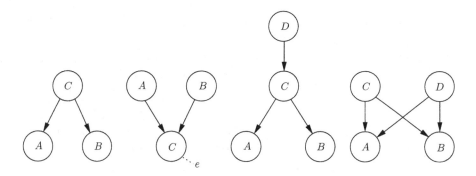

Fig. 3.9. Examples in which an intermediate variable C "resolves" undirected dependencies. In examples (a) and (b), A and B are not independent, whereas A and B are not independent given D in examples (c) and (d).

3.1.4 A Simplified Poker Game

In this poker game, each player receives three cards and is allowed two rounds of changing cards. In the first round, you may discard any number of cards from your hand and get replacements from the pack of cards. In the second round, you may discard at most two cards. After the two rounds of card changing, I am interested in an estimate of my opponent's hand.

The hypothesis events are the various types of hands in the game. They may be classified in the following way (in increasing rank): nothing special, 1 ace, 2 of the same value, 2 aces, flush (3 of a suit), straight (3 of consecutive value), 3 of the same value, straight flush. Ambiguities are resolved according to rank. This is, of course, a simplification, but it is often necessary to do so in modeling. The hypothesis events are collected into one hypothesis variable OH (opponent's hand) with the preceding classes as states.

The only information to acquire is the number of cards the player discards in the two rounds. Therefore, the information variables are FC (first change) with states $0, 1, 2, 3$ and SC (second change) with states $0, 1, 2$. By saying this, we are making an approximation again. The information on the cards you have seen is relevant for your opponent's hand. If, for example, you have seen three aces, then he cannot have two aces.

A causal structure for the information variables and the hypothesis variable could be as in Figure 3.10. However, this structure will leave us with no clue as to how to specify the probabilities.

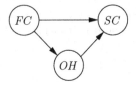

Fig. 3.10. An oversimplified structure for the poker game. The variables are *FC* (first change), *SC* (second change), and *OH* (opponent's hand).

What we need are mediating variables describing the opponent's hands in the process: the initial hand *OH0* and the hand *OH1* after the first change of cards. The causal structure will then be as in Figure 3.11.

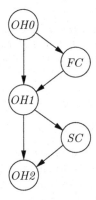

Fig. 3.11. A structure for the poker game. The two mediating variables *OH0* and *OH1* are introduced. *OH2* is the variable for my opponent's final hand.

To determine the states of *OH0* and *OH1*, we must produce a classification that is relevant for determining the states of the children (*FC* and *OH1*, say). We may let *OH0* and *OH1* have the states *nothing special, 1 ace, 2 of consecutive value, 2 of a suit, 2 of the same value, 2 of a suit and 2 of consecutive value, 2 of a suit and 2 of the same value, 2 of consecutive value and 2 of the same value, flush, straight, 3 of the same value, straight flush.*

We defer further discussion of the classification to the section on specifying the probabilities (Section 3.2).

3.1.5 Naive Bayes Models

In the previous sections we saw examples of Bayesian networks that were designed to capture the independence properties in the domains being modeled. However, the first Bayesian diagnostic systems were actually constructed based on much simpler models, namely so-called *naive Bayes models*. In a naive Bayes model the information variables are assumed to be independent given the hypothesis variable (see Figure 3.12).

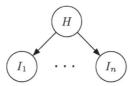

Fig. 3.12. A naive Bayes model.

Using this assumption, the conditional probability distribution for the hypothesis variable given the information variables is very easy to calculate, and the overall process (from model specification to probability updating) can be summarized as follows:

- Let the possible diseases be collected into one hypothesis variable H with prior probability $P(H)$.
- For all information variables I, acquire the conditional probability distribution $P(I \mid H)$ (the *likelihood* of H given I).
- For any set of observations $f_1, \ldots, f_n$ on the variables $I_1, \ldots, I_n$, calculate the product $P(f_1, \ldots, f_n \mid H) = P(f_1 \mid H) \cdot P(f_2 \mid H) \cdots P(f_n \mid H)$. This product is also called the likelihood for H given $f_1, \ldots, f_n$. The posterior probability for H is then calculated as

$$P(H \mid f_1, \ldots, f_n) = \mu P(H) P(f_1, \ldots, f_n \mid H)$$
$$= \mu P(H) \prod_{i=1}^{n} P(f_i \mid H), \tag{3.1}$$

where $\mu = 1/P(f_1, \ldots, f_n)$ is a normalization constant.

What is particularly attractive with the calculation in equation (3.1) is that the time complexity is linear in the number of information variables, and that each term in the product involves only two numbers (assuming that the hypothesis variable is binary), one for $P(f_i \mid H = y)$ and one for $P(f_i \mid H = n)$. On the other hand, as we also saw from the insemination example, the independence assumption need not hold, and if the model is used anyway, the conclusions may be misleading. However, in certain application areas (such as diagnosis) the naive Bayes model has been shown to provide very good performance, even when the independence assumption is violated. This is partly due to the fact that for many diagnostic problems we are interested only in identifying the most probable disease. In other words, if the conditional independence assumption does not change which state has the highest probability, then the naive Bayes model can be used without affecting the performance of the system. We shall return to these models in Section 8.1.

3.1.6 Causality

In the examples presented in the previous section, there was no problem in establishing the links and their directions. However, you cannot expect this part of the modeling always to go smoothly.

First, causal relations are not always obvious – recall the debates on whether smoking causes lung cancer or whether a person's sex has an impact on his/her ability in the technical sciences. Furthermore, causality is not a well-understood concept. Is a causal relation a property of the real world, or rather, is it a concept in our minds helping us to organize our perception of the world? For now, we make only one point about this issue, namely that in some situations you may be able to infer information about causality based on actions that change the state of the world. For example, assume that you are confronted with two correlated variables A and B, but you cannot determine a direction. If you observe the state of A, you will change your belief of B and vice versa. A good test then is to imagine that some outside agent *fixes* the state of A. If this does not make you change your belief of B, then A is not a cause of B. On the other hand, if this imagined test indicates no causal arrow in any direction, then you should look for an event that has a causal impact on both A and B. If C is such a candidate, then check whether A and B become independent given C (see Figure 3.9). We shall briefly return to the issue of discovering causal relations in Section 7.1, where we discuss methods for learning Bayesian networks from data.

3.2 Determining the Conditional Probabilities

The numbers (conditional probabilities) that you need to specify for a Bayesian network are called the *parameters* of the network. The basis for the conditional probabilities can have an epistemological status ranging from well-founded theory over frequencies in a database to subjective estimates. We will give examples of each type.

3.2.1 Milk Test

For the milk test in Figure 3.1, we need $P(Infected?)$ and $P(Test | Infected?)$. The retailer of the test should provide $P(Test | Infected?)$. Any producer of such kinds of tests is supposed to have performed a series of tests yielding the relevant numbers, namely the frequency of *false positives*, $P(Test = pos | Infected? = no)$, and the frequency of *false negatives*, $P(Test = neg | Infected? = yes)$. Let both numbers be 0.01.

The numbers provided by the retailer are not sufficient for the user of the test. In the case of a positive test result, the milk may still be clean, and to come up with a probability we need the prior probabilities $P(Infected?)$.

An estimate of the prior probability would in this case be the daily frequency λ of infected milk for each cow at the particular farm. Estimating λ may be a bit tricky because the farmer may have no experience with actually testing the milk from each specific cow with a perfect test. Assume that this particular farm has 50 cows, and that the milk from all cows is poured into a container and transported to the dairy, which tests the milk with a very precise test. The farmer's experience is that on average the dairy reports his milk to be infected once a month.

Now we must make various assumptions. The first assumption could be that the daily λ is the same for all cows. The next assumption could be that outbreaks of infected milk for the cows in the farm are independent. This yields a coin-tossing model with $P(Infected? = yes) = \lambda$. The information we have is that if we toss fifty coins at the same time, the frequency of at least one of them coming up with $Infected? = yes$ is 1 out of 30. That is, in 29 days out of 30, none of the cows are infected and the probability that all the cows are clean on a given day is therefore $29/30$. Moreover, from the assumption of the outbreaks being independent we also have that the probability of all 50 cows being clean on a given day is $(1 - \lambda)^{50}$:

$$P(Inf_1, \ldots, Inf_{50}) = (1 - \lambda_1) \cdots (1 - \lambda_{50}) = (1 - \lambda)^{50}.$$

Combining all this, we now have

$$(1 - \lambda)^{50} = \frac{29}{30},$$

which yields the estimate

$$\lambda = 1 - \left(\frac{29}{30}\right)^{0.02} \approx 0.0007.$$

This completes the model, and next you can use a computer system to calculate posterior probabilities. The interesting question for this situation is, if we get a positive test result, what is the probability that the milk is infected? This is left as an exercise (see Exercise 3.5).

For the seven-day model in Figure 3.2, we also need $P(Inf_{i+1} \mid Inf_i)$. There are two numbers to estimate: the risk of becoming infected and the chance of being cured. These numbers must be based on experience. For the sake of the example, let the risk of becoming infected be 0.0002 and the chance of being cured 0.3. This gives the numbers in Table 3.1.

For the seven-day model with a two-day memory of infection (Figure 3.3), we need $P(Inf_{i+1} \mid Inf_i, Inf_{i-1})$. If we assume that the risk of being infected is the same as before, that the infection always lasts at least two days, and that after this the chance of being cured is 0.4 each of the following days, then the numbers are as in Table 3.2 (see Exercise 3.10).

For the seven-day model with two-day memory of infection as well as correctness of test (Figure 3.4), we furthermore need $P(Test_{i+1} \mid Inf_i, Inf_{i+1},$

		Inf_i	
		yes	no
Inf_{i+1}	yes	0.7	0.0002
	no	0.3	0.9998

Table 3.1. $P(Inf_{i+1} \mid Inf_i)$.

		Inf_{i-1}	
		yes	no
Inf_i	yes	0.6	1
	no	0.0002	0.0002

Table 3.2. $P(Inf_{i+1} = yes \mid Inf_i, Inf_{i-1})$.

$Test_i$). If we assume that a correct test has a 99.9% chance of being correct next time, and an incorrect test has a 90% risk of also being incorrect next time, we can calculate all required numbers for the four-dimensional table. However, by introducing mediating variables, Cor_i, the specification of numbers could be easier, and the tables would be smaller. Figure 3.13 shows how the model could be simplified.

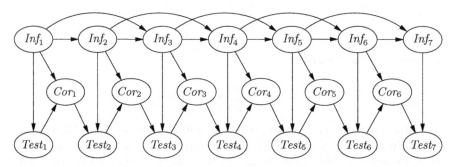

Fig. 3.13. A seven-day model with a two-day memory for infection and a one-day memory of correctness of test.

With the preceding assumptions, the required tables are as in Table 3.3.

3.2.2 Stud Farm

The stallion Brian has sired Dorothy on the mare Ann and sired Eric on the mare Cecily. Dorothy and Fred are the parents of Henry, and Eric has sired Irene on Gwenn. Ann is the mother of both Fred and Gwenn, but their fathers are in no way related. The colt John with

		Inf_i	
		yes	no
$Test_i$	pos	1	0
	neg	0	1

		Cor_{i-1}	
		yes	no
Inf_i	yes	0.999	0.1
	no	0.001	0.9

Table 3.3. The conditional probability distributions $P(Cor_i = yes \mid Inf_i, Test_i)$ and $P(Test_i = pos \mid Inf_i, Cor_{i-1})$.

the parents Henry and Irene has been born recently; unfortunately, it turns out that John suffers from a life-threatening hereditary disease carried by a recessive gene. The disease is so serious that John is displaced instantly, and since the stud farm wants the gene out of production, Henry and Irene are taken out of breeding. What are the probabilities for the remaining horses to be carriers of the unwanted gene?

The genealogical structure for the horses is given in Figure 3.14.

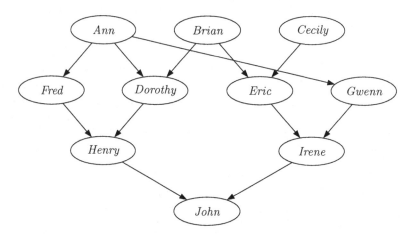

Fig. 3.14. Genealogical structure for the horses in the stud farm.

The only information variable is John. Before the information on John is acquired, he may have three genotypes: he may be sick (aa), a carrier (aA), or he may be pure (AA). The hypothesis events are the genotypes of all other horses in the stud farm.

The conditional probabilities for inheritance are both empirically and theoretically wellstudied, and the probabilities are as shown in Table 3.4.

The inheritance tables could be as in Table 3.4. However, for all horses except John, we have additional knowledge. Since they are in production, they cannot be of type aa. A way to incorporate this would be to build a

	aa	aA	AA
aa	(1, 0, 0)	(0.5, 0.5, 0)	(0, 1, 0)
aA	(0.5, 0.5, 0)	(0.25, 0.5, 0.25)	(0, 0.5, 0.5)
AA	(0, 1, 0)	(0, 0.5, 0.5)	(0, 0, 1)

Table 3.4. $P(Child\,|\,Father, Mother)$ for genetic inheritance. The numbers (α, β, γ) are the child's probabilities for (aa, aA, AA).

Bayesian network in which all inheritance is modeled in the same way and afterward enter the findings that all horses but John are not aa. It is also possible to calculate the conditional probabilities directly. If we first consider inheritance from parents that may be only of genotype AA or aA, we get Table 3.5.

	aA	AA
aA	(0.25, 0.5, 0.25)	(0, 0.5, 0.5)
AA	(0, 0.5, 0.5)	(0, 0, 1)

Table 3.5. $P(Child\,|\,Father, Mother)$ when the parents are not sick.

The table for John is as in Table 3.5. For the other horses, we know that aa is impossible. This is taken care of by removing the state aa from the distribution and normalizing the remaining distribution. For example, $P(Child\,|\,aA, aA) = (0.25, 0.5, 0.25)$, but since aa is impossible, we get the distribution $(0, 0.5, 0.25)$, which is normalized to $(0, 0.67, 0.33)$. The final result is shown in Table 3.6.

	aA	AA
aA	(0.67, 0.33)	(0.5, 0.5)
AA	(0.5, 0.5)	(0, 1)

Table 3.6. $P(Child\,|\,Father, Mother)$ with aa removed.

In order to deal with Fred and Gwenn, we introduce the two unknown fathers I and K as mediating variables and assume that they are not sick. For the horses at the top of the network, we specify prior probabilities. This will be an estimate of the frequency of the unwanted gene, and there is no theoretical way to derive it. Let us assume that the frequency is such that the prior belief of a horse being a carrier is 0.01.

In Figure 3.15, the final model with initial probabilities is shown; Figure 3.16 gives the posterior probabilities given that John is aa; and in Fig-

ure 3.17 you can see the posterior probabilities with the prior beliefs at the
top changed to 0.0001. Note that the sensitivity to the prior beliefs is very
small for the horses whose posterior probability for *carrier* is much greater
than 0, for instance in the cases of Ann and Brian.

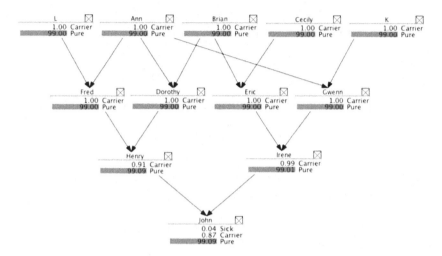

Fig. 3.15. The stud farm model with initial probabilities.

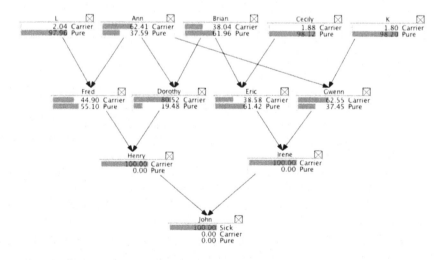

Fig. 3.16. Stud farm probabilities given that John is sick.

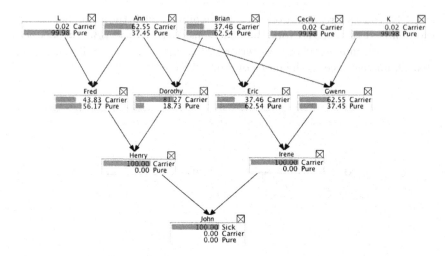

Fig. 3.17. Stud farm probabilities with prior probabilities for top variables changed to (0.0001, 0.9999).

3.2.3 Poker Game

In the stud farm example, the conditional probabilities were established mainly through theoretical considerations. This should also be attempted for the model of the poker game developed in Section 3.1.4, but it cannot be carried through entirely.

Consider for example $P(FC|OH0)$. It is not possible to give probabilities that are valid for any opponent. It is heavily dependent on the opponent's insight, psychology, and game strategy. We will assume the following strategy:

- If nothing special (*no*), then change 3.
- If 1 ace (*1 a*), then keep the ace.
- If 2 of consecutive value (*2 cons*), 2 of a suit (*2 s*), or 2 of the same value (*2 v*), then discard the third card.
- If 2 of a suit and 2 of consecutive value, then keep 2 of a suit (this strategy could be substituted by a random strategy for keeping either 2 of a suit or 2 of consecutive value).
- If 2 of a suit and 2 of the same value or 2 of consecutive value and 2 of the same value, then keep the 2 of the same value.
- If flush (*fl*), straight (*st*), 3 of the same value (*3 v*), or straight flush (*sfl*), then keep it.

Based on the preceding strategy, a logical link between *FC* and *OH0* is established. Note that the strategy makes the states for combined hands redundant. They play no role, and therefore we remove them.

The strategy for $P(SC|OH1)$ is the same except that in the case of *no*, only 2 cards are discarded.

These strategies seem to be the most rational. However, deterministic strategies in games do not always work, since they give your opponent valuable information about your hand. A good strategy should therefore be random rather than deterministic. Sometimes you may, for example, change nothing although you have a weak hand. Some people call it bluff, but it is really a way of increasing your opponent's uncertainty no matter what you do.

The remaining probabilities to specify are $P(OH0), P(OH1 \mid OH0, FC)$, and $P(OH2 \mid OH1, SC)$.

The Probability Distribution $P(OH0)$

The states are $(no, 1\ a, 2\ cons, 2\ s, 2\ v, fl, st, 3\ v, sfl)$, and through various (approximated) combinatorial calculations, the prior probability distribution is found to be $P(OH0) = (0.1569, 0.0765, 0.0635, 0.4447, 0.1694, 0.0494, 0.0353, 0.0024, 0.0024)$. For example, in order to determine the probability $P(OH0 = st)$ we first calculate the number of different ways in which we can obtain a straight: by disregarding permutations of the three cards, we get $52 \cdot 4 \cdot 4$ by letting $ka2$ be a straight. However, since we do not want to include straight flushes, we subtract the number of ways (52) in which we can obtain a straight flush (again disregarding permutations), and finally we divide by the number of ways to draw three cards out of 52 cards (the latter is equal to the binomial coefficient $\binom{52}{3}$):

$$P(OH0 = st) = \frac{52 \cdot 4 \cdot 4 - 52}{\binom{52}{3}} = 0.0353.$$

The Probability Distribution $P(OH1 \mid OH0, FC)$

Due to the logical links between $OH0$ and FC, it is sufficient to consider only nine out of the possible 36 parent configurations, namely $(no, 3)$, $(1\ a, 2)$, $(2\ cons, 1)$, $(2\ s, 1)$, $(2\ v, 1)$, $(fl, 0)$, $(st, 0)$, $(3\ v, 0)$, $(sfl, 0)$. The last four are obvious. In Table 3.7, the results of the approximate combinatorial calculations are given.

The probabilities for the remaining parent configurations may be whatever is convenient, so put, for example, $P(OH1 \mid 3\ v, 1) = (1, 0, \ldots, 0)$.

The Probability Distribution $P(OH2 \mid OH1, SC)$

First, a table $P(OH2' \mid OH1, SC)$ similar (but not identical in the numbers) to Table 3.7 can be calculated. However, the states of $OH2'$ are not the ones we are interested in. We are interested in the *value* of the hand, and a state such as 2 cons is of no value unless one of them is an ace. Therefore, the probabilities for the states of $OH2'$ are transformed to probabilities for $OH2$. For the transformation, the following rules are used:

	(OH0, FC)				
	(no, 3)	(1 a, 2)	(2 cons, 1)	(2 s, 1)	(2 v, 1)
no	0.1583	0	0	0	0
1 a	0.0534	0.1814	0	0	0
2 cons	0.0635	0.0681	0.3470	0	0
2 s	0.4659	0.4796	0.3674	0.6224	0
OH1 2 v	0.1694	0.1738	0.1224	0.1224	0.9592
fl	0.0494	0.0536	0	0.2143	0
st	0.0353	0.0383	0.1632	0.0307	0
3 v	0.0024	0.0026	0	0	0.0408
sfl	0.0024	0.0026	0	0.0102	0

Table 3.7. $P(OH1 | OH0, FC)$ for the nonobvious parent configurations.

$$1\ a = 1\ a + \frac{1}{6}(2\ cons + 2\ s),$$

$$no = no + \frac{5}{6}(2\ cons + 2\ s).$$

The probabilities of $2\ a$ are calculated specifically. The resulting probabilities are given in Table 3.8.

	(OH1, Sc)				
	(no, 2)	(1 a, 2)	(2 cons, 1)	(2 s, 1)	(2 v, 1)
no	0.5613	0	0.5903	0.5121	0
1 a	0.1570	0.2425	0.1181	0.1024	0
2 v	0.1757	0.0667	0.1154	0.1154	0.8838
OH2 2 a	0.0055	0.1145	0.0096	0.0096	0.0736
fl	0.0559	0.0559	0	0.2188	0
st	0.0392	0.0392	0.1666	0.0313	0
3 v	0.0027	0.0027	0	0	0.0426
sfl	0.0027	0.0027	0	0.0104	0

Table 3.8. $P(OH2 | OH1, SC)$ for the nonobvious configurations.

Using a model such as the one in Figure 3.11 and with the conditional probability tables specified in this section, we have established a model for assisting a (novice) poker player. However, if my opponent knows that I use the system, he can change cards in such a way that affects my estimate of his hand.

3.2.4 Transmission of Symbol Strings

A language L over 2 symbols $(\mathbf{a}, \mathbf{b})$ is transmitted through a channel. Each word is surrounded by the delimiter symbol c. In the transmission some characters may be corrupted by noise and be confused with others.

A five-letter word is transmitted. Give a model that can determine the probabilities for the transmitted symbols given the received symbols.

There are five hypothesis variables $T_1, \ldots, T_5$ with states a, b and five information variables $R_1, \ldots, R_5$ with states a, b, c. There is a causal relation from T_i to R_i. Furthermore, there may also be a relation from T_i to $T_{i+1} (i = 1, \ldots, 4)$ encoding that certain pairs of symbols are more likely to occur than others. You could also consider more-involved relations from pairs of symbols to symbols, but for now we refrain from doing that. The structure is given in Figure 3.18.

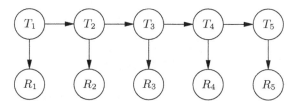

Fig. 3.18. A model for symbol transmission. T_i are the symbols transmitted; R_i are the symbols received.

The conditional probabilities can be established through experience. The probabilities $P(R_i \mid T_i)$ will be based on statistics describing the frequencies of confusion. Let Table 3.9 be the result.

	$T = a$	$T = b$
$R = a$	0.80	0.15
$R = b$	0.10	0.80
$R = c$	0.10	0.05

Table 3.9. $P(R \mid T)$ under transmission.

You may obtain the probabilities $P(T_{i+1} \mid T_i)$ by investigating the five-letter words in L. What is the frequency of the first letter? What is the frequency of the second letter given that the first letter is **a**? You continue to do this for each letter. You can refine this frequency analysis by also taking the frequencies of the words into consideration. Let Table 3.10 be the result of a frequency analysis.

You can calculate the required probabilities from Table 3.10 using the fundamental rule. The prior probabilities for T_1 are $(0.5, 0.5)$, and $P(T_2, T_1)$ is

First 2	Last 3 letters							
letters	aaa	aab	aba	abb	baa	bab	bba	bbb
aa	0.017	0.021	0.019	0.019	0.045	0.068	0.045	0.068
ab	0.033	0.040	0.037	0.038	0.011	0.016	0.010	0.015
ba	0.011	0.014	0.010	0.010	0.031	0.046	0.031	0.045
bb	0.050	0.060	0.056	0.057	0.016	0.023	0.015	0.023

Table 3.10. Frequencies of five-letter words in L. The word **abaab**, for example, has frequency 0.040.

achieved by adding the elements in each row. Table 3.11 gives two conditional probabilities.

	a	b		a	b
a	0.6	0.4	a	0.24	0.74
b	0.4	0.6	b	0.76	0.26

$$P(T_2 \mid T_1) \qquad P(T_3 \mid T_2)$$

Table 3.11. Two conditional probabilities for five-letter words in L.

An alternative model would be to have a hypothesis variable, *Word*, with 32 states and with Table 3.10 as prior probabilities (see Figure 3.19).

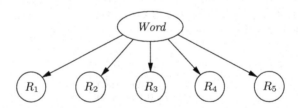

Fig. 3.19. An alternative model for symbol transmission. *Word* is the set of possible transmitted words.

This is manageable because of the small number of five-letter words over $\{a, b\}$; but if the alphabet had 24 symbols, and if six-letter words were considered, the number of states in *Word* would become intractably large. On the other hand, the model of Figure 3.18 may be too simple to catch the dependencies in Table 3.10, so the task really is to analyze the table in order to find the simplest structure describing it. There are methods for doing this, and we return to this topic in Chapter 7.

3.2.5 Cold or Angina?

The estimation of the conditional probabilities for the example introduced in Section 3.1.2 has a very subjective flavor based on my own experience with colds and anginas. I estimate the following probabilities: $P(Cold?)$, $P(Angina?)$, $P(See\ Spots?\,|\,Angina?)$, $P(Fever?\,|\,Cold?, Angina?)$, $P(Sore\ Th\ roat?\,|\,Cold?, Angina?)$.

Because in the morning I do not recall having been chilly yesterday, the prior probabilities $P(Cold?)$ and $P(Angina?)$ are my subjective recollections of how often I wake up in the morning with a cold or with an angina. Because cold is more frequent than angina, I put $P(Cold?) = (0.97, 0.03)$ and $P(Angina?) = (0.993, 0.005, 0.002)$; the order of the states are taken from Section 3.1.2.

Without angina or with mild angina, I will not see spots. With severe angina, I would expect to see spots, but I may not. I put $P(See\ Spots?\,|\,An\ gina? = severe) = (0.1, 0.9)$.

The Probability Distribution $P(Sore\ Throat?\,|\,Cold?, Angina?)$

If I suffer from neither a cold nor angina, I have a background probability of 0.05 of having a sore throat in the morning; this background probability covers everything other than cold and angina that may result in a sore throat. A cold as well as angina may give me a sore throat. If I only have a cold, the probability of a sore throat is 0.4. If I have mild angina, the probability of a sore throat is 0.7, and in the case of severe angina, I will certainly have a sore throat. What if I have both a cold and mild angina? I do not have sufficient experience to come up with a reliable estimate. Instead, I can use the two conditional probabilities from before: out of 100 mornings, I will wake up five mornings with a "background produced" sore throat. Out of the remaining 95 mornings, the cold yields a sore throat in 40% of them, that is, 38 mornings. Out of the remaining 57 mornings, mild angina will cause a sore throat in 70% of them: 39.9 mornings. In total, if I have both mild angina and a cold, I will have a sore throat in 82.9 mornings out of 100. The number 82.9 indicates an unjustified precision, and for psychological reasons we set the probability to 0.85. In Section 3.3.2 on "noisy-or," we give a systematic treatment of this method of estimating probabilities. The full table for $P(Sore\,Throat?\,|\,Cold?, Angina?)$ is given in Table 3.12. It is left as an exercise to complete the model.

	Angina? = no	Angina? = mild	Angina? = severe
Cold? = no	0.05	0.7	1
Cold? = yes	0.4	0.85	1

Table 3.12. $P(Sore\ Throat? = yes\,|\,Cold?, Angina?)$.

3.2.6 Why Causal Networks?

As mentioned previously, the structure of a Bayesian network need not reflect cause–effect relations. The only requirement is that the d-separation properties of the network hold for the domain modeled. There are, however, good reasons to strive for causal networks. The model in Figure 3.20 can be used to illustrate some of the points. We have a disease *Dis* and two tests, *Ts* and *Tt*.

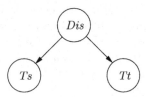

Fig. 3.20. A model for a disease with two tests.

When diagnosing, you usually reason opposite to the directions of the arrows in Figure 3.20, and trained physicians are usually inclined to provide conditional probabilities in the diagnostic direction. A model reflecting this might look like the one in Figure 3.21 a).

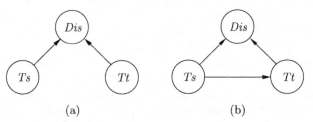

(a)   (b)

Fig. 3.21. Diagnostic models for the situation in Figure 3.20: (a) with a wrong independence, (b) with no (conditional) independence.

The model in Figure 3.21(a) is not correct. According to this model, *Ts* and *Tt* are independent (which is not the case in Figure 3.20), and there is no way to correct it by specifying the potentials in a sophisticated manner. To correct the model, you must add some extra structure making *Ts* and *Tt* dependent. You may, for example, introduce a link from *Ts* to *Tt*, as is done in Figure 3.21(b). Therefore, to get a correct model, it is not sufficient to acquire $P(Dis \mid Ts, Tt)$ together with the "priors" $P(Ts)$ and $P(Tt)$. This also illustrates another point, namely that a correct model of a causal domain is minimal with respect to links. In other words, if for some reason you wish to represent a causal relation with a link directed opposite to the causal direction, then the total number of links can not decrease, and most likely it will increase.

The model in Figure 3.20 has another advantage over the models in Figure 3.21, namely that the conditional probabilities $P(Ts \mid Dis)$ and $P(Tt \mid Dis)$ are more stable than the conditional probabilities specified for the models in Figure 3.21. The conditional probabilities for Figure 3.20 reflect general properties of the relation between diseases and tests, and they are the ones that a manufacturer of tests can publish, whereas the conditional probabilities for Figure 3.21 are a mixture of disease–test relations and prior frequencies of the disease.

It may happen that it is not possible to acquire the conditional probabilities for a correct model, but instead, other types of conditional probabilities are available. Assume, for example, that for the model in Figure 3.20, we can acquire only the potentials $P(Dis \mid Ts)$, $P(Dis \mid Tt)$, $P(Ts)$, and $P(Tt)$. Using Bayes' rule on $P(Dis \mid Ts)$ and $P(Ts)$, we get $P(Dis)$ and $P(Ts \mid Dis)$. The same can be done with $P(Dis \mid Tt)$ and $P(Tt)$. If the two calculations of $P(Dis)$ give the same result, we have the required potentials. If, on the other hand, the two calculations disagree, there is no safe way to solve the conflict. It can happen in many different situations that you have a set of potentials, but the model requires another set and there is no safe way of inferring the needed potentials. It is a lively area of research to construct engineering methods for getting the best out of what you have.

In Chapter 9, we deal with *interventions*. They provide another good reason for constructing causal models. An intervention is an action that has an impact on the state of certain variables. The impact of an intervention will spread in the causal direction, but not opposite to the causal direction. If the model does not reflect causal directions, it cannot be used to simulate the impact of interventions.

3.3 Modeling Methods

Much skepticism of Bayesian networks stems from the question of where the numbers come from. As shown in the previous section, they come from many different sources. If you are building a model over a domain in which experts actually *do* take decisions based on estimates, why should you not be able to make your Bayesian network estimate at least as well as the experts? You can, for example, use the technique described in Section 1.1 to acquire the probabilities from the experts. The acquisition of numbers is, of course, not without problems, and in this section we give some methods that can help you in this job. Also, we provide some modeling tricks.

3.3.1 Undirected Relations

It may happen that the model must contain dependence relations among variables A, B, C, say, but it is neither desirable nor possible to attach directions to them.[1] The relation may, for example, be a description of possible configurations. This difficulty may be overcome by using conditional dependence as described in Section 2.2.1 (converging influence).

Let $R(A, B, C)$ describe the relation using the values 0 and 1; $R(A, B, C) = 1$ for all valid configurations of A, B, and C. Add a new variable D with two states y and n and let A, B, and C be parents of D (see Figure 3.22). Assign D the deterministic conditional probability table given as $P(D = y \mid A, B, C) = R(A, B, C)$ (and $P(D = n \mid A, B, C) = 1 - R(A, B, C)$) and enter the evidence $D = y$. The variable D is called a *constraint variable*, and by entering $D = y$ we are basically forcing the relation/constraint to hold.

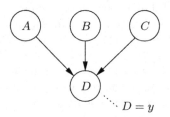

Fig. 3.22. A way to introduce undirected relations among A, B, and C.

Example 3.1. If we want to model that A, B, and C are always in the same state, then we can assign D the conditional probability table given in Table 3.13 (assuming that A, B, and C are binary).

		$C = y$		$C = n$	
		$B = y$	$B = n$	$B = y$	$B = n$
A	y	1	0	0	0
	n	0	0	0	1

Table 3.13. The conditional probability distribution $P(D = y \mid A, B, C)$ for the constraint variable D modeling that A, B, and C are always in the same state.

Example 3.2. I have washed two pairs of socks in the washing machine. The washing has been rather hard on them, so they are now difficult to distinguish. However, it is important for me to pair them correctly. To classify the socks, I have pattern and color. A classification model may be like the one in Figure 3.23. The variables S_i have states t_1 and t_2 for the two types, the

---

[1] In that case, the model is called a *chain graph*. A chain graph is an acyclic graph with both directed and nondirected links, where *acyclic* means that all cycles consist of only nondirected links.

variables P_i have two pattern types, and the variables C_i have two color types. The constraint that there are exactly two socks of each type is described in Table 3.14.

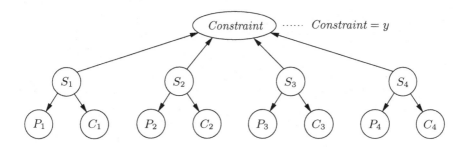

Fig. 3.23. A model for classifying pairs of socks.

S_1	t_1	t_1	t_1	t_1	t_1	t_1	t_1	t_1	t_2	t_2	t_2	t_2	t_2	t_2	t_2	t_2
S_2	t_1	t_1	t_1	t_1	t_2	t_2	t_2	t_2	t_1	t_1	t_1	t_1	t_2	t_2	t_2	t_2
S_3	t_1	t_1	t_2	t_2	t_1	t_1	t_2	t_2	t_1	t_1	t_2	t_2	t_1	t_1	t_2	t_2
S_4	t_1	t_2	t_1	t_2	t_1	t_2	t_1	t_2	t_1	t_2	t_1	t_2	t_1	t_2	t_1	t_2
P	0	0	0	1	0	1	1	0	0	1	1	0	1	0	0	0

Table 3.14. The table for $P = P(\text{Constraint} = y \mid S_1, S_2, S_3, S_4)$; t_1 and t_2 are the two states of S_1, S_2, S_3, S_4.

The situation is more subtle if the relation $R(A, B, C)$ is of probabilistic nature. If $A, B,$ and C have no parents, $R(A, B, C)$ can be a joint probability table. On the other hand, if A has a parent, then $R(A, B, C)$ may be considered as representing a feedback cycle. We shall not deal with this problem but refer the reader to the literature on chain graphs.

3.3.2 Noisy-Or

When a variable A has several parents, you must specify $P(A \mid \mathbf{c})$ for each configuration $\mathbf{c}$ of the parents. If you take the distributions from a database, the number of cases for each configuration may become too small. Also, the configurations may be too specific for any expert. You may also be in the situation that you have reasonable estimates of $P(A \mid B)$ and $P(A \mid C)$, but you require $P(A \mid B, C)$. Then, you should look for assumptions that reduce the number of distributions to specify.

Consider in Section 3.2.5 the conditional probability table for $P(\text{Sore Th roat? } \mid \text{Cold?, Angina?})$. It was possible to get estimates of $P(\text{Sore Throat?} \mid$

Cold?) and *P*(*Sore Throat?* | *Angina?*), but is there a general way to describe how they then combine into *P*(*Sore Throat?* | *Cold?*, *Angina?*)? The following is a way of describing it.

There are three events causing me to have a sore throat in the morning:

- the "background event," which in 5% of the mornings yields a sore throat;
- *cold*, which causes a sore throat with probability 0.4;
- *angina*, which when *mild* causes a sore throat with probability 0.7, and when it is *severe* it certainly causes a sore throat.

The preceding uncertainty can be interpreted as follows. If any of the causes are present, then I have a sore throat unless something has prevented it. In other words, if I have *mild* angina, then I have a sore throat unless some other circumstances prevent it, and there is a 30% chance that it is prevented. In the same way, there is a 60% chance that some inhibitor prevents me from having a sore throat although I have a cold, and the background event is prevented with probability 0.95.

Now, if we assume that the preventing factors are independent, then the combined probabilities are easy to calculate as one minus the product of the appropriate probabilities for the inhibitors (note that the background event is always a fact). The probabilities are given in Table 3.15.

	Angina? = no	*Angina? = mild*	*Angina? = severe*
Cold? = no	0.05	$1 - 0.95 \cdot 0.3$	1
Cold? = yes	$1 - 0.95 \cdot 0.6$	$1 - 0.95 \cdot 0.3 \cdot 0.6$	1

Table 3.15. Calculation of *P*(*Sore Throat? = yes* | *Cold?*, *Angina?*). Note that some numbers are slightly different from the corresponding numbers in Table 3.12.

Another way to view the calculations above is to make the independence assumptions explicit in the model. Consider the model shown in Figure 3.24(a) and introduce an intermediate node ST_C between *Sore Throat* (*ST*) and *Cold?* (*C*) as well as an intermediate node ST_A between *Sore Throat?* and *Angina?* (*A*). The node ST_C captures the effect that *Cold?* has on *Sore Throat?* (i.e., it represents a "cold-induced" sore throat), whereas ST_A represent an "angina-induced" sore throat. In order to model the "background event" we introduce two additional nodes B and ST_B, where B represent the "background event," and ST_B plays the same role as ST_C and ST_A above. The three nodes ST_A, ST_B, and ST_C also represent the inhibitors, and they are assigned the conditional probability tables shown in Table 3.16; the numbers have been deduced from the itemized list above. Finally, since we will have a sore throat no matter whether it is induced by cold, angina, or something else, we assign *ST* a conditional probability distribution that corresponds to a logical-or. The resulting model is shown in Figure 3.24(b), where the variables ST_A, ST_B, and ST_C are independent, reflecting the assumption that the inhibitors are independent. Moreover, if we marginalize out the variables ST_A, ST_B, and ST_C, we end up with the conditional probability table in Table 3.15 (see also Exercise 3.20).

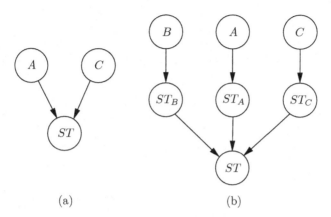

(a)   (b)

Fig. 3.24. Figure (a) shows the model structure for $P(ST \mid C, A)$, and figure (b) shows the model structure that explicitly represent the independence assumption about the inhibitors.

		A		
		no	mild	severe
ST_A	yes	0	$1 - 0.3$	1
	no	1	0.3	0

$P(ST_A \mid A)$

		B
		yes
ST_B	yes	$1 - 0.95$
	no	0.95

$P(ST_B \mid B)$

		C	
		no	yes
ST_C	yes	0	$1 - 0.6$
	no	1	0.6

$P(ST_C \mid C)$

Table 3.16. The conditional probability tables $P(ST_A \mid A)$, $P(ST_B \mid B)$, and $P(ST_C \mid C)$.

The preceding construction is an example of the simplifying assumption called a *noisy-or*. In what follows we put this assumption into a more general context, albeit only with binary variables.

Let $A_1, \ldots, A_n$ be binary variables listing all the causes of the binary variable B. Each event $A_i = y$ causes $B = y$ unless an *inhibitor* prevents it, and the probability for that is q_i (see Figure 3.25).

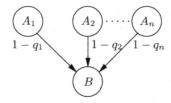

Fig. 3.25. The general situation for noisy-or. Here q_i is the probability that the impact of A_i is inhibited.

In other words, $P(B = n \mid A_i = y) = q_i$. We assume that *all inhibitors are independent*. Then $P(B = n \mid A_1, A_2, \ldots, A_n) = \prod_{j \in Y} q_j$, where Y is the set of indices for variables in the state y. For example,

$$P(B = y \mid A_1 = y, A_2 = y, A_3 = \cdots = A_n = n)$$
$$= 1 - P(B = n \mid A_1 = y, A_2 = y, A_3 = \cdots = A_n = n)$$
$$= 1 - q_1 \cdot q_2.$$

By assuming "noisy-or," the number of probabilities to estimate grows linearly with the number of parents.

Note 1. We require $P(B = y \mid A_1 = \cdots = A_n = n)$ to be 0. This may seem to restrict the applicability of the approach. However, as in the preceding example, if $P(B = y) > 0$ when none of the causal events in the model are on, then introduce a background event that is always on.

Note 2. The complementary construction to noisy-or is called *noisy-and*. A set of causes should all be "on" in order to have an effect. However, the causes have random inhibitors, which are mutually independent.

Note 3. As in Figure 3.24(b), noisy-or can be modeled directly without performing the calculations (see Figure 3.26). This highlights the assumptions behind the noisy-or gate. If a cause is on, then its effect may be prevented by an inhibitor, and the probabilities for the inhibitors to be present are independent.

Note 4. The noisy-or model has been generalized to variables having more than two states, and in this form it is called a *noisy-max*; in this model we assume that the states of B are ordered.

3.3.3 Divorcing

Let $A_1, \ldots, A_n$ be a list of variables all of which are causes of B. If you wish to specify $P(B \mid A_1, \ldots, A_n)$, you might have a very large knowledge acquisition

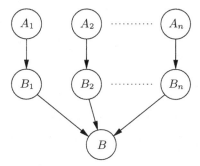

Fig. 3.26. Direct modeling of a noisy-or gate. Here $P(B_i \mid A_i)$ is the original $P(B \mid A_i)$, and $P(B \mid B_1, \ldots, B_n)$ is logical or.

task ahead of you. Either you need to ask the experts on the distribution of B given very specific parent configurations or, if the table must be extracted from a database, you need a very large set of cases. The following example illustrates the problem.

Example 3.3 (Granting a loan). A bank will decide on a mortgage loan for a customer who wishes to purchase a house. The customer is asked to fill in a form giving information on various financial and personal matters together with various key information on the house. The answers are used to estimate the probability that the bank will get its money back.

The information can be the following: type of job, yearly income, other financial commitments, number and types of cars in the family, number of previous addresses during the last five years, number of children in the family, number of divorces, size and age of the house, price of the house, and type of environment.

In principle, each slot in the form represents a variable with a causal impact on the variable *Money back?*. If we assume that each parent variable has five states, we have already listed a parent space with $5^{11} \approx 5,000,000$ configurations. For each configuration, we request a distribution for A. No person can estimate that number of distributions, nor can he or she estimate a distribution for a divorced businesswoman with a yearly income of $50,000, having loans of $70,000 already, one car, three previous addresses, two children, wanting to purchase a twenty-year-old house of 150 m^2 at the price of $200,000 in a farming area. Also, if the distributions are to be taken from a database, the bank will need at least 50,000,000 cases that may not be more than 10 years old.

To handle this kind of task, we *divorce* the parents. The set of parents $A_1, \ldots, A_i$ for B is divorced from the parents $A_{i+1}, \ldots, A_n$ by introducing a mediating variable C, making C a child of $A_1, \ldots, A_i$ and a parent of B (see Figure 3.27).

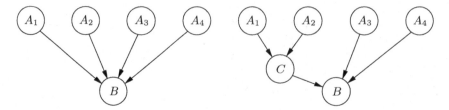

Fig. 3.27. Parents A_1 and A_2 are divorced from A_3 and A_4 by introducing the variable C.

The assumption behind divorcing is the following (with reference to Figure 3.27).

The set of configurations (A_1, A_2) can be partitioned into the sets $c_1, \ldots, c_m$ such that whenever two configurations (a_1, a_2) and (a'_1, a'_2) are elements in the same c_i, then $P(B \mid a'_1, a'_2, A_3, A_4) = P(B \mid a_1, a_2, A_3, A_4)$. The divorcing variable then has $c_1, \ldots, c_m$ as states.

In the example of granting a loan, it is impossible to perform an analysis as before, and you will group the variables based on another type of insight into the domain. For example, the variables about the house can be grouped and given a common child variable describing how safe the mortgage will be, the financial variables may be grouped for a variable describing the applicant's financial abilities; and the remaining variables may describe the applicant's stability.

In connection to the example of granting a loan, it should be noted that if we only want to perform a classification, then we need not build a Bayesian network. Other techniques such as statistical classifiers and classification trees (see Section 8.4) may be more adequate. However, if we also wish to calculate decision recommendations, we will need the posterior probabilities provided by a Bayesian network. We will deal further with this in Chapter 9.

3.3.4 Noisy Functional Dependence

There are ways of directing the divorcing. "Noisy-or" and "noisy-and" are examples of a general method called *noisy functional dependence*.

Example 3.4 (Headache). Headache (Ha) may be caused by fever (Fe), hangover (Ho), fibrositis (Fb), brain tumor (Bt), and other causes (Ot), and you may choose to soothe it with aspirin (As) (we ignore the effect aspirin has on fever). Let Ha have the states *no, mild, moderate, severe*. The various causes support each other in the effect. If, for example, $Ho = y$ or $Fe = y$ is present, then it may yield a *mild* Ha, but if both are present, then the Ha would be *moderate*. Furthermore, if also $As = y$, then Ha may drop to *no* or *mild*. Although the various parents of Ha combine in a rather involved manner, we still have the feeling that the impacts of the causes are independent. This kind of independence can be described as follows: if the headache is at level l, and we add an extra cause for headache, then the result is a headache at level q independent of how the initial state has been caused.

Assume that we can estimate conditional probabilities of type $P(Ha \mid C)$, and we want to combine the effects of the various causes. For this, we can imagine that we attach a number to the states of Ha: $no \mapsto 0$, $mild \mapsto 1$, $moderate \mapsto 2$, $severe \mapsto 4$, and the "adding up" of the effects consists in adding the numbers. A model could be similar to the one in Figure 3.28.

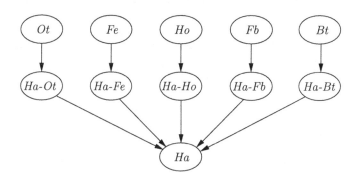

Fig. 3.28. A model for causes of headache. The bottom node adds up the effects.

The hidden assumption behind this method of adding up is that the effect from any cause is independent of the current state of headache, and it is faithfully reflected in the numbers attached to the headache states. To make it explicit in the model, we can give each headache node a child with numbers as states, these nodes are given a common child that adds the numbers, and a new node translates the numbers to Ha states (see Figure 3.29).

Now, for $P(Nu\text{-}Ha \mid Nu\text{-}Ot, Nu\text{-}Fe, Nu\text{-}Ho, Nu\text{-}Fb, Nu\text{-}Bt)$ we can perform divorcing, we can add one number at a time (see Figures 3.30 and 3.31), or we can represent the function in any other kind of compact way.

The effect of aspirin can be included in two different ways. Either it subtracts a number from the sum or it has a direct effect on the headache state.

3.3.5 Expert Disagreements

It may happen that we are in a situation in which the experts disagree on the conditional probabilities for a model. Consider the model in Figure 3.32, and assume that we have three experts who agree on $P(B)$ and $P(C \mid A)$, but they disagree on $P(A)$ and $P(D \mid B, C)$. For the three experts, we have $P(A = y) = (0.1, 0.3, 0.4)$, and the table for $P(D \mid B, C)$ can be seen in Table 3.17.

If you have equal confidence in the three experts, you can take the mean of the three numbers. If your confidence in the experts varies, you may incorporate this and calculate a weighted average. For example, you may give the first

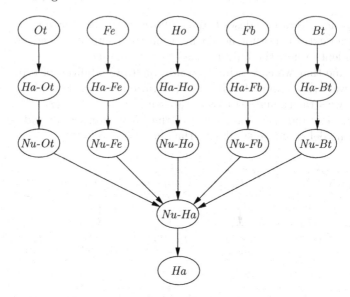

Fig. 3.29. A model that adds the headache states by transforming to numbers, adding, and transforming back to headache states again.

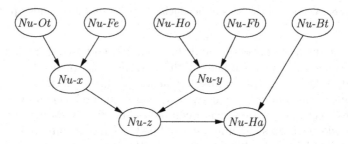

Fig. 3.30. The adder represented through divorcing.

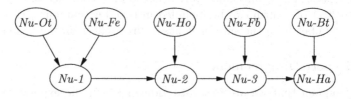

Fig. 3.31. The adder represented through adding one number at a time.

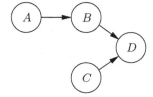

Fig. 3.32. A model with expert disagreements. All variables are binary.

		B	
		y	n
C	y	(0.4, 0.4, 0.6)	(0.7, 0.9, 0.7)
	n	(0.6, 0.4, 0.5)	(0.9, 0.7, 0.9)

Table 3.17. $P(D = y \mid B, C)$ for the three different experts s_1, s_2, s_3.

two experts a confidence weight 1 and the third expert a confidence weight 2. Because the total confidence weight is 4, you get a confidence distribution $(0.25, 0.25, 0.5)$, and for A you have $P(A = y) = 0.25 \cdot 0.1 + 0.25 \cdot 0.3 + 0.5 \cdot 0.4 = 0.3$. The probability $P(D \mid B, C)$ is shown in Table 3.18.

		B	
		y	n
C	y	0.5	0.75
	n	0.5	0.85

Table 3.18. $P(D = y \mid B, C)$ weighted with confidence distribution $(0.25, 0.25, 0.5)$.

The experts can be represented explicitly in the model by introducing a variable S with states s_1, s_2, and s_3. The variable S has a link to the nodes, about whose tables the three experts disagree (see Figure 3.33).

The variable S is given the confidence distribution $(0.25, 0.25, 0.5)$ as before, and the child variables have a conditional probability table for each expert. The table $P(D = y \mid B, C, S)$ is as in Table 3.17.

By modeling the different expert opinions explicitly, you have prepared the model for *adaptation*. Whenever you have a case with evidence e entered into the model, you will get $P(S \mid e)$, which is an updated indication of which expert to believe. That is, you get a new confidence distribution that can be used for the next case, see also Section 6.3.

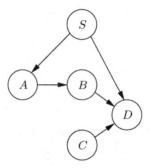

Fig. 3.33. The model from Figure 3.32 with the experts represented explicitly by the node S.

3.3.6 Object-Oriented Bayesian Networks

Complex Bayesian network models often include copies of almost-identical network fragments. Consider, for example, the Bayesian network shown in Figure 3.34, and assume that X_1 and X_2 have the same state space ($\text{sp}(X_1) = \text{sp}(X_2)$), and that the conditional probability tables associated with the nodes labeled A are identical; similarly for the nodes labeled B, C, D, and E. Given these two assumptions we see that the network contains four identical copies of the same *network fragment* defined by the five nodes A, B, C, D, E.

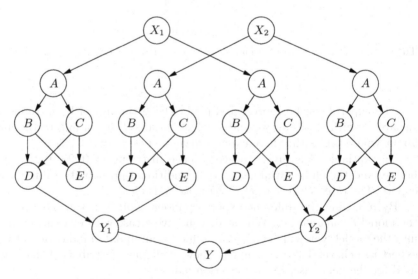

Fig. 3.34. A Bayesian network containing repetitive substructures.

The occurrence of such repetitive structures can be exploited during model construction. For example, instead of explicitly specifying the same network fragment multiple times, we could instead construct a generic network fragment that can be *instantiated* the required number of times. By borrowing terminology from the object-oriented programming paradigm, we call such a generic network fragment a *class*, and each network fragment that is produced by instantiating the class is called an *object*. Figure 3.35 shows a class description (called Class-name) for the duplicated network fragment in Figure 3.34. In order for the class to support the specification of the conditional probability distribution for A, the class includes an artificial node X (drawn as a dashed node) having the same state space as X_1 and X_2. Note that this node does *not* correspond to an actual variable, but should rather be seen as a "placeholder" that simply allows us to specify the probability distribution for A. The shaded nodes in Figure 3.35 indicate the part of the class/object that is accessible outside the object; they may be parents of nodes outside the object. Nodes that are neither dashed nor shaded are encapsulated within the object, and they may therefore be considered invisible to the rest of the model.

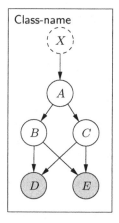

Fig. 3.35. A class model for the duplicated network fragment in Figure 3.34. Class-name is the name of the class.

Given such a class description, we can make an equivalent representation of the model in Figure 3.34 by instantiating the class four times and connecting X_1, X_2, Y_1, and Y_2 to the objects (labeled Inst. 1, Inst. 2, Inst. 3, Inst. 4) as appropriate. The resulting model is shown in Figure 3.36 and is called an *object-oriented Bayesian network model* (OOBN). The dashed arcs indicate which node X is a placeholder for in the various objects.

As implied by the discussion above, an object (or a class) can be seen as a function that given a certain input provides a probability distribution over a

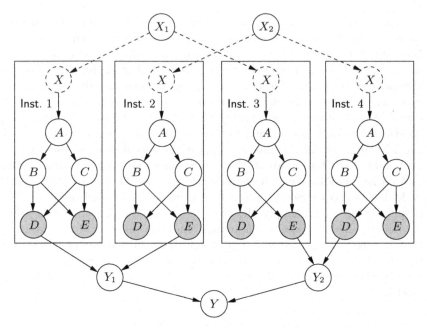

Fig. 3.36. An object-oriented Bayesian network representation of Figure 3.34.

set of variables. For example, the class shown in Figure 3.35 specifies a probability distribution over D and E given a state for X. Based on this perspective,
we can partition the elements in an object into three sets: *input attributes*,
output attributes, and *encapsulated attributes*. In the example above, X is an
input attribute, D and E are output attributes, and A, B, and C are encapsulated attributes. Following standard programming terminology, the input
attributes in the class description can be seen as the *formal parameters* of the
corresponding function, whereas the *actual parameters* passed to an object
are identified as the parents of the input attributes in the surrounding model.
Thus, X can be considered a formal parameter, and X_1 is the actual parameter passed to the left-most object in Figure 3.36. In general, we also allow
encapsulated attributes and output attributes to be objects themselves. However, input attributes must correspond to variables, since they serve as the
parameters passed to the object. Note that the simplest type of class/object
consists of a single variable, where the input attributes correspond to the
parents of that variable.

The specification of encapsulated attributes is closely related to the concept of *information hiding* in the object-oriented programming paradigm. By
taking this idea one step further, we obtain a straightforward mechanism for
simplifying the visual representation of a model by abstracting away irrelevant
details. For example, by abstracting away the encapsulated attributes in Figure 3.36 we obtain the OOBN shown in Figure 3.37. In general, when objects

are encapsulated within other objects this approach provides us a method for obtaining a hierarchical representation of the model; each level corresponds to a particular level of abstraction revealing the encapsulated attributes for the current layer of objects.

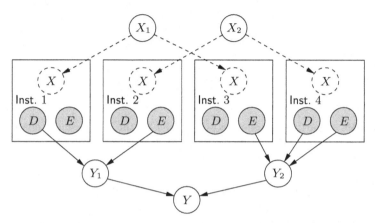

Fig. 3.37. An object-oriented Bayesian network model corresponding to the model shown in Figure 3.36. The encapsulated attributes have been hidden to simplify the representation.

Top-Down Construction of OOBNs

The input attributes and the output attributes are also referred to as the *interface* of the object, since instantiating these nodes will d-separate the internal part of the object (the encapsulated attributes) from the rest of the network (the proof is left as an exercise. This property supports a top-down model construction process: you may start constructing the model at a high level of abstraction by including only the interfaces of the objects without specifying their internal details. Later you can change the abstraction level and start specifying/refining the internal class description.

For example, assume that you should construct a Bayesian network model for the safety characteristics of a car. We know that the type of car and its maintenance level influence both the general steering characteristics of the car as well as its braking capabilities. In turn, these two aspects influence the steering safety and the braking power of the car.

We also know that the steering safety and the braking power are influenced by the grip of the car, and the grip is mainly determined by the tire type and the tire mileage. However, it may happen that at the time of model specification we do not know (or do not want to specify) the relationship between the grip of the car and tire type and mileage. See Figure 3.38 for a

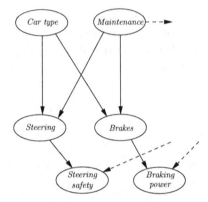

Fig. 3.38. A partial Bayesian network model for the safety characteristics of a car. The dashed arrows indicate unspecified parent and child relations.

partial Bayesian network representation. We could instead construct a class representing the grip of the car with a rudimentary internal structure and simply include the interface of the class in the model. An example is shown in Figure 3.39. Figure 3.40 shows two possible specifications of a class modeling the tire grip. The leftmost class could serve as an initial approximation to the more detailed specification shown at the right-hand side of Figure 3.40.

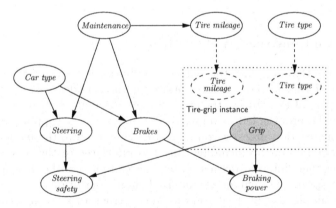

Fig. 3.39. An object-oriented Bayesian network model of the driving characteristics of a car.

Subclassing and Inheritance

A powerful property of object-oriented modeling is the use of subclassing (or inheritance) between classes. When a class C' is a *subclass* of another class

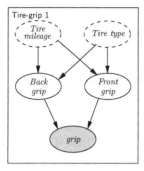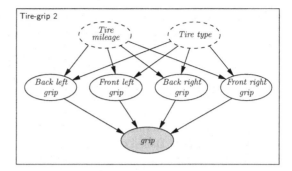

Fig. 3.40. Two possible refinements of the interface for the grip class illustrated in Figure 3.39. In the rightmost refinement, we model the grip on each of the tires.

C (also called the *superclass* for C'), then an instance of C can always be substituted with an instance of class C'. For example, consider again the two classes in Figure 3.40. We wish for the class Tire grip 2 to be viewed as a subclass of Tire-grip 1, which means that any instance of Tire-grip 1 can be substituted with an instance of Tire-grip 2. This example is quite obvious, since the two classes have the same interface connecting them to the rest of the model. However, suppose now that we should refine our grip model so that it also covers the car type; we assume that for a car with front-wheel drive there is a tendency for the front tires to be more worn than for a car with rear-wheel drive (conversely for cars with rear-wheel drive). One way to include these considerations into the model is to construct a class as in Figure 3.41.

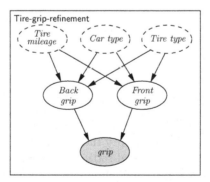

Fig. 3.41. The class Tire-grip-refinement taking the car type into account.

We would now like to be able to replace the instance in Figure 3.39 with an instance of class Tire-grip-refinement. However, this raises a technical question:

If we simply replace the instance in Figure 3.39 without connecting the input node *Car type* to an actual node in the model, then both *Back Grip* and *Front Grip* would have a parent with an unspecified probability distribution (see Figure 3.42). In order to avoid this problem, we associate a so-called *default potential* with each input node in the class; a default potential is simply a probability distribution that will be used when an input node is not connected to a node in the surrounding model. For the example above, we could specify the default potential $P(Car\ type) = (0.5, 0.5)$, assuming that the node is binary. Based on these considerations we require that if a class C' should be a *subclass* of another class C, then it should hold that:

- the set of input variables for C is a subset of the input variables for C', and
- the set of output variables for C is a subset of the output variables for C'.

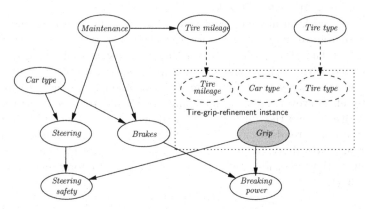

Fig. 3.42. An object-oriented Bayesian network model of the driving characteristics of a car. The input node *Car type* is associated with the default potential $P(Car\ type) = (0.5, 0.5)$.

We can construct additional subclasses of Tire-Grip representing different aspects of the grip of the car. The classes can be organized in a hierarchy according to their subclass/superclass relationship. In turn we can view this class hierarchy as a model repository that facilitates a quick top-down model construction, and for more general settings, we can construct generic repositories of classes representing common modeling problems.

When we subsequently use the object-oriented Bayesian network model for answering queries (i.e., doing belief updating), we first observe that an object-oriented Bayesian network can be seen as a standard Bayesian network with some extra features for simplifying the model specification. This also implies that inference in an OOBN can be performed by first transforming the model into a standard Bayesian network, and then applying any inference algorithm on the produced network (see Chapter 4). Transforming an OOBN into a BN is basically a matter of recursively merging each input node with its parent in the surrounding model. Methods have also been developed whereby you keep the OOBN structure and respect the privacy of the encapsulated attributes. The inference method transmits probability distributions only over the interface nodes between the objects.

3.3.7 Dynamic Bayesian Networks

When working with domains that evolve over time, you can introduce a discrete time stamp and have a model for each unit of time. We call such a local model a *time slice*. Consider, for example, the model for infected milk in Figure 3.43.

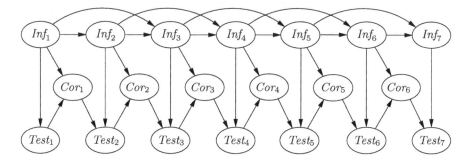

Fig. 3.43. A seven-day model with a two-day memory for infection as well as correctness of test.

For each time slice i, you have three variables Inf_i, $Test_i$, and Cor_i. The three variables are connected in a time slice, as shown in Figure 3.44.

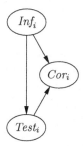

Fig. 3.44. A time slice for infected milk.

The time slices are connected through *temporal links* to constitute a full model. If the structures of the time slices are identical, and if the temporal links are the same, we say that the model is a *repetitive temporal model*. If the conditional probabilities are also identical, we call the model a *dynamic Bayesian network model*.

The model for transmission of symbols in Section 3.2.4 can be considered a temporal repetitive model, but it is not a dynamic Bayesian network because the conditional probabilities are not identical. On the other hand, the seven day model in Figure 3.2 is a dynamic Bayesian network.

A special category of time-stamped model is that of the *hidden Markov models*. They are strictly repetitive models with an extra assumption (the Markov property): the past has no impact on the future given the present. The model in Figure 3.2 is an example of a hidden Markov model, but in Figure 3.43 influence from Inf_{i-1} may flow to Inf_{i+1} regardless of our knowledge of time slice i. The latter model can, however, be transformed to a hidden Markov model by introducing a copy Inf_i^* of Inf_{i-1} in the ith time slice (see Figure 3.45).

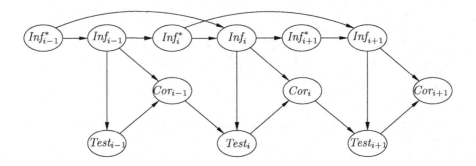

Fig. 3.45. The model of Figure 3.43 transformed into a hidden Markov model.

The reason for the term *hidden* Markov model is that under the surface (the test results) there is a hidden activity that cannot be observed (the infections).

A *Kalman filter* is a hidden Markov model in which exactly one variable has relatives outside the time slice. The model in Figure 3.2 is a Kalman filter. A *Markov chain* is a Kalman filter consisting of exactly one variable in each time slice. Note that a hidden Markov model can be transformed to a Markov chain by taking the cross product of all variables in each time slice.

In modeling domains that are evolving over time, there is a distinction between *finite-horizon* and *infinite-horizon* domains. The infected milk problem is an infinite-horizon domain, and a typical finite-horizon domain is a cornfield from sowing to harvest.

Specifying a repetitive temporal model can be eased by introducing a couple of new features to the specification language. Apart from the structure of a time slice, you must specify the number of time slices and the temporal links. The number of slices can be written in a special box, and you can introduce a special kind of arrow to specify temporal links. A number attached to a temporal link can specify the number of time steps to jump (if no number is specified, the link goes from slice i to slice $i + 1$). In Figure 3.46, we have used an extended specification language for the model in Figure 3.43.

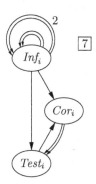

Fig. 3.46. A compact specification of the model in Figure 3.43 (an extension of Figure 3.44). The $\rightrightarrows$ indicates a temporal link. The number "2" attached to one of them specifies that it jumps two time steps (no number attached means a jump from slice i to slice $i + 1$).

Dynamic Bayesian networks are easily modeled through the object-oriented approach: the output variables are the variables with a child in later time slices, and the input variables are parents from earlier time slices. In Figure 3.46 the output variables are Inf_i and Cor_i, and the input variables are Inf_{i-1}, Inf_{i-2}, and Cor_{i-1}.

So from a modeling point of view, it is quite straightforward to work with time-stamped models. However, they will often yield calculational problems (see Exercise 3.25 and Chapter 4).

3.3.8 How to Deal with Continuous Variables

Consider the *Cold or Angina?* example from Section 3.1.2, in which the variable *Fever?* was given a discrete state space with three states (chosen a bit arbitrarily). A more natural way of representing fever would be to use a continuous variable (typically drawn using a double circle as in Figure 3.47(a)).

With a continuous variable we can no longer encode the uncertainty using a conditional probability table. Instead we will have to specify a *density function* for each combination of states for the parent variables for *Fever?*. A typical

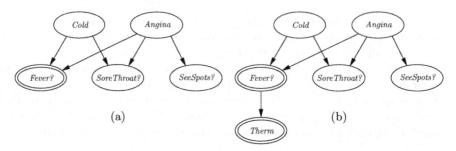

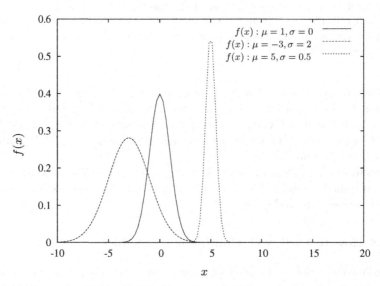

Fig. 3.47. Figure (a) shows the cold and angina model in which *Fever?* is represented by a continuous variable (drawn as a double circle). In Figure (b) the model is extended with another continuous variable *Therm* that models the accuracy of the thermometer.

density function is the normal distribution (or Gaussian distribution), which is defined by a mean μ and a variance σ^2 (see Figure 3.48 for examples):

$$f(x) = \frac{1}{\sqrt{2\pi\sigma^2}} \exp\left(-\frac{(x-\mu)^2}{2\sigma^2}\right).$$

Fig. 3.48. Example of normal distributions with different values for the mean and the variance.

For the example above, we should therefore specify a μ and a σ^2 for each state combination of the variables *Cold* and *Angina* (the resulting function is also called a *conditional Gaussian distribution*). A possible specification could be as in Table 3.19.

		Cold?	
		no	*yes*
	no	$(37°C, 0.25)$	$(37.5°C, 0.75)$
Angina?	*mild*	$(38°C, 0.5)$	$(38.5°C, 1)$
	severe	$(39°C, 0.75)$	$(39.5°C, 1.25)$

Table 3.19. Means and variances for the *Fever?* variable.

The model in Figure 3.47(a) can be extended to also represent the accuracy of the thermometer. Specifically, the thermometer that I use is rather old with an accuracy corresponding to a variance of 0.25. In addition to this it has a peculiar tendency of showing 1°C plus 5% more than the actual temperature. This situation is modeled in Figure 3.47(b). The continuous variable *Therm* represents the thermometer, and it is assigned a conditional Gaussian distribution, where the variance is set to 0.25 and the mean is specified as a linear function of *Fever?*:

$$\mu_{Therm} = 1.0 + 1.05 \cdot x_{Fever?}.$$

Given this model, we can now answer queries such as $P(Cold \mid Therm = 39.2°C, SoreThroat? = yes, SeeSpots? = no)$ and $f(Fever \mid Therm = 39.2°C, SoreThroat? = yes, SeeSpots? = no)$; the latter density is a linear combination of conditional Gaussian distributions. For example, if we use the probabilities specified in Section 3.2.5 together with the conditional Gaussian distributions described above we get $P(Cold \mid Therm = 39.2°C, SoreThroat? = yes, SeeSpots? = no) = (0.13(y), 0, 87(n))$, and for $f(Fever \mid Therm = 39.2°C, SoreThroat? = yes, SeeSpots? = no)$ we get a mean and a variance of 36.67°C and 0.127, respectively. We will not present the methods for calculating posterior probabilities in networks with continuous variables.

Bayesian networks containing both discrete and continuous variables are also called *hybrid Bayesian networkshybrid*. Unfortunately, in order to perform exact probability updating in these types of networks we need to put some rather severe constraints on the networks. In general, we require that:

- Each continuous variable be assigned a (linear) conditional Gaussian distribution. That is, for each configuration **c** of the discrete parents, the variance $\sigma_{\mathbf{c}}^2$ is a constant (independent of the continuous parents) and the mean $\mu_{\mathbf{c}}$ is a linear function of the continuous parents $Y_1, \ldots, Y_m$:

$$\mu_{\mathbf{c}} = a_{\mathbf{c}} + \sum_{i=1}^{m} a_{\mathbf{c}}^i y_i.$$

- No discrete variable have continuous parents.

Note that if a continuous variable does not have any parents, then it is assigned an unconditional normal distribution.

Whether these two constraints can be met is strongly dependent on the domain being modeled. For example, you may argue that it is inappropriate to assign a conditional Gaussian distribution to the *Fever?* variable, since the distribution is defined over the entire real line and it will therefore also assign a nonzero probability mass to impossible temperature intervals. On the other hand, when specifying probabilities you are almost always making some kinds of approximations, and the question is then whether the specified Gaussian distribution is within an acceptable distance from what you deem the "correct" distribution. If it is not, you have to look for other ways of specifying the probabilities (an example of this is given below). The second constraint is more serious, since it puts restrictions on the structure of the domains that can be modeled. For instance, if we were to extend the model with a child, *Headache?* (having states *yes* and *no*), of *Fever?*, then the structural constraint would be violated.

If it is not possible to meet the two constraints above, then one possibility would be to approximate by discretizing the continuous variables. Assume that we have the specification in Table 3.19, and we should now specify intervals for a finite set of states. For the three states *no*, *low*, and *high*, it would be natural to use knowledge of fever. In other situations, you would try to determine intervals such that for each parent configuration most of the probability mass is concentrated in a few intervals. This may not be possible, and it will often be a delicate matter to establish a good set of intervals. In the current situation, we define low fever to be in the interval $(37.5°C, 38.5°C)$. Consequently, *no* is $(-\infty, 37.5°C)$ and *high* is $(38.5°C, \infty)$. Next, you use Table 3.19 to calculate the probability mass for each interval. The result is given in Table 3.20.

		Cold?	
		no	yes
	no	(0.834, 0.165, 0.01)	(0.5, 0.376, 0.124)
Angina?	mild	(0.24, 0.52, 0.24)	(0.159, 0.341, 0.5)
	severe	(0.042, 0.24, 0.718)	(0.037, 0.149, 0.814)

Table 3.20. The result of sampling Table 3.19 to the intervals for *no*, *low*, and *high*.

3.3.9 Interventions

You may wish to incorporate actions that change the state of some variables. You may, for example, wish to model the result of cleaning the spark plugs in the car start problem. If you use the model in Figure 2.16 directly and enter your cleaning of the spark plugs by entering $SP = yes$, you get incorrect results. The problem is that you may no longer have a start problem, and the state of St may be changed due to your action. The problem is called *persistence*. You may extend the model in Figure 2.16 with a variable *Clean?*, but then you also must introduce new nodes for the variables that may change state. Because you have a causal model, the nonpersistent nodes are the descendants of the nodes affected by the intervention (see Fig. 3.49). The variable *Clean?* has a special status in the model. It is not meaningful to give it prior probabilities, and the descendants of the nodes have no meaning before a decision on *Clean?* has been taken. Therefore, it is customary to give this kind of node a rectangular shape.

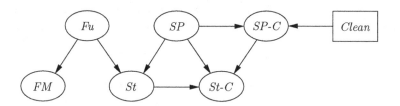

Fig. 3.49. A network modeling the effect of cleaning the spark plugs.

The conditional probabilities for new nodes are natural. If *Clean?* = *no*, then *SP-C* is in the same state as *SP*, and if *Clean?* = *yes* and *SP* = *yes*, then the probability that *SP-C* = *no* is the probability that you can clean the spark plugs properly. For *St-C*, you still have a start problem unless it was due to dirty spark plugs and they have been properly cleaned.

3.4 Special Features

A Bayesian network model is primarily used for belief updating. However, you may request other kinds of information from a model. This section outlines some types of requests. Chapter 5 gives a more detailed presentation. To illustrate the features in this section, we use the sore throat example from Section 3.1.2 (see Figure 3.50). However, we change the potentials slightly: when I suffer from mild angina, I will see yellow spots with probability 0.01, and it also happens with probability 0.001 that I have severe angina without a sore throat, provided that I do not have a cold. The rest of the potentials can be found in Sections 3.2.5 and Section 3.3.8.

We use the evidence $e = \{$ *Fever?* = *no*, *SoreThroat?* = *no*, *See Spots?* = *yes*$\}$ (do not ask why I looked down my throat).

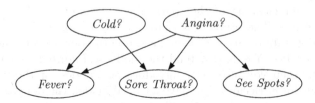

Fig. 3.50. The sore throat model.

3.4.1 Joint Probability Tables

Because it is not unusual to suffer from both cold and angina, it may be of interest to use the model in Figure 3.50 to calculate the joint probability table $P(Angina?, Cold? | e)$. This can be done by use of the fundamental rule

$$P(Angina?,\ Cold? | e) = P(Angina? | Cold?, e)P(Cold? | e).$$

Read $P(Cold? | e)$ from the system; then enter and propagate first $Cold? = yes$ and then $Cold? = no$ to get $P(Angina? | Cold?, e)$.

This method is conceptually easy, but if you request the joint table for many variables, it is computationally very time-consuming. Other methods are presented in Chapter 5.

3.4.2 Most-Probable Explanation

Instead of requesting the full joint probability table, I may request the most-probable configuration of $Cold?$ and $Angina?$. This can be achieved much faster than by calculating $P(Cold?, Angina? | e)$ and picking the state with highest probability.

In general, you have a set of instantiated variables and you request the most-probable configuration of the remaining variables. This is also called the *most-probable explanation*, MPE. MPE can be calculated similarly to probability updating (see Section 2.3.4 and Chapter 4). The only difference is that instead of marginalizing by summing out, you take the maximum. The distributive law for max reads $\max(ab, ac) = a\max(b, c)$. In the general form, it says

$$\text{If } A \notin \text{dom}(\phi_1), \text{ then } \max_A \phi_1\phi_2 = \phi_1 \max_A \phi_2.$$

Most Bayesian network systems have a special feature for calculating MPE.

3.4.3 Data Conflict

Although the evidence e yields posterior probabilities for $Cold?$ as well as for $Angina?$, it is more likely that I have misinterpreted what I saw in the throat.

In other words, in the light of neither fever nor sore throat, it is very likely that the evidence *See Spots? = yes* is faulty. It would be nice if the system by itself could raise a flag indicating that the evidence does not seem coherent.

To investigate coherence of the evidence, a *conflict measure* is defined. The idea behind the measure is that correct findings from a coherent case covered by the model support each other, and therefore we will expect them to be positively correlated. For example, if e_1 and e_2 are two pieces of evidence, then we would expect $P(e_1 \mid e_2) > P(e_1)$ and therefore $P(e_1, e_2) = P(e_1 \mid e_2)P(e_2) > P(e_1)P(e_2)$. Let $e = \{e_1, \ldots, e_m\}$ be a set of findings. Based on the intuition above, the conflict measure on e is defined as

$$\mathrm{conf}(e) = \log_2 \frac{P(e_1) \cdots P(e_m)}{P(e)}.$$

The conflict measure is easy to calculate because $P(e)$ is communicated by the system (see Example 3.9) and $P(e_i)$ can be read from the model in its initial state. If $\mathrm{conf}(e)$ is positive, the findings are not positively correlated, and we can take this as an indication that the evidence is conflicting. To be quite accurate, a high conflict measure is an indication that there is discrepancy between model and evidence. This may be due to flawed findings, it may be because we are faced with a very rare case, or the situation may not be covered by the model. This is discussed in more detail in Section 5.5.

3.4.4 Sensitivity Analysis

Sensitivity analysis refers to analyzing how sensitive the conclusions (the probabilities of the hypothesis variables) are to minor changes. The changes may be variations of the parameters of the model or may be changes of the evidence (*SE analysis*). In general, sensitivity analysis is rather technical and in this section we only give some hints. It is treated in more detail in Chapter 5.

Consider the angina example. The conclusion is $P(Angina? \mid e) = (0, 0.98, 0.02)$. SE analysis consists in answering questions such as, "what are the crucial findings?", "what if one of the findings was changed or removed?" or "what set of findings would be sufficient for the conclusion?" If we consider the conclusion to be that I suffer from mild angina, we see that the finding *See Spots? = yes* is not sufficient in itself because it indicates severe angina, nor is any of the other findings. Instead, *See Spots? = yes* together with *SoreThroat = no* is sufficient, and with these two findings fixed, the conclusion is insensitive to any finding on *Fever?*.

Now consider the parameters $t = P(SoreThroat? = no \mid Angina? = severe, Cold? = no)$ and $s = P(See\ Spots)yes \mid Angina? = mild)$. The initial values of t and s are 0.001 and 0.01, respectively. What we might look for is a functional expression for $P(Angina? = mild \mid e)(t)$ and $P(Angina? = mild \mid e)(s)$. This is called one-way sensitivity analysis. We might also request two-way sensitivity analysis by establishing $P(Angina? = mild \mid e)(t, s)$.

It follows from a general theorem that $P(e)(t)$ as well as $P(Angina? = mild, e)(t)$ are linear expressions (see Section 5.7), and hence $P(Angina? = mild \mid e)(t)$ is a quotient of two linear expressions. From the initial propagation, we can acquire $P(e)(0.001)$ and $P(Angina? = mild \mid e)(0.001)$. By changing t to 0.002 and propagating, we get $P(e)(0.002)$ and $P(Angina? = mild \mid e)(0.002)$. These four values are sufficient for determining the four constants in the functional expression for $P(Angina? = mild \mid e)(t)$.

3.5 Summary

Types of Variables in Building a Bayesian Network Model

Hypothesis variables: Variables with a state that is asked for. They are, however, either impossible or too costly to observe directly.

Information variables: Variables that can be observed.

Mediating variables: Variables introduced for a special purpose. It may be to properly reflect the independence properties in the domain, to facilitate the acquisition of conditional probabilities, to reduce the number of distributions to acquire for the network, or for other purposes.

Warning: It is tempting to introduce mediating variables in order to have a more refined model of the domain; however, if they do not serve any other purpose you should get rid of them. They jeopardize performance.

Acquiring Conditional Probabilities

Theoretically well founded probabilities as well as frequencies and purely subjective estimates can be used in the same network.

If the number of distributions is too large for a reasonable estimation, a simplifying assumption can reduce it.

Noisy-or: Let B have the parents $A_1, \ldots, A_n$ (all variables binary). Suppose that $A_i = y$ causes $B = y$ unless it is inhibited by an inhibitor Q_i that is active with probability q_i. Assume that the inhibitors are independent. Then,

$$P(B = n \mid a_1, \ldots, a_n) = \prod_{j \in Y} q_j,$$

where Y is the set of indices for the states y.

Divorcing: Let B have the parents $A_1, \ldots, A_n$. Assume that the set of configurations of $(A_1, \ldots, A_i)$ can be partitioned into the sets $c_1, \ldots, c_m$ such that whenever two configurations a_1^* and a_2^* of $(A_1, \ldots, A_i)$ are elements in the same c_j, then

$$P(B \mid a_1^*, A_{i+1}, \ldots, A_n) = P(B \mid a_2^*, A_{i+1}, \ldots, A_n).$$

Then, $A_1, \ldots, A_i$ can be divorced from $A_{i+1}, \ldots, A_n$ by introducing a mediating variable C with states $c_1, \ldots, c_m$, making C a child of $A_1, \ldots, A_i$ and a parent of B.

Other Tricks

Undirected relations – in particular, logical constraints – can be modeled by introducing a dummy child of the constrained variables and letting its potential reflect the relation.

For a specification language for *repeating structures*, see Figure 3.51.

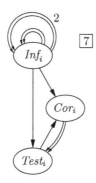

Fig. 3.51. A compact specification of a repeating structure with 7 slices. The $\rightrightarrows$ indicates a temporal link. The number "2" attached to one of them specifies that it jumps two time steps (no number attached means a jump from slice i to slice $i+1$).

Expert disagreements on potentials can be represented in a model by introducing a node representing the experts.

Continuous variables can be represented in the model if:

- they do not have any discrete children, and
- they are assigned a linear conditional Gaussian distribution.

If these two conditions cannot be met, an alternative is to transform them into variables with a finite number of states.

3.6 Bibliographical Notes

Naive Bayes was used by de Dombal *et al.* (1972) and can be traced back at least to Minsky (1963). Noisy-or was first described by Pearl (1986); divorcing was used in MUNIN (Andreassen *et al.*, 1989). Exercise 3.27 is based on (Cooper, 1990). Chain graphs are treated in depth in (Lauritzen, 1996). Dynamic Bayesian networks are described in (Kjærulff, 1992). The compact representation of repetitive structures was suggested by Bangsø and Wuillemin (2000). Andreassen (1992) discusses various ways of transforming conditional Gaussian variables into finite variables. A method not described in this chapter is *similarity networks* (Heckerman, 1990). The method helps in eliciting the conditional probabilities. Other elicitation methods can be found in (Druzdzel and van der Gaag, 1995). Object oriented Bayesian networks were introduced in (Koller and Pfeffer, 1997); the version presented here is the one from (Bangsø and Wuillemin, 2000). References for the special features in Section 3.4 are given in Section 5.9.

3.7 Exercises

Exercise 3.1. Peter is currently taking three courses on the topics of probability theory, linguistics, and algorithmics. At the end of the term he has to take an exam in two of the courses, but he has yet to be told which ones. Previously he has passed a mathematics and an English course, with good grades in the mathematics course and outstanding grades in the English course. At the moment, the workload from all three courses combined is getting too big, so Peter is considering dropping one of the courses, but he is unsure how this will affect his chances of getting good grades in the remaining ones. What are reasonable variables of interest in assessing Peter's situation? How do they group into information, hypothesis, and mediating variables?

Exercise 3.2. Assume that three mornings in a row I wonder whether my sore throat is due to cold or angina. Construct a model.

Exercise 3.3. Construct a model extending the model in Section 3.1.3 with a scanning test.

Exercise 3.4. Consider the following variables relating to a single household consisting of a couple and possibly some children:

- *Illness at the moment*, with states *severe illness*, *minor illness*, and *no illness*.
- *History of illness*, with states *cases of severe illness*, *often minor illnesses*, and *rarely minor illness*.
- *Number of children*, with states *none*, *one*, *two*, *three*, and *four and up*,
- *Working parents*, with states *both*, *father*, *mother*, and *none*.
- *Religion*, with states *Christianity*, *Judaism*, *Islam*, *Buddhism*, *Atheism*, and *other*.
- *Household income*, with states *$0–$50000*, *$50000–$100000*, and *$100000– and up*.
- *Fish-eating habits*, with states *often fish* and *rarely fish*.

- *Fiber-eating habits*, with states *lots of fiber* and *not much fiber*.
- *Drinking habits*, with states *never alcohol*, *wine once in a while*, *often wine*, and *wine every day*.

Try to construct a Bayesian network incorporating the above variables according to your perception of the world. What are the d-separation properties of the network you constructed?

Exercise 3.5. [E] Construct a model for a single milk test (Section 3.2.1). What is the probability of infected milk given a positive test result?

Exercise 3.6. [E] Ground meat purchased in the supermarket may be infected. On average, it happens once out of 600 times. A test with results *positive* and *negative* can be used. If the meat is *clean*, the test result will be *negative* in 499 out of 500 cases, and if the meat is *infected*, the test result will be *positive* in 499 out of 500 cases.

Construct a Bayesian network and use a software system to calculate the probability of *infected* for meat with a positive test result.

Exercise 3.7. [E] Complete the Bayesian network for Cold or angina? and perform a self-diagnosis.

Exercise 3.8. [E] Consider the insemination example from Section 3.1.3. Let the probabilities be as in Table 3.21 ($Ho = y$ means that hormonal changes have taken place) $P(Pr) = (0.87, 0.13)$.

	$Pr = y$	$Pr = n$		$Ho = y$	$Ho = n$
$Ho = y$	0.9	0.01	$BT = y$	0.7	0.1
$Ho = n$	0.1	0.99	$BT = n$	0.3	0.9

	$Ho = y$	$Ho = n$
$UT = y$	0.8	0.1
$UT = n$	0.2	0.9

Table 3.21. Tables for Exercise 3.8.

(i) What is $P(Pr \mid BT = n, UT = n)$?
(ii) Construct a naive Bayes model. Determine the conditional probabilities for the model using the model above. What is $P(Pr \mid BT = n, UT = n)$ in this model?

Exercise 3.9. [E] Use the model from Exercise 3.8 to calculate $P(UT = y, BT = y)$. Enter the two pieces of evidence into the model and prompt your system to update probabilities. As a side effect, the system computes $P(e)$, the probability of the evidence entered. Find out how your system provides it.

Exercise 3.10. [E]

(i) Implement the seven-day model in Figure 3.13. Are the initial probabilities stable over time?
(ii) Consider the conditional probability tables $P(Inf_2 \mid Inf_1)$ and $P(Inf_1) = (0.0007, 0.9993)$ and assume that the risk of becoming infected is 0.0002. We require that the initial probabilities be stable: $P(Inf_2) = P(Inf_1) = (0.0007, 0.9993)$. Show that the chance of being cured must be 2/7.
(iii) Consider the conditional probabilities $P(Inf_{i+2} \mid Inf_i, Inf_{i+1})$, and assume that the risk of being infected is the same as above. We require stable initial probabilities. Show that the chance of being cured for a more than one day infection must be 0.4.

Exercise 3.11. Show that the procedure described in Section 3.1.5 is equivalent to updating in the model in Figure 3.12.

Exercise 3.12. [E] Consider the stud farm example in Section 3.2.2 and let the prior probability for aA be 0.005.

(i) Enter the model into your Bayesian network system.
(ii) Add to the model the frequency 0.001 for mutation of the gene from A to a.
(iii) Construct a model for the situation in part (ii), but for a recessive gene borne by the female sex chromosome. (Note that horses with the disease are taken out of production.)

Exercise 3.13. [E] Consider the transmission example from Section 3.2.4.

(i) From Table 3.10, calculate the remaining conditional probabilities for the model in Figure 3.18.
(ii) Implement the model.
(iii) The sequence *baaca* is received. What is the most-probable symbol transmitted according to the model in Figure 3.18? What is the most-probable word?
(iv) What is the most-probable word according to the model in Figure 3.19?

Exercise 3.14. [E] Consider the simplified poker game in Sections 3.1.4 and 3.2.3.

(i) Implement the system.
(ii) Extend the system with a facility giving the chances that your hand is better than your opponent's hand.

Exercise 3.15. [E] Construct a naive Bayes model of the simplified poker game example in Sections 3.1.4 and 3.2.3 with *OH2* being the class variable. Use your implemented model from Exercise 3.14 to calculate the needed probabilities for the naive Bayes model. What is $P(OH2 \mid FC1 = 1, FC2 = 2)$ using the model from Exercise 3.14? What is $P(OH2 \mid FC1 = 1, FC2 = 2)$ using the naive Bayes model?

Exercise 3.16. You are confronted with three doors, A, B, and C. Behind exactly one of the doors there is $10,000. When you have pointed at a door, an official will open another door with nothing behind it. After he has done so, you are allowed to alter your choice. Should you do that?

Exercise 3.17. Extend the model in Figure 3.23 to incorporate constraints on color and pattern for the same sock.

Exercise 3.18. The *drive* in golf is the first shot in playing a hole. If you drive with a *3-wood* (a particular type of golf club), there is a 2% risk of a miss (a bad drive), and $\frac{1}{4}$ of the good drives have a length of 180 m, $\frac{1}{2}$ are 200 m, and $\frac{1}{4}$ have a length of 220 m. You may also use a *driver* (another type of golf club). This will on average increase the length by 10%, but you will also have 3 times as high a risk of a miss. Both wind and the slope of the hole may affect the result of the drive. Wind doubles the risk of a miss, and the length is affected by 10% (longer if the wind is from behind and shorter otherwise). A downhill slope yields 10% longer drives, and an uphill slope decreases the length of the drive by 10%.

Estimate the probabilities for miss and length given the various factors.

Exercise 3.19. The *putt* is (usually) the last shot on a golf hole. My ball is lying 1 m away from the hole, and under normal circumstances I will miss 1 putt out of 10. However, when it rains, I miss 1 out of 7; if it is windy, I miss 1 out of 4; if the green is curved, I miss 1 out of 3; and if I am putting for a birdie (one under par), I miss 1 out of 2.

Estimate the probabilities for success and failure given the various factors.

Exercise 3.20. Show that the model in Figure 3.26 corresponds to the one in Figure 3.25.

Exercise 3.21. [E] Show that noisy or may be modeled as described in Figures 3.30 and 3.31. Apply this model to the putting problem of Exercise 3.19, and compare the number of quantities to specify.

Exercise 3.22.

(i) Complete the model in Section 3.3.4.

$$P(Ha) = P(Ha \mid Ot = y) = (0.93, 0.04, 0.02, 0.01),$$
$$P(Ha \mid Fe = y) = P(Ha \mid Ho = y) = P(Ha \mid Fb = y) = (0.1, 0.8, 0.1, 0),$$
$$P(Ha \mid Bt = y) = (0.3, 0.2, 0.2, 0.3).$$

$As \setminus Ha_1$	no	mild	moderate	severe
y	(1, 0, 0, 0)	(0.7, 0.3, 0, 0)	(0.1, 0.7, 0.2, 0)	(0, 0.1, 0.7, 0.2)
n	(1, 0, 0, 0)	(0, 1, 0, 0)	(0, 0, 1, 0)	(0, 0, 0, 1)

Table 3.22. $P(Ha \mid Ha_1, As)$ for Exercise 3.22.

(ii) Include aspirin in the basis of Table 3.22.

Exercise 3.23. Specify the model in Figure 3.4 as an OOBN.

Exercise 3.24. Construct an OOBN model for the stud farm in Section 3.2.2. Use default potentials for horses with parents outside the model.

Exercise 3.25. [E] Consider the model in Figure 3.52. All variables have ten states.

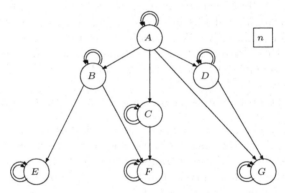

Fig. 3.52. A compact representation of a dynamic Bayesian network for Exercise 3.25.

(i) Implement one time slice (with any set of potentials).
(ii) Implement three time slices.
(iii) How many time slices can you implement before your system reports that it requires extra memory?

Exercise 3.26. [E] Consider a soccer tournament with 8 teams. Teams 1 to 4 are poor ones, and Teams 5 to 8 are good ones. Each match is between two teams drawn at random from those that have played the same number of matches previously in the tournament. The loser of each match is eliminated from the tournament. The probability of a good team winning a match against another team is 0.8 if the other team is a poor one, and 0.5 if the other team is a good one. The probability of a poor team winning a match against another poor team is 0.5. What is the probability of a poor team making it to the final? (Hint: For each match, generate a variable that represents the winner (with states *poor team* and *good team*), and variables that represent each contestant in the opening matches (with states *poor team* and *good team*). Finally, use constraint nodes to ensure compliance with the restrictions in the exercise.)

Exercise 3.27. [E] The following relations hold for the Boolean variables A, B, C, D, E, and F:

$(A \vee \neg B \vee C) \wedge (B \vee C \vee \neg D) \wedge (\neg C \vee E \vee \neg F) \wedge (\neg A \vee D \vee F) \wedge$
$(A \vee B \vee \neg C) \wedge (\neg B \vee \neg C \vee D) \wedge (C \vee \neg E \vee \neg F) \wedge (A \vee \neg D \vee F)$.

 (i) Is there a truth value assignment to the variables making the expression true? (Hint: Represent the expression as a Bayesian network.)
 (ii) We receive the evidence that A is false and B is true. Is there a truth value assignment to the other variables making the expression true?
 (iii) The *satisfiability problem* for propositional calculus is, given a Boolean expression $\mathbb{E}$ (over n Boolean variables), is there a truth-value assignment to the variables that makes $\mathbb{E}$ true?
 Show that a method for calculation of probabilities in Bayesian networks yields a method for solving the satisfiability problem for propositional calculus. (Hint: Assume that $\mathbb{E}$ is in conjunctive normal form.)
 (iv) Show that probability calculation in Bayesian networks is NP-hard.

Exercise 3.28. You have the model $A \rightarrow B$ and $P(A) = (0.7, 0.3)$. Two experts give the tables in Table 3.23, and you have no reason to believe more in one expert than in the other.

You receive the evidence $A = y$. What are the posterior probabilities for B and the experts?

$B \setminus A$	y	n
y	0.9	0.4
n	0.1	0.6

$P_1(B \mid A)$

$B \setminus A$	y	n
y	0.6	0.4
n	0.4	0.6

$P_2(B \mid A)$

Table 3.23. Table for Exercise 3.28.

Exercise 3.29. [E]

(i) Take your model from Exercise 3.7 and enter the evidence $e = \{Fever? = no, Sore\ Throat? = no, See\ Spots? = yes\}$. How does your system react? Change the potentials such that $P(Sore\ Throat? = no \mid Angina? = severe, Cold? = no) = 0.001$, and $P(See\ Spots? \mid Angina? = mild) = 0.01$.

(ii) Calculate $P(Cold?, Angina? \mid e)$.

(iii) Calculate MPE(e).

(iv) Calculate conf(e).

(v) Determine $P(Angina? = mild \mid e)(s)$, where $s = P(See\ Spots? = yes \mid Angina? = mild)$.

4

Belief Updating in Bayesian Networks

In this chapter, we present algorithms for probability updating. An efficient updating algorithm is fundamental to the applicability of Bayesian networks. As shown in Chapter 2, access to $P(\mathcal{U}, e)$ is sufficient for the calculations. However, because the joint probability table increases exponentially with the number of variables, we look for more-efficient methods. Unfortunately, no method guarantees a tractable calculational task. However, the method presented here represents a substantial improvement, and it is among the most-efficient methods known.

We shall use the framework of potentials. A conditional probability table $P(A \mid \mathrm{pa}(A))$ is a function $\phi : \mathrm{pa}(A) \cup \{A\} \to [0 : 1]$, and we call it a potential. For the algebra of probability tables we shall for notational convenience use functional notation. That is, the product $P(A \mid \mathrm{pa}(A)) \cdot P(B \mid \mathrm{pa}(B))$ is considered as a product of two functions $\phi_1(A, \mathrm{pa}(A))\phi_2(B, \mathrm{pa}(B))$. The reader is expected to be familiar with Section 1.4.

Sections 4.1–4.6 present the junction tree algorithm, a version of the variable elimination method. Section 4.7 presents an alternative method with any-space properties, recursive conditioning, and in Sections 4.8 and 4.9 we outline different approximation methods.

4.1 Introductory Examples

To repeat the fundamentals from Chapter 2 and for pinpointing the issues in belief updating for Bayesian networks, we consider in this section a simple example. Consider the Bayesian network in Figure 4.1 over the universe $\mathcal{U}$. The potentials specified for BN are $\phi_1 = P(A_1), \phi_2 = P(A_2 \mid A_1), \phi_3 = P(A_3 \mid A_1), \phi_4 = P(A_4 \mid A_2), \phi_5 = P(A_5 \mid A_2, A_3)$, and $\phi_6 = P(A_6 \mid A_3)$.

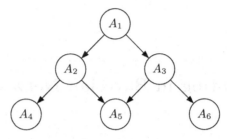

Fig. 4.1. A simple Bayesian network, BN.

4.1.1 A Single Marginal

Let us first assume that we wish to calculate $P(A_4)$. From the chain rule, we have

$$P(\mathcal{U}) = \phi_1\phi_2\phi_3\phi_4\phi_5\phi_6 \text{ and } P(A_4) = \sum_{A_1,A_2,A_3,A_5,A_6} \phi_1\phi_2\phi_3\phi_4\phi_5\phi_6.$$

To avoid calculating $P(U)$, we use the distributive law (Section 1.4):

$$P(A_4) = \sum_{A1} \phi_1(A_1) \sum_{A_2} \phi_2(A_2, A_1)\phi_4(A_4, A_2) \sum_{A_3} \phi_3(A_3, A_1)$$

$$\sum_{A_5} \phi_5(A_5, A_2, A_3) \sum_{A_6} \phi_6(A_6, A_3).$$

First, calculate $\phi_6'(A_3) = \sum_{A_6} \phi_6(A_6, A_3)$, then multiply $\phi_6'(A_3)$ by $\phi_5(A_5, A_2, A_3)$ and calculate $\phi_5'(A_2, A_3) = \sum_{A_5} \phi_5(A_5, A_2, A_3)\,\phi_6'(A_3)$; $\phi_5'(A_2, A_3)$ is multiplied by $\phi_3(A_3, A_1)$, and so forth. Notice that in the calculation of $\phi_5'(A_2, A_3)$ you can apply the distributive law again; that is, you need not multiply by $\phi_6'(A_3)$ before you marginalize A_3 out. The calculation is sketched graphically in Figure 4.2.

The reason for using the distributive law is to reduce the size of the tables to handle. The full joint, $P(\mathcal{U})$, requires a space incorporating all six variables. For the process illustrated in Figure 4.2, the largest potential to handle contains three variables. In Figure 4.3, the structure is repeated, but in each bucket (drawn as an ellipse) we have indicated the variables to handle, and the variables in a mailbox (drawn as a rectangle) indicate the domain of the potential communicated.

In the preceding calculations, we performed the marginalizations in a particular order, namely A_6, A_5, A_3, A_2, A_1, and this is reflected in the structure of Figure 4.2. Because marginalization is commutative (Section 1.4), it can be done in any order. It is standard to use the term *elimination order* rather than marginalization order. If we use the reversed elimination order, we get the structure in Figure 4.4.

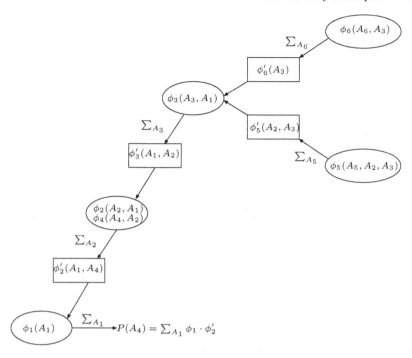

Fig. 4.2. An illustration of the process of marginalizing down to A_4. The elliptic nodes are buckets containing potentials. In a bucket, the potentials are multiplied by the incoming potentials, a variable is marginalized out, and the result is placed in a mailbox (a rectangular node) for a neighboring bucket.

Figure 4.5 illustrates the domains to handle for the last elimination order. As can be seen, the domains for the first order are smaller than the domains for the last order.

Because the size of the domains to handle is a good measure of complexity, we will address the task of finding an elimination order yielding the smallest domains to handle.

4.1.2 Different Evidence Scenarios

In the preceding calculations, we assumed that no evidence was entered into the network. By analyzing the process illustrated in Figure 4.2, we realize some simplifications. Because $\phi_5 = P(A_5 \mid A_2, A_3)$ and $\phi_6 = P(A_6 \mid A_3)$, we have that $\phi'_5 = \sum_{A_5} P(A_5 \mid A_2, A_3) = \mathbf{1}$ and $\phi'_6 = \sum_{A_6} P(A_6 \mid A_3) = \mathbf{1}$, where $\mathbf{1}$ is the unit potential. Also,

$$\phi'_3 = \sum_{A_3} \phi_3 \phi'_5 \phi'_6 = \sum_{A_3} \phi_3 \mathbf{1} \cdot \mathbf{1} = \sum_{A_3} \phi_3 = \sum_{A_3} P(A_3 \mid A_1) = \mathbf{1}.$$

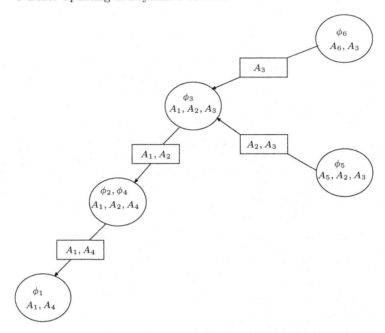

Fig. 4.3. A structure indicating the domains of the various potentials to handle.

We note that ϕ_3' is void, and the entire process is reduced to calculating $\sum_{A_1} P(A_1) \sum_{A_2} P(A_2 \mid A_1) P(A_4 \mid A_2)$.

The nodes A_3, A_5, and A_6 are examples of so-called *barren nodes*.[1] A node A is barren if neither A nor any of A's descendants have received evidence. The conditional probability potential attached to a barren node has an impact only on descendant nodes.

If we have the evidence $A_5 = a_5$ and $A_6 = a_6$, the evidence is represented as two 0-1 findings, $\mathbf{e}_5$ and $\mathbf{e}_6$ (Section 2.3.3). The formula is

$$P(A_4, e) = \sum_{A_1, A_2, A_3, A_5, A_6} \phi_1 \phi_2 \phi_3 \phi_4 \phi_5 \phi_6 \mathbf{e}_5 \mathbf{e}_6,$$

and we have (Section 2.3.3)

$$P(A_4 \mid e) = \frac{P(A_4, e)}{\sum_{A_4} P(A_4, e)}.$$

To calculate $P(A_4, e)$, the effect on the frame in Figure 4.3 is that the two evidence potentials are added in the buckets with ϕ_5 and ϕ_6 attached to them (see Figure 4.6).

---

[1] This term was first used in connection with influence diagrams (Section 9.4), where barren nodes have no influence on the decisions.

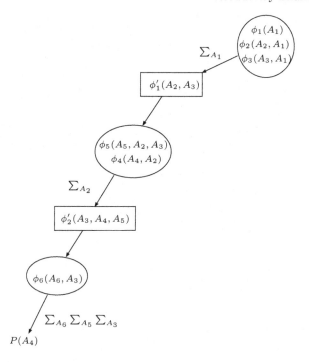

Fig. 4.4. The structure resulting from eliminating in an order that is the reverse of that from Figure 4.2.

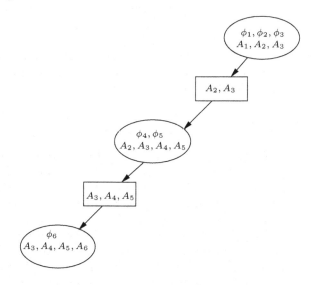

Fig. 4.5. The domains for the elimination order A_1, A_2, A_3, A_5, A_6.

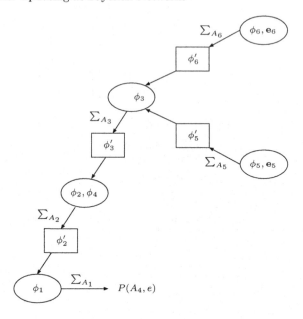

Fig. 4.6. The frame from Figure 4.2 incorporating the evidence $e_5 : A_5 = a_5$ and $e_6 : A_6 = a_6$.

The effect of e is that the variables A_5 and A_6 are instantiated in the potentials ϕ_5 and ϕ_6, and the marginalizations of A_5 and A_6 are redundant, that is, $\phi_5' = P(A_5 = a_5 \mid A_2, A_3)$ and $\phi_6' = P(A_6 = a_6 \mid A_3)$.

The process in Section 4.1.1 is sufficiently general to encompass all types of evidence scenarios. The task is to supplement this general process with methods taking advantage of simplifications due to the particular evidence scenario, such as identification of barren nodes.

4.1.3 All Marginals

Assume that we wish to compute $P(A_i, e)$ for all i. Without taking advantage of the special evidence scenario, we can for each node use the method from Section 4.1.1. Assume that we calculate $P(A_2, e)$ through the elimination order A_6, A_5, A_3, A_1, A_2. Then, the frame of potentials looks as in Figure 4.7.

As can be seen, the frame in Figure 4.7 is very similar to the frame in Figure 4.3. Only one arrow is reversed, and many calculations from the calculation of $P(A_4, e)$ can be reused. In this chapter, we present a systematic way of exploiting reuse in calculating all marginals. The resulting method has a complexity equivalent to two single-variable marginalizations.

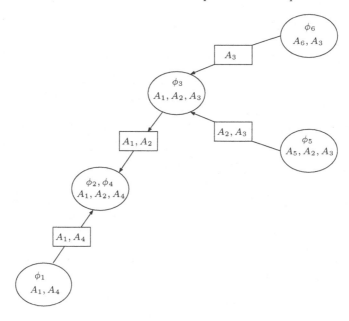

Fig. 4.7. A frame for computing $P(A_2, e)$ through the elimination order A_6, A_5, A_3, A_1, A_2.

4.2 Graph-Theoretic Representation

As illustrated in Section 4.1, belief updating for Bayesian networks consists basically in calculating sums of products. In this section, we deal systematically with this task without explicit reference to Bayesian networks. The methods presented are general and can be applied to a large variety of tasks.

4.2.1 Task and Notation

We will work with a set of real-valued potentials $\Phi = \{\phi_1, \ldots, \phi_m\}$ over finite variables from the universe $\mathcal{U} = \{A_1, \ldots, A_n\}$.

Let Ψ be any set of potentials. The product of all potentials ψ in Ψ is denoted by $\prod \psi$. We will also use the notation $\prod_{i=1}^{k} \psi_i$ for the product $\psi_1 \cdots \psi_k$, and if the boundaries are apparent from the context, we write $\prod \psi_i$.

The potential $\sum_X \phi(X, Y, \ldots, Z)$ is the sum $\phi(x_1, Y, \ldots, Z) + \cdots + \phi(x_k, Y, \ldots, Z)$, and it is defined over $(Y, \ldots, Z)$. We say that X has been *marginalized* out of $\phi(X, Y, \ldots, Z)$. If $\mathcal{V}$ is a set of variables, then $\sum_{\mathcal{V}}$ is a notation for marginalizing out all variables in $\mathcal{V}$. Because marginalization is commutative (Section 1.4), this notation is unambiguous.

Instead of sum notation, we may also use projection notation. We let $\phi^{\downarrow X}(X, Y, \ldots, Z)$ denote the potential resulting from marginalizing out $(Y, \ldots, Z)$; the potential is projected down to X. If $\mathcal{W}$ is a set of variables, then $\phi^{\downarrow \mathcal{W}}$

denotes the result of marginalizing out all variables except the members of $\mathcal{W}$.

Task: Compute $(\prod \Phi)^{\downarrow A_i}$ for all A_i.

Definition 4.1. *Let Φ be a set of potentials, and let X be a variable. Then X is eliminated from Φ through the following procedure:*

    1. Remove all potentials in Φ with X in their domains. Call the set of removed potentials Φ_X.
    2. Calculate $\phi^{-X} = \sum_X \prod \Phi_X$.
    3. Add ϕ^{-X} to Φ. Call the result Φ^{-X}; $\Phi^{-X} = \{\Phi \setminus \Phi_X, \phi^{-X}\}$.

Note that elimination of the variable X corresponds to using the distributive law on the product. Instead of calculating the product, we keep the factors in a bucket and do not multiply before we are forced to do so.

Proposition 4.1. *The task $(\prod \Phi)^{\downarrow X}$ is solved by repeatedly eliminating the variables except for X.*

It remains to establish an elimination order.

4.2.2 Domain Graphs

To get an overview of the consequences of various elimination orders, the task is represented graphically.

Definition 4.2. *Let $\Phi = \{\phi_1, \ldots, \phi_m\}$ be potentials over $\mathcal{U} = \{A_1, \ldots, A_n\}$ with $\mathrm{dom}(\phi_i) = \mathcal{D}_i$. The domain graph for Φ is the undirected graph with the variables of $\mathcal{U}$ as nodes and with a link between each pair of variables that are members of the same $\mathcal{D}_i$.*

For the sake of exposition, we assume throughout the chapter that the graphs considered are connected.

Example 4.1. In Section 4.1.1, we dealt with a Bayesian network over the potentials $\Phi = \{\phi_1(A_1), \phi_2(A_2, A_1), \phi_4(A_4, A_2), \phi_3(A_3, A_1), \phi_5(A_5, A_2, A_3), \phi_6(A_6, A_3)\}$. The domain graph for Φ is given in Figure 4.8.

Compared to the initial Bayesian network in Figure 4.1, we see that directions on the links have been dropped and that a new link (A_2, A_3) has been inserted. It is often called a *moral link* because it connects two nodes with a common child. The domain graph for a Bayesian network is called the *moral graph*.

When we eliminate a variable X, we work with the product of all potentials with X in the domain. The domain of this product consists of X and its neighbors in the domain graph, and when X is eliminated, the resulting potential has all X's neighbors in its domain. This means that in the domain

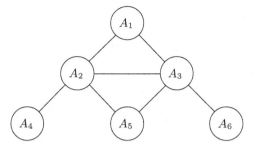

Fig. 4.8. The domain graph for $\Phi = \{\phi_1(A_1),\ \phi_2(A_2, A_1),\ \phi_3(A_3, A_1),\ \phi_4(A_4, A_2),$ $\phi_5(A_5, A_2, A_3),\ \phi_6(A_6, A_3)\}$.

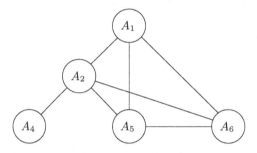

Fig. 4.9. The domain graph for Φ^{-A_3} from Figure 4.8.

graph for Φ^{-X} all neighbors of X are pairwise linked. In Figure 4.9, we show the domain graph for the example in Figure 4.8 with A_3 eliminated.

Note that the graph in Figure 4.9 has several new links. These new links are called *fill-ins*. The introduction of fill-ins highlights the fact that when eliminating A_3 you work with a potential over a domain that was not present initially. In order to avoid working with new domains, you try to avoid fill-ins. To put it another way, an elimination sequence that does not introduce fill-ins requires less space than an elimination sequence that introduces fill-ins.

In Section 4.1, we considered calculation of $P(A_4)$. In the graph-theoretic framework, it corresponds to constructing an elimination sequence ending with A_4. For the domain graph in Figure 4.8, it is possible to eliminate down to A_4 without introducing fill-ins: $A_6, A_5, A_3, A_1, A_2, A_4$. Such a sequence is called a *perfect elimination sequence*. There are several perfect elimination sequences ending with A_4, and an optimal elimination sequence will be found among them. In Figure 4.8, we see that the sequence $A_5, A_6, A_3, A_1, A_2, A_4$ as well as $A_1, A_5, A_6, A_3, A_2, A_4$ and $A_6, A_1, A_3, A_5, A_2, A_4$ are perfect elimination sequences.

Proposition 4.2. *Let* $X_1, \ldots, X_k$ *be a perfect elimination sequence, and let* X_j *be a node with a complete neighbor set.*[1] *Then, the sequence* $X_j, X_1, \ldots,$ $X_{j-1}, X_{j+1}, \ldots, X_k$ *is also a perfect elimination sequence.*

Proof. If you start by eliminating X_j, you do not introduce fill-ins. Consider variable X_i. When you eliminate X_i, you look at the uneliminated neighbors, and if a pair of them is not linked, you introduce a fill-in. Eliminating X_j before X_i does not give X_i new neighbors, and it will not enforce new fill-ins when X_i is eliminated.   □

The complexity of using a particular elimination sequence is characterized by the set of domains for the potentials used. For the elimination order $A_6, A_5, A_3, A_1, A_2, A_4$, the set of domains is $\{\{A_6, A_3\}, \{A_2, A_3, A_5\}, \{A_1, A_2, A_3\}, \{A_1, A_2\}, \{A_2, A_4\}\}$. If a domain is a subset of another domain, then it does not require extra space and we need not consider it. For example, the set $\{A_1, A_2\}$ is removed from the preceding domain set.

Definition 4.3. *The* domain set *of an elimination sequence is the set of domains of potentials produced during the elimination in which potentials that are subsets of other potentials are removed.*

Unfortunately, it does not hold that if you eliminate without introducing fill-ins, then the domain set consists only of domains from the initial set of potentials. For the preceding perfect elimination sequence, we have that when A_3 is eliminated, you work with a potential with domain $\{A_1, A_2, A_3\}$, which is not one of the initial domains. However, there is no way to avoid this. No matter which of the three variables you eliminate first, you will produce a potential with all three variables in the domain. In general, it holds that if the set $\mathcal{V}$ of variables is a complete set in the domain graph, then any elimination sequence will contain a potential with a domain including $\mathcal{V}$.

Proposition 4.3. *All perfect elimination sequences produce the same domain set, namely the set of cliques of the domain graph; a complete set is a* clique *if it is not a subset of another complete set (a maximal complete set).*

Proof. First we show that a clique V corresponds to the domain of a potential produced during the elimination. Let X be the first variable from V to be eliminated. When X is eliminated, we produce a domain D consisting of X and all its neighbors. Because all elements of V are neighbors of X, D must contain V. Let Y be a member of D. After elimination of X, there is a link between Y and all members of V. The elimination does not produce fill-ins, so the links must have been present initially, and because V is a maximal complete set, Y must be a member of V. Hence, the cliques must be members of the domain set.

---

[1] A set of nodes is *complete* if all nodes are pairwise linked.

Finally we show that each member W of the domain set is a clique. Because the elimination does not produce fill-ins, W must be a complete set in the domain graph. If W is not maximal, it is a subset of a clique V, and V is a member of the domain set, so W cannot be a member. $\qquad\square$

From Proposition 4.3, we can conclude that any perfect elimination sequence ending with the variable A is optimal with respect to calculating $P(A)$. The full task is to compute the marginals down to each variable, so the task can be solved by establishing an optimal elimination sequence for each variable.

4.3 Triangulated Graphs and Join Trees

Before continuing with the belief-updating task, we deal in detail with some purely graph-theoretic concepts. They will be used for the belief updating task in the next section.

Definition 4.4. *An undirected graph with a perfect elimination sequence is called a* triangulated graph.

Note that the term "triangulated" may be misleading. The graph (b) in Figure 4.10 is not triangulated.

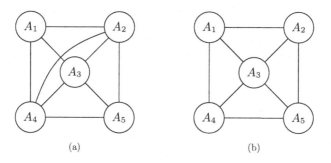

Fig. 4.10. (a) A triangulated graph; (b) a nontriangulated graph.

Notation: Let X be a node in an undirected graph. The set of neighbors of X we denote by $\mathrm{nb}(X)$, and the set of neighbors plus X we denote by $\mathrm{fa}(X)$, the family of X. If the nodes of the graph are enumerated, we use the index to write N_i rather than N_{X_i}. Nodes with a complete neighbor set are called *simplicial nodes*. A neighbor to a node X is said to be *adjacent* to X. Note that X is simplicial if and only if $\mathrm{fa}(X)$ is a clique.

Proposition 4.4. *Let G be a triangulated graph, and let X be a simplicial node. Let G' be the graph resulting from eliminating X from G (see Figure 4.11). Then G' is a triangulated graph.*

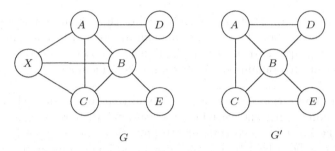

Fig. 4.11. If fa(X) is a complete set, you eliminate X from G by simply removing X together with its links.

Proof.   Follows from Proposition 4.2.   □

Note that a triangulated graph always has at least one simplicial node, namely the first one in the elimination sequence. Actually, there are at least two.

Theorem 4.1. *A triangulated graph with at least two nodes has at least two simplicial nodes.*

Proof.   We prove by induction a slightly stronger statement: let G be an incomplete triangulated graph with at least three nodes. Then, it has at least two nonadjacent simplicial nodes.

Certainly, any incomplete triangulated graph with three nodes has two nonadjacent simplicial nodes (see Figure 4.12).

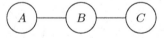

Fig. 4.12. A connected incomplete triangulated graph with three nodes.

Assume the statement to be true for all graphs with fewer than n nodes, and let G be an incomplete triangulated graph with n nodes. The first node, X, in the elimination sequence is simplicial, and we must find another one not adjacent to X. Let G' be the graph resulting from removing X from G.

The graph G' is triangulated, and any simplicial node in G' is either simplicial in G or a member of nb(X).

Because G is not complete, it must contain nodes that are not members of nb(X). If G' is complete, any of these nodes can do. If G' is not complete, we know from the induction hypothesis that it has at least two nonadjacent simplicial nodes. If both were neighbors of X, they would be adjacent.   □

Corollary 4.1. *In a triangulated graph, each variable A has a perfect elimination sequence ending with A.*

Proof. Let A be any node in the triangulated graph G. Eliminate a simplicial node $X(X \neq A)$; Theorem 4.1 ensures that such a node exists. Proposition 4.4 yields that the resulting graph is triangulated, and you can repeatedly apply Theorem 4.1 until only A is left.   □

From Corollary 4.1, we see that if you have established one perfect elimination sequence, then you can easily establish a perfect elimination sequence down to any variable. In other words, you can for each variable A establish an optimal sequence of marginalizations for calculating $P(A)$. We give the details in Section 4.4.

Unfortunately, it does not hold that all domain graphs are triangulated. The following theorem gives an easy way of checking whether a graph is triangulated, and if it is, it also gives a simple way of establishing an elimination sequence.

Theorem 4.2. *An undirected graph is triangulated if and only if all nodes can be eliminated by successively eliminating a simplicial node X.*

Proof. If all nodes can be eliminated by successively eliminating simplicial nodes, then we produce a perfect elimination sequence, and the graph is triangulated.

Now assume that the undirected graph is triangulated. Let us eliminate any simplicial node. Proposition 4.4 yields that the resulting graph is triangulated, and we can continue the procedure.   □

To check whether a graph is triangulated, you repeatedly eliminate simplicial nodes. At some stage, you run into a situation in which you cannot eliminate more nodes. If the node set is empty, then the graph is triangulated; if not, then the graph is not triangulated.

In general, it is NP-hard to determine the set of cliques in a graph. For triangulated graphs, Proposition 4.3 and Theorem 4.2 yield an easy procedure.

Algorithm 4.1 *To determine the set of cliques in a triangulated graph, you can do as follows*

1. *Eliminate a simplicial node X; $\mathrm{fa}(X)$ is a clique candidate.*
2. *If $\mathrm{fa}(X)$ does not include all remaining nodes, go to 1.*
3. *Prune the set of clique candidates by removing sets that are subsets of other clique candidates.*

□

4.3.1 Join Trees

Definition 4.5. *Let $\mathcal{G}$ be the set of cliques from an undirected graph, and let the cliques of $\mathcal{G}$ be organized in a tree T. Then T is a* join tree *if for any pair of nodes V, W all nodes on the path between V and W contain the intersection $V \cap W$.*

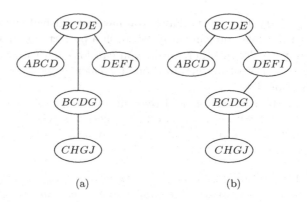

Fig. 4.13. (a) A join tree; (b) not a join tree.

Theorem 4.3. *If the cliques of an undirected graph G can be organized into a join tree, then G is triangulated.*

Proof. Let the cliques be organized in a join tree, and let V be a leaf clique with unique neighbor clique W. Any member of V that is a member of another clique must – due to the join tree condition – also be a member of W. Therefore, V must contain at least one variable X not contained in any other clique (otherwise V would be a subset of W). Then fa(X) must be complete, and X can be eliminated without creating fill-ins. We can repeat eliminating variables that are only members of V, and when all these have been eliminated, we have a graph G' with the same cliques as G except for V. Then, the join tree for G with the node V removed is a join tree for G', and we can continue by eliminating a variable from a leaf in G'. □

Theorem 4.4. *If the undirected graph G is triangulated, then the cliques of G can be organized into a join tree.*

The proof is a construction of a join tree from a triangulated graph. To illustrate the construction, we use the graph in Figure 4.14.

Construction: Establish an elimination sequence in the following way. Start with a simplicial node X. Then fa(X) is a clique. Continue eliminating nodes

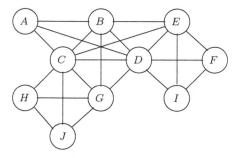

Fig. 4.14. A triangulated graph.

from fa(X) that have neighbors only in fa(X). Give fa(X) an index i according to the number of nodes eliminated, and denote the set of the remaining nodes by S_i. This set is called a *separator*. Choose a new clique in the graph G' with the eliminated nodes removed, and repeat the process with the index starting at i. Continue to do so until all cliques have been eliminated. Figure 4.15 shows the result of this process on the graph in Figure 4.14.

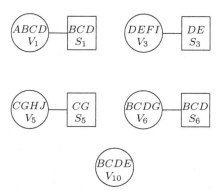

Fig. 4.15. The cliques, separators, and indices resulting from the graph in Figure 4.14. The elimination sequence used is $A, F, I, H, J, G, B, C, D, E$.

When the parts have been established as indicated in Figure 4.15, each separator S_i is connected to a clique V_j ($j > i$) such that $S_i \subset V_j$ (see Figure 4.16). This is always possible because S_i is a complete set, and when the first node from S_i is eliminated, it must be when dealing with a clique of higher index than i, and it must contain all of S_i. For convenience, we talk of the direction from V_i over S_i to V_j as upward, and we call V_j a parent of V_i.

We must prove that the structure constructed is a tree and has the join tree property. Each clique has at most one parent, so there cannot be multiple paths, and the structure is a tree.

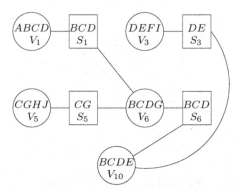

Fig. 4.16. A join tree (expanded with separators) resulting from the construction applied to the graph in Figure 4.14.

To prove the join tree condition, consider the cliques V_i and V_j ($i < j$), and let X be a member of both. There is a unique path between V_i and V_j, and we will prove that X is a member of all cliques on that path. Because X is not eliminated when dealing with V_i, it must be a member of S_i, and from the construction, X must be a member of V_i's parent V_k. If $k = j$, we are finished; otherwise we continue the argument for the smallest of the two.

Remark: The separators are so called because any separator S divides the graph into two parts, and all paths connecting the two parts must pass through S. If the join tree is constructed from a Bayesian network, the two parts are d-separated given S.

A join tree provides the framework for constructing perfect elimination sequences. Namely, notice that the simplicial nodes are those with all uneliminated neighbors in one clique, and two nodes are neighbors if they are members of the same clique. Hence, all perfect elimination sequences can be constructed from a join tree by repeatedly eliminating simplicial nodes.

4.4 Propagation in Junction Trees

In the literature you often see the terms "join tree" and "junction tree" used interchangeably. In this book we introduce a distinction.

Definition 4.6. *Let Φ be a set of potentials with a triangulated domain graph, G. A junction tree for Φ is a join tree for G with the following addition: each potential ϕ in Φ is attached to a clique containing $\mathrm{dom}(\phi)$; each link has the appropriate separator attached; each separator contains two mailboxes, one for each direction.*

If Φ is a set of conditional probabilities for a Bayesian network BN together with evidence potentials for the evidence e, we say that the junction tree represents BN with evidence e.

Notation: The propagation algorithm presented here deals with sets of potentials. A set of potentials is a representation of the product of the member potentials. Let Φ be a set of potentials whose domains are subsets of V, and let W be a subset of V. Then, $\Phi^{\downarrow W}$ is a set of potentials resulting from successively eliminating the variables in $V \setminus W$ as described in Definition 4.1. Because the elimination order is arbitrary, this notation seems to introduce some ambiguity with respect to the functions in the resulting set. Because we treat the sets as representations of products, and the product is independent of the elimination order, we will not deal with this apparent ambiguity.

Example 4.2. Consider the set $\psi = \{\phi_1(A),\ \phi_2(A,B),\ \phi_3(A,C),\ \phi_4(C,D),\ \phi_5(C)\}$, and let $W = \{B,C\}$. Then, $\psi^{\downarrow W} = \{\sum_A \phi_1(A)\phi_2(A,B)\phi_3(A,C),\ \sum_D \phi_4(C,D), \phi_5(C)\}$.

Before giving a general description of the propagation algorithm, we will go through an example.

Example 4.3. Consider the Bayesian network in Section 4.1 with potentials $\phi_1 = P(A_1), \phi_2 = P(A_2 \mid A_1), \phi_3 = P(A_3 \mid A_1), \phi_4 = P(A_4 \mid A_2), \phi_5 = P(A_5 \mid A_2, A_3), \phi_6 = P(A_6 \mid A_3)$ and with the domain graph in Figure 4.8. We know that the elimination sequence $A_6, A_5, A_3, A_1, A_2, A_4$ is perfect. The domain graph has a join tree over the cliques $V_1 = \{A_3, A_6\}$, $V_2 = \{A_2, A_3, A_5\}$, $V_4 = \{A_1, A_2, A_3\}$, $V_6 = \{A_2, A_4\}$ and the separators $S_1 = \{A_3\}$, $S_2 = \{A_2, A_3\}$, $S_4 = \{A_2\}$. The junction tree is shown in Figure 4.17.

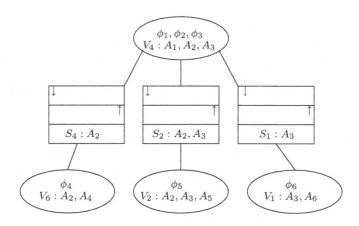

Fig. 4.17. A junction tree for the Bayesian network in Figure 4.8.

To calculate $P(A_4)$, we find a clique containing $A_4(V_6)$. It is made a temporary root, and we send messages in the direction of V_6 starting from the leaf cliques. The message $\psi_1 = \phi_6^{\downarrow A_3} = \phi_6^{\downarrow S_1}$ is placed in the appropriate S_1 mailbox, and the message $\psi_2 = \phi_5^{\downarrow\{A_2,A_3\}} = \phi_5^{\downarrow S_2}$ is placed in the appropriate S_2 mailbox. Next, V_4 assembles the incoming messages and the potentials held form the set $\Phi_4 = \{\psi_1, \psi_2, \phi_1, \phi_2, \phi_3\}$. The variables A_1 and A_3 are eliminated from Φ_4, and the result, $\psi^4 = (\phi_1\phi_2(\phi_3\psi_2\psi_1)^{\downarrow\{A_1,A_2\}})^{\downarrow A_2} = \sum_{A_1} \phi_1\phi_2 \sum_{A_3} \phi_3\psi_2\psi_1$, is placed in the appropriate mailbox (see Figure 4.18).

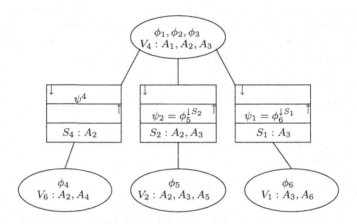

Fig. 4.18. The cliques V_1 and V_2 have sent messages to their separators, and V_4 has sent the message $\sum_{A_1} \phi_1\phi_2 \sum_{A_3} \phi_3\psi_2\psi_1$ to S_4.

Now V_6 can collect its message, multiply it by ϕ_4, and marginalize A_2 out to get $P(A_4)$.

The process just described is called *collecting evidence* to V_6. To calculate the marginal for another variable X, we can collect to a clique containing X. If, for example, we wish to calculate $P(A_6)$, we can collect to V_1. We can also prepare the junction tree for the calculation of all marginals: send messages in the direction away from V_6. This process is called *distributing evidence*. First, V_6 sends the message $\psi_4 = \phi_4^{\downarrow A_2}$ to S_4, and V_4 sends a message to S_2 as well as S_1 (see Figure 4.19). When the message for S_2 is calculated, the set $\{\psi_4, \phi_1, \phi_2, \phi_3, \psi_1\}$ is assembled, and A_1 is marginalized out. Here, we multiply only the potentials that have A_1 in the domain, and the message becomes a set of potentials: $\{\psi_4, \sum_{A_1} \phi_1\phi_2\phi_3, \psi_1\}$.

When both collecting and distributing evidence have been performed, we have performed a full propagation, and to calculate a marginal $P(X)$ we find a clique V containing X. Take, for example, A_3. The clique V_1 contains A_3. The incoming message to V_1 is the message for collecting evidence to V_1, and therefore it corresponds to a perfect elimination sequence ending with the nodes A_6 and A_3. This means that the product $\phi_6\psi^1$ is the projection of the

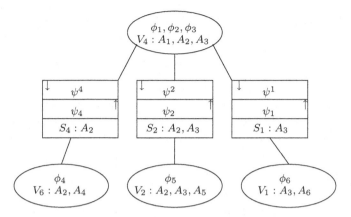

Fig. 4.19. The junction tree after a full propagation: $\psi^2 = \{\psi_4, \sum_{A_1} \phi_1\phi_2\phi_3, \psi_1\}, \psi^1 = \sum_{A_2} \psi_2\psi_4 \sum_{A_1} \phi_1\phi_2\phi_3$.

entire product down to $\{A_3, A_6\}$, and we can easily calculate $P(A_3)$ as well as $P(A_6)$.

There is a slightly easier way of calculating A_3. Consider the separator S_1. It consists of A_3 alone. For the product of the two messages of S_1, we have

$$\psi^1\psi_1 = (\sum_{A_2} \psi_2\psi_4 \sum_{A_1} \phi_1\phi_2\phi_3)(\sum_{A_6} \phi_6)$$

$$= \sum_{A_2}\sum_{A_5} \phi_5 \sum_{A_4} \phi_4 \sum_{A_1} \phi_1\phi_2\phi_3 \sum_{A_6} \phi_6$$

$$= (\phi_5\phi_4\phi_1\phi_2\phi_3\phi_6)^{\downarrow A_3} = P(A_3).$$

Next, assume that you have the evidence $e = \{e_5 : A_5 = a^5, e_6 : A_6 = a^6\}$. The evidence e is represented as two $0-1$ potentials $\mathbf{e}_5$ and $\mathbf{e}_6$. To calculate the probabilities $P(X, e)$, you place the two evidence potentials in appropriate cliques (V_2 and V_1) and perform a full propagation.

4.4.1 Lazy Propagation in Junction Trees

Each clique V holds a set of potentials denoted by Φ_V. Each separator has two mailboxes, one for each direction of the link. The messages stored in the mailboxes are sets of potentials. The messages are denoted by ψ_S or ψ^S, depending on the direction.

The basic operation in the lazy propagation procedure is *message passing*.

Definition 4.7. *Let V be a clique with set of potentials Φ_V, and let S be a neighboring separator. Let $S_1, \ldots, S_k$ be the other neighboring separators of V. Assume that each S_i has received a message Ψ_i for V.*

Then V can pass the message $(\Phi_V \cup \Psi_1 \cup \cdots \cup \Psi_k)^{\downarrow S}$ to S, and we say that the direction V to S is triggered.

The propagation method consists in repeatedly passing messages along triggered directions.

Proposition 4.5. *If you repeatedly pass messages along triggered directions in a junction tree, then you need not stop before a message has been passed in both directions over each link. In that situation, we say that the junction tree is* full.

Proof. See Exercise 4.27.   □

As shown in Example 5.3, you can start off by directing all messages toward a chosen temporary root R. In other words, the junction tree is given a direction from R and outward, and the messages are passed in the opposite direction from leaves and inward (see Figure 4.20). This procedure is called CollectEvidence(R).

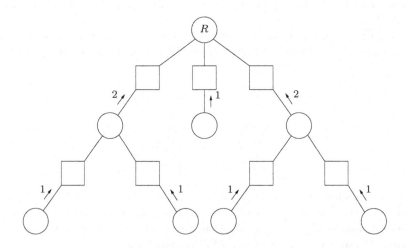

Fig. 4.20. The message passing in CollectEvidence(R).

Notice that the message passing in CollectEvidence(R) corresponds to a perfect elimination sequence ending with the nodes of R.

To fill the junction tree after a CollectEvidence(R), you need only to place messages in the opposite directions. First, R passes a message to its neighbors, they in turn pass messages further outward, and so forth out to the leaves (see Figure 4.21). This procedure is called DistributeEvidence(R). Note that messages are passed along triggered directions only if DistributeEvidence(R) is performed after CollectEvidence(R) has been performed.

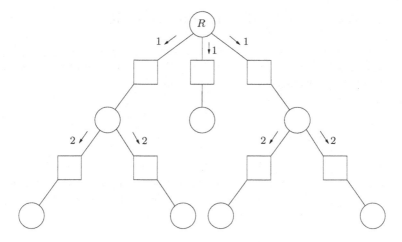

Fig. 4.21. The messages passing in DISTRIBUTEEVIDENCE(R).

Theorem 4.5. *Let the junction tree T represent the Bayesian network BN over the universe U and with evidence e. Assume that T is full.*

1. *Let V be a clique with set of potentials Φ_V, and let $S_1, \ldots, S_k$ be V's neighboring separators and with V-directed messages $\Psi_1, \ldots, \Psi_k$. Then*

$$P(V, e) = \prod \Phi_V \prod \Psi_1 \cdot \ldots \cdot \prod \Psi_k.$$

2. *Let S be a separator with the sets Ψ_S and Ψ^S in the mailboxes. Then*

$$P(S, e) = \prod \Psi_S \prod \Psi^S.$$

Proof.

1. Consider the messages passed in the direction of V. They correspond to a COLLECTEVIDENCE(V), and the message passing corresponds to a perfect elimination sequence ending with the nodes of V. Therefore,

$$P(V, e) = P(U, e)^{\downarrow V} = \prod \Phi_V \prod \Psi_1 \cdots \prod \Psi_k.$$

2. Consider S_k as before. Because

$$\prod \Psi^k = \left(\prod \Phi_V \prod \Psi_1 \cdots \prod \Psi_{k-1} \right)^{\downarrow S_k},$$

we have

$$
\begin{aligned}
P(S_k, e) = P(V, e)^{\downarrow S_k} &= \left(\prod \Phi_V \prod \Psi_1 \cdots \prod \Psi_k \right)^{\downarrow S_k} \\
&= \left(\prod \Phi_V \prod \Psi_1 \cdots \prod \Psi_{k-1} \right)^{\downarrow S_k} \prod \Psi_k \\
&= \prod \Psi^k \prod \Psi_k.
\end{aligned}
$$

$\square$

4.5 Exploiting the Information Scenario

As mentioned at the beginning of this chapter, the actual information scenarios can provide simplifications of the calculations. This is one of the reasons why we let lazy propagation work with sets of potentials rather than multiplied potentials.

4.5.1 Barren Nodes

Barren nodes (see Section 4.1.2) do not contribute to the probabilities of nonbarren nodes, and therefore we need not take their potentials into account when calculating marginals of nonbarren nodes. This is illustrated in Figure 4.22.

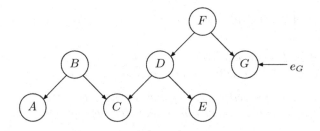

Fig. 4.22. The nodes A, B, C, D, and E are barren.

In the calculation of $P(F \mid e_G)$, the part of the network with barren nodes can be discarded. Figure 4.23 shows a junction tree for the network.

To calculate $P(F \mid e_G)$, you can collect to the clique (F, G). We see that all marginalizations to perform are of the form $\sum_X P(X \mid pa(X))$, and from the unit potential property (Section 1.4) they are all **1**.

Now assume that there is also evidence e_A for the variable A. Because A is d-separated from F, e_A does not affect $P(F \mid e_G)$. In the junction tree propagation, the message $\psi(B)$ from the clique (A, B) is no longer **1**. When the clique (B, C, D) produces a message for (D, F), the calculation is $\{P(C \mid B, D), P(B), P(a \mid B)\}^{\downarrow D}$. If we start marginalizing C out, we apply the unit potential property, and marginalizing B will result in a constant.

The handling of barren nodes can be taken care of using the following rule.

Barren node rule: Let Ψ be a set of potentials, and assume that we calculate $\Psi^{\downarrow V}$. If $A \notin V$, and the only potential in Ψ with A in the domain is of the form $P(A \mid W)$, then A is marginalized by discarding $P(A \mid W)$.

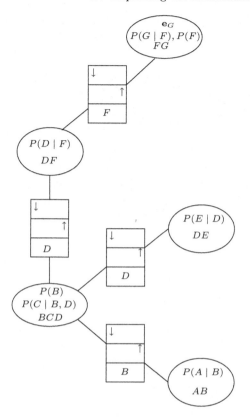

Fig. 4.23. A junction tree for the network in Figure 4.22.

4.5.2 d-Separation

When evidence is of the form that it instantiates a variable (hard evidence), then the domains to handle will be reduced with this variable. There are other simplifications due to instantiation: new pairs of variables may become d-separated, reducing the domains of the messages to communicate. We illustrate this with the example in Figure 4.24.

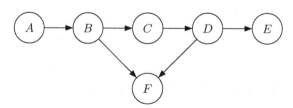

Fig. 4.24. A Bayesian network.

We will be interested in $P(E\,|\,e)$, and therefore we only consider collecting evidence to the clique (D, E). A junction tree for the network is shown in Figure 4.25.

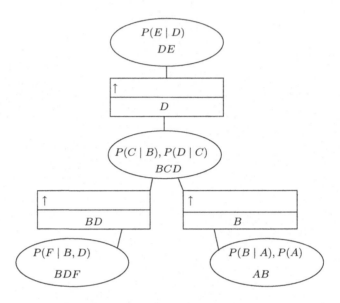

Fig. 4.25. A junction tree for the Bayesian network in Figure 4.24. Only the upward mailboxes are indicated.

First, let A be instantiated to a. The messages are given in Figure 4.26, and we see that the evidence has an impact on $P(E, e)$ through the message $\psi_1(D)$: $P(E\,|\,a) = \sum_D P(E\,|\,D)P(D\,|\,a)$.

Next, let C be instantiated to c. Then A and E are d-separated. Figure 4.27 shows how this is reflected in the messages: $P(E\,|\,c) = \sum_D P(E\,|\,D)\,P(D\,|\,c)$.

Finally, let F be instantiated to f. Then A and E are no longer d-separated. This is shown in Figure 4.28.

Note: In the examples, we have entered evidence on a variable X by instantiating the potentials including X. In general, evidence can be entered by adding the corresponding evidence potential to a clique containing X, and the instantiation is effected when X must be marginalized. This means that the evidence potential is passed to separators containing X.

4.6 Nontriangulated Domain Graphs

So far, we have considered propagation methods only for potentials with a triangulated graph. For these methods, we know that the junction tree is a

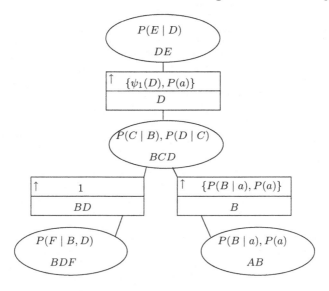

Fig. 4.26. The messages on collecting to (D, E) for A instantiated. Here $\psi_1 = \sum_C P(D \mid C) \sum_B P(C \mid B) P(B \mid a) = \sum_C P(D \mid C) P(C \mid a) = P(D \mid a)$.

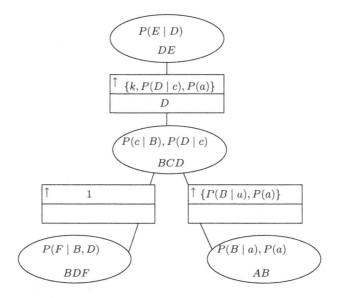

Fig. 4.27. The messages on collecting to (D, E) for A and C instantiated. Here $k = \sum_B P(c \mid B) P(B \mid a) = P(c \mid a)$.

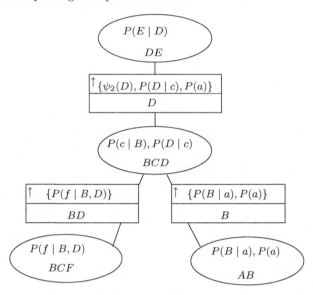

Fig. 4.28. The messages on collecting to (D, E) for A, C, and F instantiated. Here $\psi_2(D) = \sum_B P(f \mid B, D)(c \mid B)P(B \mid a)$.

propagation framework having the smallest possible domains with which to work.

If the domain graph is not triangulated, we embed it in a triangulated graph and use its junction tree. In fact, we did so in Section 4.5.2 when we handled evidence.

Example 4.4. Consider the Bayesian network in Figure 4.29. After having eliminated the variables A, C, H, I, and J, we cannot eliminate any node without adding fill-ins, and the graph is not triangulated.

The graph in Figure 4.30 is a triangulated graph extending the moral graph in Figure 4.29. We can use a junction tree for that graph (see Figure 4.31).

4.6.1 Triangulation of Graphs

It is quite easy to find a triangulated graph extending a graph G. You eliminate the variables in some order, and if you wish to eliminate a node with an incomplete neighbor set, you make it complete by adding fill-ins (the graph in Figure 4.30 is the result of eliminating in the order $A, C, H, I, J, B, G, D, E, F$). The resulting graph has a perfect elimination sequence, and it is therefore triangulated.

There are several different elimination orders, and many of them produce different triangulated graphs. We aim to work with the one yielding the smallest domains.

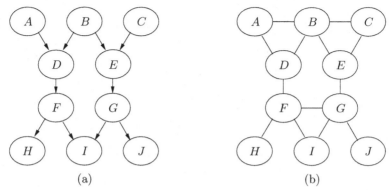

(a)   (b)

Fig. 4.29. A Bayesian network (a) with a nontriangulated moral graph (b).

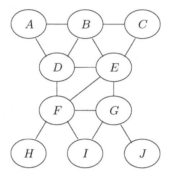

Fig. 4.30. A triangulated graph extending the moral graph in Figure 4.29.

Definition 4.8. *Let V be a set of variables. For $X \in V$, $|\mathrm{sp}(X)|$ denotes the number of states of X. The size of V, $\mathrm{sz}(V)$, is the product $\prod_{X \in V} |\mathrm{sp}(X)|$. Let BN be a Bayesian network, let G be a triangulated graph extending BN's moral graph, and let $V_1, \ldots, V_n$ be the cliques of G. The size of G is the sum $\mathrm{size}(G) = \sum_i \mathrm{sz}(V_i)$.*

Unfortunately, it is NP-hard to determine an elimination sequence yielding a triangulation of minimal size. However, there are heuristic algorithms that have proven to give fairly good results. One example is the following:

Heuristic: Repeatedly eliminate a simplicial node, and if this is not possible, eliminate a node X of minimal $\mathrm{sz}(\mathrm{fa}(X))$.

Example 4.5. Let the number of states for the variables in Figure 4.29 be as follows: A, B, C, H, I, and J have two states, D has four states, E has five states, F has six states, and G has seven states. After having eliminated the variables A, C, H, I, and J, we eliminate a nonsimplicial node. We have $\mathrm{sz}(\mathrm{fa}(B)) = 40, \mathrm{sz}(\mathrm{fa}(D)) = 48, \mathrm{sz}(\mathrm{fa}(E)) = 70, \mathrm{sz}(\mathrm{fa}(F)) = 168$, and

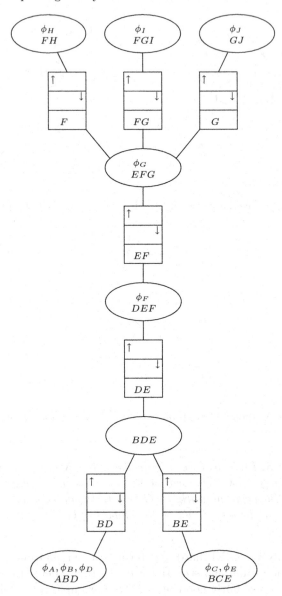

Fig. 4.31. A junction tree with potentials from the Bayesian network in Figure 4.29. Notation: $\phi_X = P(X \mid pa(X))$.

$sz(fa(G)) = 210$. We choose to eliminate B, creating the fill-in (D, E). With this new link, we have new sizes $sz(fa(D)) = 120$ and $sz(fa(E)) = 140$. We eliminate D and add the fill-in (E, F). Now the graph is triangulated. However, in this case the triangulation is not optimal (see Exercise 4.32).

For later use, we establish the following proposition.

Proposition 4.6. *Let $A_1, \ldots, A_n$ be an elimination sequence triangulating the graph G, and let A_i and A_j be two nonneighbors in $G(i < j)$. Then the elimination sequence introduces the fill-in (A_i, A_j) if and only if there is a path $A_i - X - \cdots - A_j$ such that all intermediate nodes are eliminated before A_i.*

Proof. Assume that fill-ins may be introduced that violate the proposition, and let (A_i, A_j) be such a fill-in with i as small as possible. Let the link be introduced on eliminating the node A_k. Because new fill-ins cannot be attached to A_i when it has been eliminated, we must have $k < i$. One of the links (A_k, A_j) and (A_i, A_k) on eliminating A_k must be a fill-in (if not, the (A_i, A_j) fill-in does not violate the proposition). Let it be (A_i, A_k). Due to the choice of (A_i, A_j) the proposition holds for (A_i, A_k), hence there is a path $A_k - X - \cdots - A_i$ such that all intermediate nodes are eliminated before A_k (see Figure 4.32). If also (A_k, A_j) is a fill-in, the same must hold. Connecting these two paths yields a path $A_i - X - \cdots - A_j$ such that all intermediate nodes are eliminated before A_i, a contradiction.

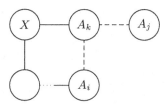

Fig. 4.32. A path connecting A_i and A_j through nodes eliminated before A_i.

Next, assume that we have a path $A_i - X - \ldots - A_j$ such that all intermediate nodes are eliminated before A_i. Let A_k be any node on the path to be eliminated, and let Y and Z be the neighbors on the path. After the elimination of A_k, there is a link (Y, Z), and there is still a path $A_i - X - \ldots - A_j$ such that all intermediate nodes are eliminated before A_i, so the property is invariant under elimination. When all the nodes before A_i are eliminated, the path must be the link (A_i, A_j).   □

4.6.2 Triangulation of Dynamic Bayesian Networks

Return to Exercise 3.25 and consider the model in Figure 3.52. In Figure 4.33, we have folded it out to three time slices.

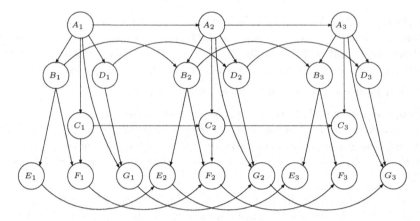

Fig. 4.33. Three time slices of the model in Figure 3.52.

As you have probably experienced when solving Exercise 3.25, your computer ran out of memory when you tried to compile the model folded out to four or five time slices. The reason is that the cliques become too large.

A conceptually simple way of considering propagation in dynamic Bayesian network models is that information is transmitted from one time slice to the next (if the task is forecasting) or to the previous time slice (if the task is to find out what happened in the past). In other words, probability potentials describing time slice i are transmitted from time slice i to time slice $i + 1$ or to time slice $i - 1$.

Let us consider forward passing from time slice i to time slice $i+1$, and let W be the set of variables with a child in slice $i+1$. We wish to pass potentials representing the joint probability of W. For the model in Figure 4.33, we pass the information from slice 1 to slice 2 by eliminating all nodes in slice 1 before any node from slice 2 is eliminated. Now consider any pair of nodes (X_2, Y_2). If there is a path in slice 1 connecting them, then they will be linked after the elimination of slice 1 (Proposition 4.6). Because the moral graph for slice 1 is connected, and all nodes in slice 2 have a parent in slice 1, the entire slice 2 will be a subset of a clique if slice 1 is eliminated before any node from slice 2. If you process only two time slices, you may avoid this clique explosion by using another elimination sequence. However, it will inevitably arrive when you extend the number of time slices to process. Some cliques will contain all variables with a child in the next slice or will contain all variables with a parent in the previous slice.

This situation is not reserved for models with connected time slices. Consider the model in Figure 4.34. If the model is folded out to four time slices, and the first three slices are eliminated before any node from slice four, then slice four becomes a complete set. Figure 4.35 shows the moral graph for four slices of the model. The reader can check that all pairs of nodes in slice four have a connecting path through the past slices.

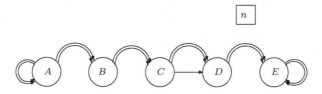

Fig. 4.34. A dynamic Bayesian network model with very sparse connection inside the time slices.

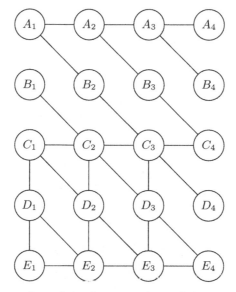

Fig. 4.35. The moral graph for four time slices of the model in Figure 4.34.

As indicated above, you may think of propagation in dynamic Bayesian networks as a way of passing probabilities of output nodes forward in time. The problem is that most often, the required probability distribution is the joint distribution over all output variables. If this is intractable, you can approximate the joint distribution by partitioning the set of output variables. If $\mathcal{O}$ is partitioned into $\{\mathcal{O}_1, \mathcal{O}_2, \mathcal{O}_3\}$, then instead of passing $P(\mathcal{O})$ you pass $\{P(\mathcal{O}_1), P(\mathcal{O}_2), P(\mathcal{O}_3)\}$. It has been proven that the error introduced does not accumulate over time, but converges to a finite error (in Kullback-Leibler divergence; see Definition 6.2).

4.7 Exact Propagation with Bounded Space

One of the biggest problems with exact propagation algorithms such as the junction tree based approach described in Section 4.4 is that the probability tables can become intractably large. In this section we will investigate an exact propagation algorithm in which space can be traded for time. For this particular propagation algorithm, we will consider calculation of probabilities only of the form $P(x, \mathbf{e})$, since $P(x \mid \mathbf{e})$ can subsequently be found by $P(x \mid \mathbf{e}) = P(x, \mathbf{e})/(\sum_x P(x, \mathbf{e}))$.

4.7.1 Recursive Conditioning

Consider the Bayesian network in Figure 4.36 and assume that we are interested in the probability $P(f)$.

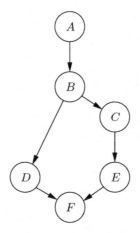

Fig. 4.36. A Bayesian network.

By calculating $P(f)$ using, for example, variable elimination (Section 2.3.4) or lazy propagation (Section 4.4.1), we basically first establish an elimination sequence and then use the distributive law. For example, by using the elimination sequence F, E, D, C, B, A we would get

$$
\begin{aligned}
P(f) \\
&= \sum_A \sum_B \sum_C \sum_D \sum_E P(A, B, C, D, E, f) \\
&= \sum_A P(A) \sum_B P(B \mid A) \sum_C P(C \mid B) \sum_D P(D \mid C) \sum_E P(E \mid D) P(f \mid D, E).
\end{aligned}
$$

$$(4.1)$$

The sequence in which the calculations are performed can be encoded in a *computation tree*. The leaf nodes represent the conditional probability distributions in the model, and for each internal node the potentials defined by the two subtrees are multiplied and the variables indicated by the label of the node are marginalized out (see Figure 4.37).

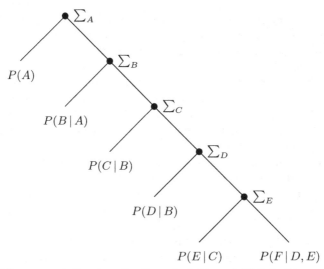

Fig. 4.37. The computation tree for the calculation of $P(f)$ in Figure 4.36 using the elimination sequence E, D, C, B, A.

Based on the computation tree in Figure 4.37 we can easily specify an algorithm that calculates $P(f)$ and performs the same operations as in equation (4.1): evaluate the computation tree from the leaves toward the root. When an internal node is reached, multiply the two potentials calculated in the two subtrees for that node and marginalize out the appropriate variables.

Another way of doing the calculations would be to start at the root $\sum_A$ and recursively evaluate the subtrees for each state of A; when the recursive calls return, the results are added up. Assuming that A is binary, for the calculations in equation (4.1) this would correspond to

$$P(f) = P(a_1) \sum_B P(B \mid a_1) \sum_C P(C \mid B) \sum_D P(D \mid C) \sum_E P(E \mid D) P(f \mid D, E)$$
$$+ P(a_2) \sum_B P(B \mid a_2) \sum_C P(C \mid B) \sum_D P(D \mid C) \sum_E P(E \mid D) P(f \mid D, E),$$
$$(4.2)$$

where, for example, the first term is the result of the recursive calls made at node $\sum_B$:

$$\sum_B P(B \mid a_1) \sum_C P(C \mid B) \sum_D P(D \mid C) \sum_E P(E \mid D)P(f \mid D, E)$$

$$= P(b_1 \mid a_1) \sum_C P(C \mid b_1) \sum_D P(D \mid C) \sum_E P(E \mid D)P(f \mid D, E)$$

$$+ P(b_2 \mid a_1) \sum_C P(C \mid b_2) \sum_D P(D \mid C) \sum_E P(E \mid D)P(f \mid D, E).$$

Compared to equation (4.1) we can say that when the computation tree is "read" from the root toward the leaves, we condition in the internal nodes, and when it is "read" from the leaves towards the root, we marginalize out in the internal nodes.

By continuing the "recursive conditioning" above, we see that the storage requirements are considerably reduced. Specifically, for handling the intermediate results we have to store only the initial conditional probability distributions together with a single number for each internal node in the computation tree, i.e., the space complexity is linear in the number of nodes. Unfortunately, this reduction in space comes at a price. In this particular example, the number of recursive calls corresponds to the size of the state space of all the variables involved. Assuming that the variables are binary, this would amount to 32 recursive calls. Note, however, that the size of the call stack is proportional to the depth of the tree.

In general, the number of recursive calls increases exponentially with the height of the computation tree, so to reduce the time complexity we should aim for a more balanced tree structure. For example, consider again the Bayesian network in Figure 4.36, but assume now that we have the elimination ordering B, A, E, C, D:

$$P(f) = \sum_B \sum_A \sum_E \sum_C \sum_D P(A, B, C, D, E, f)$$

$$= \sum_B \left[\sum_A P(A)P(B \mid A) \right] \tag{4.3}$$

$$\times \left[\sum_E \left[\sum_C P(C \mid B)P(E \mid C) \right] \left[\sum_D P(D \mid B)P(f \mid D, E) \right] \right].$$

The corresponding computation tree is shown in Figure 4.38. In this tree the calculation of $P(f)$ requires only $2 \cdot (2 + 2 \cdot (2 + 2)) = 20$ recursive calls.

In the two examples above, we condition on only one variable at a time. The reason is that both elimination sequences ensure that each time we condition on a variable, the remaining variables can be partitioned into two d-separated sets. This, however, is not the case in general. For example, for the elimination sequence D, C, E, B, A, neither D nor C can alone partition the variables into independent sets; hence a node in the tree is labeled with both variables (see Figure 4.39):

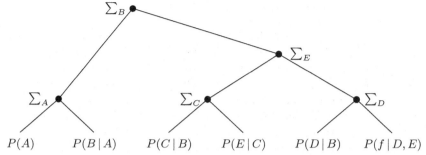

Fig. 4.38. The computation tree for the calculation of $P(f)$ in Figure 4.36 using the elimination sequence B, A, E, C, A.

$$P(f) = \sum_D \sum_C \sum_E \sum_B \sum_A P(A, B, C, D, E, f)$$

$$= \sum_D \sum_C \left[\sum_E P(E \mid C) P(f \mid D, E) \right.$$

$$\times \left[\sum_B P(C \mid B) P(D \mid B) \left[\sum_A P(A) P(B \mid A) \right] \right] \right].$$

It should also be noted that the computation graph is not required to be binary; for example, if conditioning on a variable partitions the remaining variables into three or more d-separated sets, then the corresponding node may have more than two children in the computation tree.

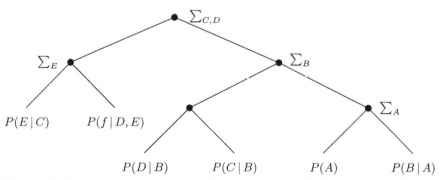

Fig. 4.39. The computation tree for the calculation of $P(f)$ in Figure 4.36 using the elimination sequence D, C, E, B, A.

In general, the set of variables attached to a node T corresponds to the set of noninstantiated variables shared by its two subtrees T_l and T_r. This set is also called the *cutset* for the node:

$$\text{cutset}(T) = (\text{dom}\,(T_l) \cap \text{dom}\,(T_r)) \setminus \text{a-cutset}(T),$$

where $\text{dom}\,(T_i)$ are the variables that appear in the conditional probability tables in the subtree T_i and a-cutset(T) is the union of the cutsets associated with the ancestral nodes for T in the tree (if T is the root node, then a-cutset(T) = $\emptyset$). Thus, the a-cutset is the set of nodes already instantiated. For example, in the tree in Figure 4.39, the cutset for the root node is $\{C, D\}$ and the a-cutset for the node labeled $\sum_A$ is $\{B, C, A\}$. In particular, the a-cutset for the unlabeled node is $\{B, C, D\}$, which covers all variables in the subtree, hence this node is given the empty cutset.

Before we present a more-formal specification of the algorithm it should be noted that in the above examples we incorporated the evidence f directly in the computation tree, indicating that a new computation tree is constructed for each piece of evidence. A more-efficient approach would be to first construct a single computation tree with no evidence inserted. Then, when evidence arrives we simply "record" the variables that should be instantiated such that no summations are performed for these variables.

Algorithm 4.2 reflects this approach for calculating the probability of a configuration $\mathbf{e}$ based on a computation tree for a Bayesian network. Observe that at each recursive call we record the corresponding instantiation and unrecord it when the call returns.

Algorithm 4.2 [RecursiveConditioning] *In order to calculate $P(\mathbf{e})$ using recursive conditioning on the tree T, do:*

1. *If T is the root, then record instantiation $\mathbf{e}$.*
2. *If T is a leaf, then:*
    a) *Return LookUp(T).*
3. *Else*
    a) *Set $p := 0$.*
    b) *For each noninstantiated configuration $\mathbf{c}$ of cutset(T) do:*
        i. *Record instantiation $\mathbf{c}$.*
        ii. *Set*

$$p := p + \prod_{i=1}^{m} RecursiveConditioning(T_i),$$

    where $T_1, \ldots, T_m$ are the children of T.
        iii. *Unrecord instantiation $\mathbf{c}$.*
    c) *Return p.*

□

Algorithm 4.3 [LookUp] *To find the value of the leaf node T under the recorded instantiations, do:*

1. *Let X be the variable associated with T and let $P(X \mid \text{pa}(X))$ be the conditional probability table assigned to X.*
2. *If X is instantiated, then:*

    a) Let x be the recorded instantiation for X and let π be the recorded
    instantiation for pa(X).
    b) Return $P(x \mid \pi)$.
3. else
    a) Return 1.

$\square$

Clearly, this algorithm requires only as much space as is needed to store the computation tree, and this is linear in the number of variables (hence for this aspect the shape of the tree is of no importance). However, the situation is different if we consider the time complexity. The time complexity can be estimated by counting the number of recursive calls, and it can be shown (see Exercise 4.38) that for a balanced tree it is $O(n^{w+1})$ and for an unbalanced tree it is $O(n \cdot \exp(w \cdot n)))$, where w is the size of the largest cutset.

This also indicates that it is important to find a good computation tree representation of the Bayesian network, and as we also indicated above this is closely connected with finding a good elimination sequence (see Section 4.6.1). In fact, given an elimination sequence that produces a maximum clique size of w, there are algorithms that will return a computation tree in which the cutset is not larger than w. The idea is to build the tree from the leaves to the root, where appropriate subtrees are joined according to the sequence in which the variables are marginalized out.

As for the tree in Figure 4.37, the algorithm above may perform redundant recursive calls to a subtree. This may happen when the a-cutset for a node/subtree includes a variable that is not in the domain of any of the probability tables associated with the subtree in question; we shall call all nonredundant nodes in a-cutset(T) the *context* for T. A way of controlling the number of redundant recursive calls is to cache previous calculations. Since we assume that we do not have enough memory to cache all values, the trick is therefore to find a good strategy for selecting the values to cache. If cache?$_T(\mathbf{x})$ is a function that determines whether to cache the value for subtree T evaluated in the context $\mathbf{x}$, we can directly control how much memory the algorithm is allowed to use. Algorithm 4.2 can easily be modified to support such a caching strategy: before a recursive call is made in context $\mathbf{x}$ we check whether a value for that context is already stored in the cache; if this is the case we simply return that value; otherwise, the call is completed and the result is cached if this is in accordance with cache?$_T(\mathbf{x})$.

4.8 Stochastic Simulation in Bayesian Networks

The junction tree based propagation methods described in the beginning of this chapter require tables for the cliques in the triangulated graph. These cliques may be very large, and it may happen that the space requirements of the tables cannot be met by the hardware available. When this is the case either you can make a tradeoff between time and space (using, for example, recursive conditioning as described in Section 4.7) or you can trade space for accuracy by using an approximate inference method.

In this section, we give a flavor of a class of approximate methods that are based on a technique called *stochastic simulation*. To illustrate the methods, consider the Bayesian network in Figure 4.40, with the conditional probabilities specified in Table 4.1, and assume that we want to estimate the probability of $E = y$. Now suppose also that we have access to a database containing configurations over the five variables and for which the distribution of the configurations follows the probability distribution specified by the Bayesian network. Given such a database, we can estimate the probability of $E = y$ by counting the number of cases that contain $E = y$ and divide it by the total number of cases:

$$P(E) \approx \frac{N(E = y)}{N}.$$

Since we (usually) do not have access to such a database, stochastic simulation instead tries to simulate such an access. This is done by drawing a large number of random configurations over (A, B, C, D, E) using the Bayesian network. There are several different algorithms for performing this type of sampling, and their main differences lie in how the samples are generated and how the probabilities are estimated from the sampled configurations.

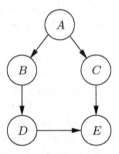

Fig. 4.40. An example network. All variables have the states y and n.

4.8.1 Probabilistic Logic Sampling

Probabilistic logic sampling is one of the simplest sampling procedures. To illustrate the approach, consider again the Bayesian network in Figure 4.40 and assume for simplicity that we have not received any evidence. A configuration can now be sampled by iteratively sampling a state of each of the variables. First a state of variable A is sampled. A random generator (with an even distribution) is asked to give a real number between 0 and 1. If the number is less than 0.4 (the prior probability of $A = y$), the state is y; otherwise, the

	A	
B	y	n
y	0.3	0.8
n	0.7	0.2

$P(B\,|\,A)$

	A	
C	y	n
y	0.7	0.4
n	0.3	0.6

$P(C\,|\,A)$

	B	
D	y	n
y	0.5	0.1
n	0.5	0.9

$P(D\,|\,B)$

	C	
D	y	n
y	(0.9, 0.1)	(0.999, 0.001)
n	(0.999, 0.001)	(0.999, 0.001)

$P(E\,|\,C, D)$

Table 4.1. The conditional probabilities for the example network. $P(A) = (0.4, 0.6)$.

state is n. Assume that the result is y. From the conditional probability table $P(B\,|\,A)$, we have that $P(B\,|\,y) = (0.3, 0.7)$. The random generator is asked again, and if the number is less than 0.3, the state of B is y. This procedure continues until we also have a state for C, D, and E. Observe that the sequence in which we generate the sample follows the topological ordering of the nodes in the network: we start at the nodes without parents and work ourselves toward the nodes without children; when visiting a variable we sample a state for that variable using its associated probability table conditioned on the configuration of the parent variables that have already been sampled.

The next configuration is sampled through the same procedure, and this is repeated until N configurations have been sampled. In Table 4.2, an example set of configurations is given.

	CDE							
AB	yyy	yyn	yny	ynn	nyy	nyn	nny	nnn
yy	4	0	5	0	1	0	2	0
yn	2	0	16	0	1	0	8	0
ny	9	1	10	0	14	0	16	0
nn	0	0	4	0	0	0	7	0

Table 4.2. A set of 100 configurations of (A, B, C, D, E) sampled from the network in Figure 4.40 and Table 4.1.

The probability distributions for the variables can now be calculated by counting in the sample set (see Exercise 4.39). For example, for 99 of the samples in Table 4.2, the state of E is y, and this gives an estimated probability:

$$P(E) \approx \left(\frac{N(E = y)}{N}, \frac{N(E = n)}{N} \right) = \left(\frac{99}{100}, \frac{1}{100} \right) = (0.99, 0.01).$$

So far, only marginal probabilities have been calculated. However, a straightforward approach to handle evidence is simply to discard the configurations that do not conform to it. In other words, a new series of stochastic simulations is started, and whenever a state of an observed variable is drawn, you stop simulating if the state drawn is not the one observed. In general, if we have evidence **e** and we are interested in estimating $P(X_k \mid \mathbf{e})$ using N samples, then probabilistic logic sampling can be performed as follows:

1. Let $(X_1, \ldots, X_n)$ be a topological ordering of the variables.
2. For $j = 1$ to N:
   a) For $i = 1$ to n:
      - Sample a state x_i for X_i using $P(X_i \mid \mathrm{pa}(X_i) = \pi)$, where π is the configuration already sampled for $\mathrm{pa}(X_i)$.
   b) If $\mathbf{x} = (x_1, \ldots, x_n)$ is consistent with **e**, then

$$N(X_k = x_k) := N(X_k = x_k) + 1,$$

   where x_k is the state that was sampled for X_k.
3. Return:

$$P(X_k = x_k \mid \mathbf{e}) \approx \frac{N(X_k = x_k)}{\sum_{x \in \mathrm{sp}(X_k)} N(X_k = x)}.$$

The preceding method does not require a triangulation of the network, nor is it necessary to store the sampled configurations (as we did in Table 4.2). It is enough to store the counts for each variable of interest. Whenever a sampled configuration has been determined, the counts of all variables are updated, and the sample can be discarded. The method therefore saves much space, and each configuration is determined in time linear in the number of variables. These benefits, however, come at the expense of accuracy and time. In particular, this method has a serious drawback when the probability of the evidence is small. For instance, assume that for the preceding example we have the observations $B = n$ and $E = n$. The probability for $(B = n, E = n)$ is 0.00282, which means that in order to get 100 configurations, you should for this tiny example expect to perform more than 35,000 stochastic simulations. In general, since the probability of the evidence drops off exponentially fast, this method can be hopelessly time-consuming even when we have only a few pieces of evidence.

4.8.2 Likelihood Weighting

You might be tempted to overcome the shortcoming of probabilistic logic sampling by simply fixing the evidence variables $\mathcal{E}$ to their observed states and sample only from the nonevidence variables; in this way no samples need to be discarded. However, since a sample is generated by going from the root nodes down to the leaves, this naive procedure would result in a sample in which the value for a given variable takes only the evidence from its ancestors into account and not the evidence pertaining to the variables further down in the network. For example, if we should try to estimate $P(A \mid B = n, E = n)$ using this modified sampling procedure we would actually estimate $P(A)$. The problem is that instead of sampling from the distribution $P(\mathcal{U}, \mathbf{e})$ specified by the evidence and the Bayesian network, we are in fact sampling from a probability distribution somewhere in between the prior distribution $P(\mathcal{U})$ and the posterior distribution $P(\mathcal{U} \mid \mathbf{e})$. To be more precise, if $\mathrm{pa}(X)''$ are the parents of X that have received evidence $(\mathrm{pa}(X) = \mathrm{pa}(X)' \cup \mathrm{pa}(X)'')$, then the joint distribution $P(\mathcal{U}, \mathbf{e})$ that we would like to sample from can be expressed as

$$
P(\mathcal{U}, \mathbf{e}) = \underbrace{\prod_{X \in \mathcal{U} \setminus \mathcal{E}} P(X \mid \mathrm{pa}(X)', \mathrm{pa}(X)'' = \mathbf{e})}_{\text{Part 1}}
$$
$$
\times \underbrace{\prod_{X \in \mathcal{E}} P(X = e \mid \mathrm{pa}(X)', \mathrm{pa}(X)'' = \mathbf{e})}_{\text{Part 2}}.
\tag{4.4}
$$

However, the distribution that we are actually sampling from is

$$
\text{Sampling distribution} = \prod_{X \in \mathcal{U} \setminus \mathcal{E}} P(X \mid \mathrm{pa}(X)', \mathrm{pa}(X)'' = \mathbf{e}),
$$

which corresponds only to Part 1 of equation (4.4).

Fortunately, this also points to a simple way of compensating for the estimation problem above: weigh each of the generated samples $\mathbf{x}$ with a weight corresponding to Part 2 of equation (4.4). That is, instead of adding 1 to the count $N(X_i)$ (as we did for probabilistic logic sampling) we add a weight $w(\mathbf{x}, \mathbf{e})$:

$$
w(\mathbf{x}, \mathbf{e}) = \prod_{E \in \mathcal{E}} P(E = e \mid \mathrm{pa}(X) = \pi),
$$

where π is the configuration of $\mathrm{pa}(X)$ specified by $\mathbf{x}$ and $\mathbf{e}$.

This updating approach, called *likelihood weighting*, ensures that we get the correct counts for estimating the probabilities. This can also be seen by combining the weight calculation and the sampling distribution, which together correspond to the distribution $P(\mathcal{U}, \mathbf{e})$.

Now consider again the example network above and assume that we want to estimate $P(A \mid B = n, E = n)$. As before, we start by sampling a state of A using a random generator (let the resulting state be y). Since B has received the evidence $B = n$, no state is sampled, and instead we continue to C and sample a state using $P(C \mid A = y) = (0.7, 0.3)$; assume that the sampled state is n. Next we sample a state for D using $P(D \mid B = n) = (0.5, 0.5)$ (assume that we get $D = y$). Since E has received evidence, $E = n$, we now have a complete configuration over all five variables and the sampling stops. Next we calculate the weight associated with the sampled configuration:

$$w((A = y, B = n, C = n, D = y, E = n), (B = n, E = n))$$
$$= P(B = n \,|\, A = y)P(E = n \,|\, C = n, D = y) = 0.7 \cdot 0.001 = 0.0007.$$

This value is then added to $N(A = y)$ (and to $N(C = n)$ and $N(D = y)$ as well if we are also interested in the probabilities for these two variables). We then continue to generate samples (and weights) as above, and when a sufficient number of samples has been generated we return the estimate $P(A \,|\, B = n, E = n) \approx N(A = y)/(N(A = y) + N(A = n))$.

In general, if we are interested in estimating $P(X_k \,|\, \mathcal{E} = \mathbf{e})$ using N samples, then the likelihood weighting algorithm can be summarized as follows:

1. Let $(X_1, \ldots, X_n)$ be a topological ordering of the variables.
2. For $j = 1$ to N:
   a) w:=1.
   b) For $i = 1$ to n:
      - Let $\mathbf{x}'$ be the configuration of $(X_1, \ldots, X_{i-1})$ specified by $\mathbf{e}$ and the previous samples.
      - If $X_i \notin \mathcal{E}$, then:
         - Sample a state x_i for X_i using $P(X_i \,|\, \mathrm{pa}(X_i) = \pi)$, where $\mathrm{pa}(X_i) = \pi$ is consistent with $\mathbf{x}'$.
      - else
         $$w := w \cdot P(X_i = e_i \,|\, \mathrm{pa}(X_i) = \pi), \text{ where } \mathrm{pa}(X_i) = \pi \text{ is consistent with } \mathbf{x}'.$$
   c) $N(X_k = x_k) := N(X_k = x_k) + w$, where x_k is the sampled state for X_k.
3. Return:
$$P(X_k = x_k \,|\, \mathbf{e}) \approx \frac{N(X_k = x_k)}{\sum_{x \in \mathrm{sp}(X_k)} N(X_k = x)}.$$

Although likelihood sampling is an improvement over probabilistic logic sampling it may still require a large number of samples. This is typically the case when there is a large difference between the sampling distribution and $P(\mathcal{U}, \mathbf{e})$ and, again, this is often the case when the evidence is unlikely.

4.8.3 Gibbs Sampling

Other methods have been constructed for dealing with this problem. A widely used method is *Gibbs sampling*. In Gibbs sampling, you start with some configuration consistent with the evidence (for example determined by probabilistic logic sampling), and then you randomly change the state of the variables in topological order. In one sweep through the variables, you determine a new configuration, and then you use this configuration for a new sweep, and so on. From this perspective, Gibbs sampling differs from the above two procedures by generating a new sample based on the current one.

Consider again the example above and let the evidence be $B = n$ and $E = n$. Assume also that we are given the starting configuration $ynyyn$. Now, to generate a sample we first calculate the probability of A given the other states of that configuration, that is, $P(A \mid B = n, C = y, D = y, E = n)$. From the network, we see that the Markov boundary for A includes only B and C; hence it is sufficient to calculate $P(A \mid B = n, C = y)$. It is easily done by Bayes' rule, which gives $(0.8, 0.2)$. We then draw a number from the random generator, and let us assume that the number is 0.456, resulting in $A = y$. The next free variable is C. We calculate

$$P(C \mid A = y, B = n, D = y, E = n) = P(C \mid A = y, D = y, E = n)$$
$$= (0.996, 0.04),$$

and draw a number from the random generator; assume that it results in $C = y$.

In general, the calculation proceeds as follows. Let A be a variable in a Bayesian network BN, let $B_1, \ldots, B_n$ be the remaining variables, and let $\mathbf{b} = (b_1, \ldots, b_n)$ be a configuration of $(B_1, \ldots, B_n)$. Then, $P(A, \mathbf{b})$ is the product of all conditional probabilities in BN with B_i instantiated to b_i. Therefore, $P(A, \mathbf{b})$ is proportional to the product of the potentials involving A, and $P(A \mid \mathbf{b})$ is the result of normalizing this product. Note that the calculation of $P(A \mid \mathbf{b})$ is a local task.

To return to the example, the next variable is D, and we follow the same procedure. Assume that the result is $D = n$. Then the configuration from the first sweep is $ynynn$. The next sweep follows the same procedure. Assume that the state of A changes to n. Then we calculate $P(C \mid A = n, D = n, E = n)$ and so forth.

In this way, a large sample of configurations consistent with the observations is produced. The question is whether the sample is representative of the probability distribution. It is not always so. It may be that the initial configuration is rather improbable, and therefore the first samples likewise are out of the mainstream. For this reason you usually discard the first 5-10% of the samples. It is called the *burn-in*. A related problem is the dependence among the samples: two successive samples will in general not be independent, since the second sample is generated by altering the first sample. In this way, these samples are also not representative of the probability distribution, and you therefore typically try to compensate for this by recording samples only at certain intervals.

Another problem is that you may be stuck in certain "areas" of the configurations. Perhaps there is a set of very likely configurations, but in order to reach them from the one you are in, a variable should change to a state that is highly improbable given the remaining configuration (see Exercise 4.43).

Finally, the method relies on an initial starting configuration. Unfortunately, it may be very hard to find such a configuration, and in fact this problem is NP-hard (see Exercise 4.44).

4.9 Loopy Belief Propagation

There is a popular approximate method that is not a version of sampling. It is called *loopy belief propagation* (LBP). LPB has been extremely successful in a setting not directly connected to Bayesian networks, namely in *error-correcting codes*; the so-called *turbo codes*.

LBP is a message passing algorithm similar to the junction tree algorithm in Section 4.4. However, instead of having cliques in a junction tree for passing messages, it uses the nodes in the Bayesian network directly.

The message passing structure consists of one node for each variable in the Bayesian network. A node representing the variable A holds the conditional probability table $P(A \mid pa(A))$, and it can process potentials over $fa(A)$ (the variables involved in the table). The neighbors of a node representing A are the neighbors of A in the Bayesian network, and the messages being passed over the links are potentials over the shared variables. We shall stick to the term *separator* for the domains of the potentials being passed over links, though these domains need not separate any variables from others. The structure is illustrated in Figure 4.41.

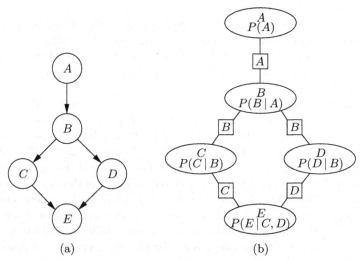

(a)   (b)

Fig. 4.41. (a) A Bayesian network. (b) The corresponding message-passing structure for LPB. Each node holds the corresponding variable's conditional probability table; the domain of a node is the variable's family. The square box on a link indicates the separator (the domain for the potentials to be passed over that link).

Note that all separators consist of one variable. If B is a child of A then the separator is A.

The processing of messages is similar to the one for junction trees: a message is sent to a neighbor by multiplying the incoming messages from all other neighbors to the potential it holds and marginalizing the result down to the separator. This is illustrated in Figure 4.42.

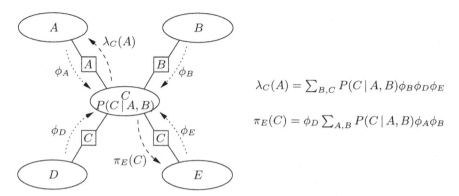

$$\lambda_C(A) = \sum_{B,C} P(C \mid A, B)\phi_B\phi_D\phi_E$$

$$\pi_E(C) = \phi_D \sum_{A,B} P(C \mid A, B)\phi_A\phi_B$$

Fig. 4.42. The node C holding $P(C \mid A, B)$ has received all messages (the ϕs). It sends a λ-message to its parent A and a π-message to its child E.

A message from a parent variable to a child variable is called a π-message (because it is in fact a probability distribution), and a message from a child to a parent is called a λ-message (for likelihood).

Since the structure may not be a tree, you cannot use the rule that a node can send to a neighbor when it has received a message from all its other neighbors. In Figure 4.41, only the node A can send a message. All other nodes wait for a message that never comes. Instead, you have a marching regime; at each step all nodes send messages to each neighbor using the messages they have received so far from the other neighbors. After each step, any node A can calculate an estimate of its own probability distribution: take the product of $P(A \mid \mathrm{pa}(A))$ and all incoming messages, marginalize it down to A, and normalize.

Now you let the method march step by step, monitor the development of the probability distributions, and use some stopping criterion. There is no guarantee that the method will converge, nor is there any guarantee that in case of convergence it will converge to the correct posterior distributions. On the other hand, very much experience has been gained, and the method converges to the correct posteriors surprisingly often.

However, sometimes the method is guaranteed to converge correctly, for example, if the network is singly connected (there are no multiple paths in the network). In that case, the junction tree will be exactly the structure for LBP (see Exercise 4.23), and when the method has marched twice the number of links in the network, the messages will be the same as the messages in the junction tree algorithm.

Unfortunately, this result is not of any use. If the Bayesian network is singly connected, the cliques are small, and exact junction tree propagation is no problem. As mentioned above, LBP does very often give good results, and much research is now directed at understanding why and characterizing situations in which you are guaranteed a result within a reasonable margin of tolerance.

4.10 Summary

Exact Belief Updating

Exact belief updating can be performed by message passing in a junction tree representation of the Bayesian network. The junction tree is obtained after triangulating the moral graph of the Bayesian network.

Moral graph: The moral graph of a Bayesian network is obtained by inserting a link between all pairs of variables with a common child, and dropping the direction on all arcs.

Triangulated graph: An undirected graph with a perfect elimination sequence is called a triangulated graph. If a graph is not triangulated, you can insert additional links (determined by, for example, node elimination), making it triangulated.

Node elimination: A node is eliminated by inserting a link between each pair of its noneliminated neighboring nodes.

Perfect elimination sequence: An elimination sequence is perfect if all nodes can be eliminated according to that sequence without inserting a link between a pair of noneliminated variables.

Clique: A complete set is a clique if it is not a subset of another complete set (a maximal complete set).

Join tree: Let $\mathcal{G}$ be the set of cliques from an undirected graph, and let the cliques of $\mathcal{G}$ be organized in a tree T. Then T is a join tree if for any pair of nodes V, W all nodes on the path between V and W contain the intersection $V \cap W$.

Junction tree: Let Φ be a set of potentials with a triangulated domain graph, G. A junction tree for Φ is a join tree for G with the following addition: each potential ϕ in Φ is attached to a clique containing dom (ϕ); each link has the appropriate separator attached; each separator contains two mailboxes, one for each direction.

Message passing: Let V be a clique with set of potentials Φ_V, and let S be a neighboring separator. Let $S_1, \ldots, S_k$ be the other neighboring separators of V. Assume that each S_i has received a message Ψ_i for V. Then V can pass the message $(\Phi_V \cup \Psi_1 \cup \cdots \cup \Psi_k)^{\downarrow S}$ to S.

Belief updating (calculating marginals): Let the junction tree T represent the Bayesian network BN over the universe U and with evidence e. Assume that each mailbox contains a message.

1. Let V be a clique with set of potentials Φ_V, and let $S_1, \ldots, S_k$ be V's neighboring separators and with V-directed messages $\Psi_1, \ldots, \Psi_k$. Then,

$$P(V, e) = \prod \Phi_V \prod \Psi_1 \cdots \prod \Psi_k.$$

2. Let S be a separator with the sets Ψ_S and Ψ^S in the mailboxes. Then,

$$P(S, e) = \prod \Psi_S \prod \Psi^S.$$

Belief Updating with Bounded Space

If there is not enough space to perform junction tree propagation, you may reduce the space complexity by applying a divide-and-conquer strategy: recursively condition on a variable (or subset of the variables) to be eliminated, solve the new smaller problems, and add up the results. A cache may be introduced to trade space for time.

Approximate Belief Updating

Stochastic simulation: Estimate $P(X \mid e)$ by sampling a large number of random configuration over the variables in the Bayesian network. Throw away the configurations that are inconsistent with e, and let N' be the resulting number of cases. Then

$$P(X \mid e) \approx \frac{N'(X)}{N'}.$$

Likelihood weighting: Estimate $P(X \mid e)$ by sampling a large number of random configurations over the noninstantiated variables in the Bayesian network. Weigh each configuration $(\mathbf{x}, \mathbf{e})$ with

$$w(\mathbf{x}, \mathbf{e}) = \prod_{E \in \mathcal{E}} P(E = e \mid \mathrm{pa}(X) = \pi),$$

where $\mathcal{E}$ are the evidence variables, and π is the configuration of $\mathrm{pa}(X)$ specified by $\mathbf{x}$ and $\mathbf{e}$.

Gibbs sampling: Estimate $P(X \mid e)$ by sampling a large number of random configurations over the noninstantiated variables in the Bayesian network. A sample is generated by starting with a configuration consistent with the evidence, and randomly changing the state of a variable by following the topological order.

Loopy belief updating (LBP): LBP is a message-passing algorithm that works directly on the Bayesian network. Messages are similar to those in junction trees, but in LBP they are passed between the families of variables in the Bayesian network.

4.11 Bibliographical Notes

Loopy belief propagation is rooted in a version of probability updating for singly connected DAGs through message passing presented by Kim and Pearl (1983). In (Pearl, 1986), cutset-conditioning was used to reduce propagation in multiply connected networks to propagation in singly connected networks. Shachter (1986) introduced arc reversal and uses it for a probability updating procedure in the bucket elimination style. Two versions of join tree propagation were presented in the late 1980s. Shafer and Shenoy (1990) proposed the method presented in this book. They did not exploit lazy evaluation but worked with multiplied potentials. Lauritzen and Spiegelhalter (1988) and Jensen *et al.* (1990b) proposed what is now called the Hugin method. It also works with multiplied potentials, but the potentials in the cliques are changed dynamically. This, together with a division operation in the separators, reduced the calculation substantially for join trees with branching higher than three. A detailed study of the similarities and differences of the two methods is reported in (Shafer, 1996). Lazy propagation (Madsen and Jensen, 1999b) dissolves the difference between Shafer-Shenoy and Hugin propagation.

The concepts of triangulated graphs and join trees have been discovered and rediscovered with various names. In (Bertele and Brioschi, 1972), they are used for dynamic programming, and Beeri *et al.* (1983) uses them for database management. A good reference on triangulated graphs is (Golumbic, 1980). The heuristic for triangulating nontriangulated domain graphs given in this chapter is due to Kjærulff (1990), and more can be found in (Cano and Moral, 1995). The problem of inference in dynamic Bayesian networks has been treated in (Boyen and Koller, 1998).

Recursive conditioning was introduced in (Darwiche, 2001). Probabilistic logic sampling was proposed by Henrion (1988), and Fung and Chang (1990) and Shachter and Peot (1990) introduced likelihood-weighted sampling for Bayesian networks. Gibbs sampling was originally introduced for image restoration by Geman and Geman (1984). Gilks *et al.* (1994) have developed a system, BUGS, for Gibbs sampling in Bayesian networks.

4.12 Exercises

Exercise 4.1. *BN* has the potentials in Table 4.3.

	A			B			C	
B	y	n	C	y	n	D	y	n
y	0.2 0.6		y	0.3 0.2		y	0.9 0.6	
n	0.8 0.4		n	0.7 0.8		n	0.1 0.4	
	$P(B\,|\,A)$			$P(C\,|\,B)$			$P(D\,|\,C)$	

Table 4.3. Potentials for Exercise 4.1. $P(A) = (0.2, 0.8)$.

(i) Calculate $P(A\,|\,D = y)$.
(ii) Calculate $P(C\,|\,D = y)$.

Exercise 4.2. *BN* has the potentials in Table 4.4.

	A			B			B	
B	y	n	C	y	n	D	y	n
y	0.2 0.6		y	0.1 0.5		y	0.7 0.4	
n	0.8 0.4		n	0.9 0.5		n	0.3 0.6	
	$P(B\,|\,A)$			$P(C\,|\,B)$			$P(D\,|\,B)$	

Table 4.4. Potentials for Exercise 4.2. $P(A) = (0.2, 0.8)$.

(i) Calculate $P(A\,|\,C = y, D = y)$.
(ii) Calculate $P(A\,|\,D = y)$.

Exercise 4.3. *BN* has the potentials in Table 4.5.

	A			A			B	
B	y	n	C	y	n	C	y	n
y	0.2 0.6		y	0.1 0.5		y	(0.3, 0.7)	(0.2, 0.8)
n	0.8 0.4		n	0.9 0.5		n	(0.9, 0.1)	(0.6, 0.4)
	$P(B\,|\,A)$			$P(C\,|\,A)$			$P(D\,|\,B, C)$	

Table 4.5. Potentials for Exercise 4.3. $P(A) = (0.2, 0.8)$.

(i) Calculate $P(A\,|\,D = y), P(B\,|\,D = y), P(C\,|\,D = y)$.

(ii) Calculate $P(B \mid C = y)$.

Exercise 4.4. Consider the Bayesian network in Figure 4.43. All variables have three states.

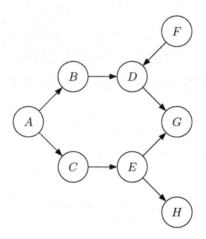

Fig. 4.43. The network for Exercise 4.4.

  (i) Calculate the size of the table $P(A, B, C, D, E, F, G = g_1, H = h_1)$.
 (ii) In the calculation of $P(A \mid G = g_1, H = h_1)$, the variables have been marginalized in the following order: B, F, D, E, C. Calculate the size of each table produced in the process, and compare the sum with the result of (i).
(iii) Determine an elimination order yielding a sum smaller than the one from (ii).

Exercise 4.5. We have the potentials $\phi_1(A_1, A_2, A_3)$, $\phi_2(A_2, A_3, A_5)$, $\phi_3(A_1, A_3, A_4)$, $\phi_4(A_5, A_6)$ over the universe $\{A_1, A_2, A_3, A_4, A_5, A_6\}$.

  (i) Determine the domain graph.
 (ii) Eliminate A_3.
(iii) Determine the domain graph for the resulting set of potentials.

Exercise 4.6. We have the potentials $\phi_1(A_1, A_2, A_3)$, $\phi_2(A_2, A_4, A_5)$, $\phi_3(A_4, A_6, A_7)$, $\phi_4(A_1, A_6, A_8)$ over the universe $\{A_1, A_2, A_3, A_4, A_5, A_6, A_7, A_8\}$.

  (i) Determine the domain graph.
 (ii) Eliminate A_1.
(iii) Determine the domain graph for the resulting set of potentials.

Exercise 4.7. Write a short algorithm that takes as input a Bayesian network over nodes $X_1, \ldots, X_n$ and an elimination sequence for all nodes but X_i, and which outputs the maximum table size that would be used during computation of $P(X_i)$ using this elimination sequence.

Exercise 4.8. Consider the Bayesian network given in Figure 4.44. What would the elimination trees (such as those in Figures 4.2 to 4.7) look like for the two elimination orders C, F, G, B, E, D and F, E, G, D, C, B?

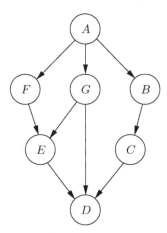

Fig. 4.44. A Bayesian network.

Exercise 4.9. Prove Proposition 4.1.

Exercise 4.10. What is $(\prod \phi)^{\downarrow A}$ for the Bayesian network in Figure 4.44?

Exercise 4.11. What are the domains encoded by the domain graph in Figure 4.45? Give an example of an elimination sequence ending with C. What do the intermediate domain graphs look like as you apply the elimination sequence? Is the sequence perfect?

Exercise 4.12. Consider the domain graph for the potentials in Exercise 4.5. Determine a perfect elimination sequence ending with A_1.

Exercise 4.13. Consider the domain graph for the potentials in Exercise 4.6. Does the graph have a perfect elimination sequence?

Exercise 4.14. Consider the Bayesian network in Figure 4.43.

(i) Determine the domain graph.
(ii) Does the domain graph have a perfect elimination sequence?

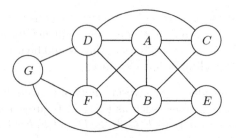

Fig. 4.45. A domain graph.

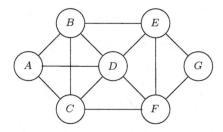

Fig. 4.46. The graph for Exercise 4.15.

Exercise 4.15. Consider the graph in Figure 4.46.

(i) Determine the simplicial nodes.
(ii) Is the graph triangulated?

Exercise 4.16. Consider the graph in Figure 4.47.

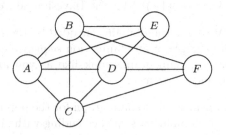

Fig. 4.47. The graph for Exercise 4.16.

(i) Determine the simplicial nodes.
(ii) Is the graph triangulated?

Exercise 4.17. Definition Let G be an undirected graph with node set U. A *path* in G is a sequence $A_1, \ldots, A_n$ of distinct nodes; where A_i and A_{i+1} are neighbors. A *cycle* is a path except $A_1 = A_n$, and all other nodes are distinct. A *chord* in a cycle $A_1, \ldots, A_n$ is a link between two nodes A_i and A_j that are not neighbors on the path. The graph G is *chord-saturated* if any cycle of length > 3 has a chord.

(i) Prove that any triangulated graph is chord-saturated. (Hint: Use induction and the fact that any cycle through a simplicial node must have a chord.)

(ii) Prove the following decomposition lemma. Let G be a incomplete chord-saturated graph with at least three nodes and with node set U. Then there is a complete subset S of U such that $G \setminus S$ is disconnected. (Hint: Let A and B be two nonadjacent nodes, and let S be a minimal set of nodes such that any path connecting A and B contains a node from S. Use chord saturation and minimality of S to prove that S is complete.)

(iii) Prove that any chord-saturated graph is triangulated. (Hint: Use (ii) to prove that any incomplete chord-saturated graph with at least two nodes has at least two simplicial nodes.)

Exercise 4.18. Prove that the moral graph of the graph in Figure 4.48 is triangulated. Give an example of a join tree for the graph.

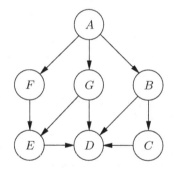

Fig. 4.48. A Bayesian network.

Exercise 4.19. Consider the domain graph from Exercise 4.5.

(i) Determine the cliques.
(ii) Construct a join tree for the graph.

Exercise 4.20. Consider the graph in Figure 4.47.

(i) Determine the cliques.

(ii) Construct a join tree for the graph.

Exercise 4.21. Consider the Bayesian network in Figure 4.49. Construct a join tree.

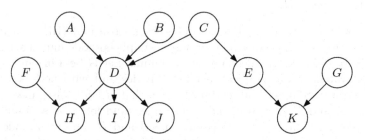

Fig. 4.49. The Bayesian network for Exercise 4.21.

Exercise 4.22. Let $\mathcal{A}$ and $\mathcal{B}$ be any two adjacent nodes in a join tree for a Bayesian network M with separator $\mathcal{S} = \mathcal{A} \cap \mathcal{B}$. Furthermore, let $\mathcal{U}_{\mathcal{A}}$ be the variables in the nodes found in the part of the join tree on $\mathcal{A}$'s side of the link, and $\mathcal{U}_{\mathcal{B}}$ those found in nodes on $\mathcal{B}$'s side of the link. Prove that for any two nodes $A \in \mathcal{U}_{\mathcal{A}} \setminus \mathcal{S}$ and $B \in \mathcal{U}_{\mathcal{B}} \setminus \mathcal{S}$, we have that A and B are d-separated by $\mathcal{S}$.

Exercise 4.23. A directed acyclic graph is *singly connected* if the graph you get by dropping the directions of the links is a tree (the graph in Figure 4.49 is singly connected).

 (i) Prove that the moral graph of a singly connected graph is triangulated. (Hint: If you successively eliminate a node with exactly one parent and no children or with no parents and exactly one child, then the result is a moral graph for a singly connected graph.)
 (ii) Prove that the separators in a join tree for a singly connected graph consist of exactly one node. (Hint: If the neighbors A and B share the neighbors C and D, then C and D are neighbors.)

Exercise 4.24. Consider the Bayesian network in Exercise 4.21.
    Indicate the potentials to communicate in a full lazy propagation with evidence $F = f$, $I = i$, $E = e$.

Exercise 4.25. Expand the join tree in Figure 4.16 to a junction tree, and add the potentials defined by the domain graph in Figure 4.14 to suitable cliques. Which messages are sent if evidence is collected to node CG?

Exercise 4.26. Consider the Bayesian network in Figure 4.50.

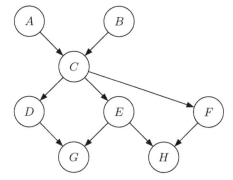

Fig. 4.50. The Bayesian network for Exercise 4.26.

(i) Construct a junction tree.
(ii) Indicate the potentials to communicate in a full lazy propagation without evidence.
(iii) Indicate the potentials to communicate with evidence $D = d$ and $H = h$.

Exercise 4.27. Prove Proposition 4.5. (Hint: Assume a deadlock (no triggered nodes).)

Exercise 4.28. Show that any asynchronous full order of message passing corresponds to a COLLECTEVIDENCE(R) followed by a DISTRIBUTEEVIDENCE(R) for some node R. (Hint: Look at the first node that receives all its messages.)

Exercise 4.29. Triangulate the domain graph from Exercise 4.6.

Exercise 4.30.

(i) Construct a junction tree for the Bayesian network in Figure 4.51 by using the elimination order F, J, B, A, I, K, E.
(ii) The numbers inside the nodes indicate the number of states. Use the heuristics from Section 4.6.1 to construct a junction tree.

Exercise 4.31. What is the moral graph of the Bayesian network in Figure 4.44? Assuming that each node has 10 states, use the heuristics following Definition 4.8 to triangulate the graph. Would the result be the same if each node had 2 states instead?

Exercise 4.32. Consider the Bayesian network in Figure 4.29, and let the number of states be as listed in Section 4.6.1. Find a better triangulation than the one obtained by using the heuristics from Section 4.6.1.

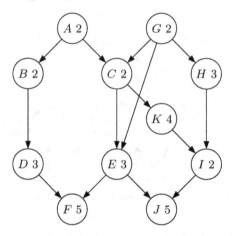

Fig. 4.51. The Bayesian network for Exercise 4.30.

Exercise 4.33. (Conditioning) Propagation methods for DAGs without multiple paths have existed for a long time. A propagation method for multiply connected DAGs consists in reducing a DAG to a set of singly connected DAGs.

(i) Consider the DAG (a) in Figure 4.52 with $P(A), P(B\,|\,A), P(C\,|\,A)$, and $P(D\,|\,B,C)$ given. Assume that $A = a$. Show that the DAG is reduced to the DAG (b) with $P(B, a), P(C, a)$, and $P(D\,|\,B,C)$ given. (Hint: Use the chain rule.. Calculate $P(B, a)$ and $P(C, a)$.

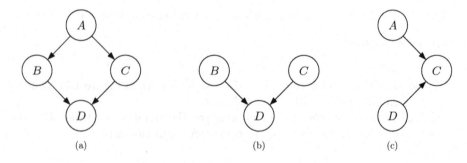

Fig. 4.52. Figures for Exercise 4.33(i)–(v).

(ii) Show that $P(D, a) = \sum_{B,C} P(D\,|\,B,C)P(B, a)P(C, a)$.
(iii) Assume that for all states a of A we have a reduced DAG as in (i). Let evidence e be entered and propagated in all the reduced DAGs, yielding $P(B, a, e), P(C, a, e), P(D, a, e)$ for all a. Calculate $P(B, e)$ and $P(A, e)$. This procedure is called *conditioning on A*.

(iv) Reduce the DAG by conditioning on B. Show that the tables are $P(A, b)$, $P(C \mid A)$, and $P(D \mid C, b)$.

(v) Show that conditioning on D does not result in a singly connected DAG. Conditioning over several variables can be performed stepwise.

(vi) Determine a minimal set of conditioning variables for the DAG in Figure 4.53 to reduce it to singly connected DAGs.

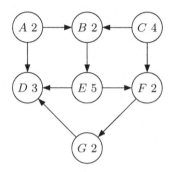

Fig. 4.53. Figure for Exercise 4.33 (vi)–(vii).

(vii) The numbers attached to the variables indicate the number of states. Determine a conditioning resulting in a minimal number of singly connected DAGs.

Exercise 4.34. Let $\mathcal{C}$ be the set of cliques from a triangulated graph. A *pre-$\mathcal{J}$-tree* is a tree over $\mathcal{C}$ with separators $S = V \cap W$ for adjacent cliques V, W. The *weight* of a pre-$\mathcal{J}$-tree is the sum of the number of variables in the separators.

(i) Prove that a join tree is a pre-$\mathcal{J}$-tree of maximal weight.
(ii) Prove that any pre-$\mathcal{J}$-tree of maximal weight is a join tree.

Exercise 4.35. (i) Consider the graph in Figure 4.35. Determine a triangulation such that no clique contains more than four nodes.
(ii) Expand the model in Figure 4.34 to six time slices. Can this model be triangulated such that no clique contains more than four nodes?

Exercise 4.36. Consider the Bayesian network in Figure 4.54, where each variable is binary, with probabilities defined as $P(A = a_1) = 0.1$, $P(B = b_1 \mid a_1) = 0.1$, $P(B = b_1 \mid a_2) = 0.9$, $P(C = c_1 \mid b_1) = 0.1$, $P(C = c_1 \mid b_2) = 0.9$, $P(D = d_1 \mid c_1) = 0.1$, and $P(D = d_1 \mid c_2) = 0.9$. Using recursive conditioning, calculate $P(a_1 \mid d_1)$.

Exercise 4.37. Construct two time slices of the model in Figure 3.52. Using recursive conditioning, what would a computation tree for calculating $P(C_2)$ look like?

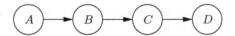

Fig. 4.54. A simple Bayesian network.

Exercise 4.38. Show that the worst case complexity of Algorithm 4.2 is $O(n \cdot \exp(wn))$, and that the complexity for a balanced tree is $O(n^{w+1})$.

Exercise 4.39. Calculate the marginals from the sample in Table 4.2 and compare the result with the exact marginals.

Exercise 4.40. From the configurations in Table 4.2, estimate the following probability distributions: $P(A)$, $P(A \mid D = n)$, and $P(C, D \mid B = y, E = n)$.

Exercise 4.41. Does your software tool allow for sampling from a Bayesian network model? Which kind of sampling technique is used?

Exercise 4.42. Using the sequence of random numbers in Table 4.41 generate as many full samples as you can for the Bayesian network model given in Figure 4.46, with conditional probabilities as defined in Table 4.1 and evidence $B = n$, using first probabilistic logic sampling, then likelihood weighting, then Gibbs sampling using sampling sequence A, C, D, E, and finally Gibbs sampling using sampling sequence A, D, E, C.

1 0.80	5 0.33	9 0.55	13 0.14
2 0.19	6 0.08	10 0.71	14 0.42
3 0.85	7 0.52	11 0.06	15 0.32
4 0.28	8 0.65	12 0.78	16 0.11

Table 4.6. A sequence of random numbers in the interval $[0, 1]$.

Exercise 4.43. The binary variables A and B are parents of the binary variable C. We have $P(A) = P(B) = (0.5, 0.5)$, and the conditional probability table is an exclusive OR table ($C = y$ if and only if exactly one of A and B is in the state y). Show that Gibbs sampling on this structure will give either $P(C = y) = 1$ or $P(C = n) = 1$.

Exercise 4.44. Given a Bayesian network over U with evidence e entered, show that it is NP-hard to find a configuration U^* such that $P(U^*, e) > 0$. (Hint: Look at Exercise 3.27.)

5

Analysis Tools for Bayesian Networks

The main reason for building a Bayesian network is to estimate the state of certain variables given some evidence. In Chapter 4, we gave methods that made it easy to access $P(A \mid e)$ for any variable A. However, this may not be sufficient. It may be crucial to establish the joint probability for a *set of variables*. Section 5.2 gives a general method for calculating $P(\mathcal{V} \mid e)$ for any set $\mathcal{V}$ of variables.

Another typical request is to ask for the most-probable configuration. We give a method for this in Section 5.3. Section 5.5 deals with methods for analyzing whether the evidence entered to the network is coherent; for example to trace flawed data.

A very important tool for a decision support system is *explanation*: a tool to explain to the user how the system came to its conclusions. A part of explanation is *sensitivity to evidence*: how sensitive is the conclusion to (small) changes in the evidence? Which parts of the evidence are crucial and/or sufficient for the conclusion? This is the subject of Section 5.6.

Finally, we present methods for analyzing how sensitive posterior probabilities are to changes in the numbers specified in the model.

The procedures in this chapter are based on lazy propagation as presented in Chapter 4, but most of them are also valid using other propagation methods. In lazy propagation, you work with sets of potentials representing the product. Often you will perform the product of the union of two sets of potentials. We shall call this operation to "take the product of the two sets" and unless necessary for the exposition, we do not bother whether this is done by taking the union of the two sets or by actually taking all potentials in the two sets and multiplying them together.

5.1 IEJ Trees

Let e_X be a finding of the form "only the states $x'_1, \ldots, x'_q$ of the variable X are possible". If you know $P(X)$ then $P(e_X)$ is easy to calculate, namely as the sum of the probabilities for the states declared possible.

We shall in several situations need $P(e)$ for a set of findings e and therefore we repeat Theorem 4.5 in condensed form.

Proposition 5.1. *Let BN be a Bayesian network and let $e = \{e_1, \ldots, e_m\}$ be evidence. When e has been entered and a full propagation has been performed, then $P(e)$ can be calculated in the following way: take any separator S, multiply the two messages in the mail boxes (to get $P(S, e)$) and marginalize all variables out of the product.*

Proposition 5.1 can be used for more than calculation of probabilities of evidence. Assume that some Bayesian network has received evidence e, and we want to calculate the probability of the configuration $\mathbf{c} = (A = a, B = b, C = c)$ given e. That is, we want $P(\mathbf{c}|e)$. Proposition 5.1 yields $P(e)$. If now we enter $\mathbf{c}$ as further evidence and perform an extra propagation, then Proposition 5.1 yields $P(\mathbf{c}, e)$, and the fundamental rule gives

$$P(\mathbf{c}|e) = \frac{P(\mathbf{c}, e)}{P(e)}.$$

This technique can for example solve the question from Section 3.2.4 with the model in Figure 3.18: the sequence *baaca* is received; what is the probability that the transmitted word is **baaba**?

Sometimes we may want to calculate $P(e')$ for various subsets $e' \subseteq e$. To do this we can work with two copies of the junction tree. In the first copy we have performed an initial propagation, and the appropriate messages are placed in the mailboxes. In the second copy we have entered and propagated evidence, and the messages from this propagation are stored in the mailboxes. To be precise, we can work with junction trees in which the separators have four mailboxes (See Figure 5.1). We call this kind of junction tree an *IEJ tree* (for Initial-Evidence Junction).

The separator S in Figure 5.1 divides the evidence into two sets: the evidence e_V entered at the left of S and e_W entered at the right of S. From Proposition 5.1 we have that $P(S)$ is the product of Φ^V and Φ^W, and $P(S, e)$ is the product of Φ^V_e and Φ^W_e. Now look at the pair (Φ^V, Φ^W_e). This pair is the pair of messages we would have, had we entered e_W only. Therefore, the product must be $P(S, e_W)$, and we can easily calculate $P(e_W)$ as well as $P(S \,|\, e_W)$. Similarly for $P(S, e_V)$ and $P(S \,|\, e_V)$.

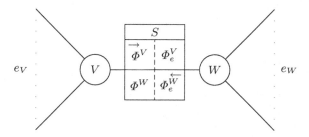

Fig. 5.1. The separators in an IEJ tree contain four mailboxes, two for each direction. One of the mailboxes contains the message from the initial propagation, and the other contains the message from a propagation of evidence.

5.2 Joint Probabilities and A-Saturated Junction Trees

When dealing with utility functions (see Chapter 9) over several variables and in various other connections, we will be faced with a request for the joint probability of several variables.

Take for example the stud farm example from Section 3.2 and the situation in Figure 3.16, and assume that the farmer has to decide on a new mating among the horses Fred, Dorothy, Eric and Gwenn. Which pair should be chosen to minimize the risk of getting a carrier as offspring?

If the set requested is a subset of a node in the junction tree, then you have the joint distribution directly. If not, the technique from Section 5.1 can be used by entering and propagating all configurations, but it is troublesome.

A better technique is to perform propagation without eliminating variables from the requested set. This technique is called *variable propagation*.

Example 5.1. Assume that we request $P(A, B, C, D, E)$ from the junction tree in Figure 5.2.

Then collect to (DEH) and in the operations, do not marginalize A, B, and C. In Figure 5.3, the functions communicated in the operation are indicated. Note that the "sending" of functions does not mean that the functions are moved. What is sent is a pointer to a table for the function, and since variable propagation involves fewer marginalizations than normal propagation, it may be faster. However, when finally the incoming messages are multiplied, we have to work with a considerably larger domain.

5.2.1 A-Saturated Junction Trees

Sometimes a variable A may be of particular interest. It may be a hypothesis variable, and you may be interested in investigating $P(A \mid X)$ for many different variables X. You may enter each state of X and propagate, but it requires one propagation for each state of each variable. Instead, you can make

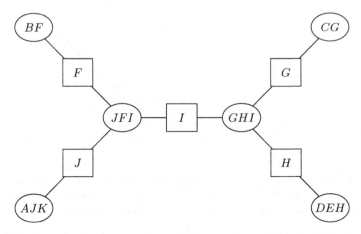

Fig. 5.2. A junction tree from which we request $P(A, B, C, D, E)$.

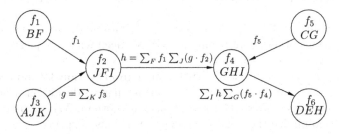

Fig. 5.3. The messages passed in performing variable propagation for the calculation of $P(A, B, C, D, E)$. We assume that each clique holds one function (over its domain).

A present in the entire junction tree: perform a full propagation but do not eliminate A. The result is called an *A-saturated junction tree*.

If $\mathcal{W}$ is a set of variables, we can do the same, and the result is a $\mathcal{W}$-saturated junction tree. The propagation in the example above is the COL-LECTEVIDENCE part of establishing an (A, B, C, D, E)-saturated junction tree. Notice that the work requested for establishing a $\mathcal{W}$-saturated junction tree does not exceed the work required for a normal propagation. More space may be required, though.

Proposition 5.2. *Let T be a $\mathcal{W}$-saturated junction tree with evidence e, and let X be any variable. Then $P(\mathcal{W} \mid X, e)$ is calculated through the following procedure*

1. *Choose any node V or separator S in T containing X.*
2. *$P(V \cup \mathcal{W}, e)$ is the product of V's set of potentials with the incoming messages ($P(S \cup \mathcal{W}, e)$ is the product of the two messages in S).*
3. *$P(\mathcal{W}, X, e) = \sum_{V \setminus (\mathcal{W} \cup \{X\})} P(V \cup \mathcal{W}, e)$.*
4. *$P(X, e) = \sum_{\mathcal{W}} P(\mathcal{W}, X, e)$.*

5. $P(\mathcal{W} \mid X, e) = \frac{P(\mathcal{W}, X, e)}{P(X, e)}$.

Note that in a $\mathcal{W}$-saturated junction tree you can get $P(\mathcal{W} \mid X, e)$ for each X through one local calculation. On the other hand, this local calculation is more complex than in the case of a normal junction tree. In the extreme, $\mathcal{W}$ may be very close to the universe, and the "local" calculation is extremely demanding.

We shall later deal with $\mathcal{W}$-saturated IEJ trees. They contain four messages in each separator, and they can be used for easy calculations of $P(\mathcal{W}|e')$ for various subsets of the evidence.

5.3 Configuration of Maximal Probability

In the example in Section 3.2.4 concerning transmission of symbol strings, the immediate task is to find out which symbol string most probably has been transmitted. Using propagation of variables, the joint probabilities for all possible strings can be calculated, and thereby the most-probable string can be found. This may require an intractably large table. There is, however, a much more efficient method.

Example 5.2. Consider a small system consisting of the variables A, B, and C with the joint probability determined by the conditional probabilities specified in Table 5.1, and suppose that we want to find out which configuration of (A, B, C) has maximal probability.

	a_1	a_2
b_1	0.6	0.2
b_2	0.4	0.8

$P(B|A)$

	b_1	b_2
c_1	0.2	0.7
c_2	0.8	0.3

$P(C|B)$

Table 5.1. Probability tables for a small system, $P(A) = (0.4, 0.6)$.

Let us start calculating the probability α of the most-probable configuration; α is the largest number in the joint probability table $P(A, B, C)$:

$$
\alpha = \max_{A,B,C} P(A, B, C) = \max_{A,B,C} P(A)P(B|A)P(C|B)
$$

$$
= \max_A \left(\max_B (\max_C (P(A)P(B|A)P(C|B))) \right)
$$

$$
= \max_A \left(\max_B (P(A)P(B|A) \max_C P(C|B)) \right)
$$

$$
= \max_A \left(P(A) \max_B (P(B|A) \max_C P(C|B)) \right).
$$

In the equations above, we first used the chain rule for Bayesian networks and next the distributive law for the max operation.

So first we determine $\max_C P(C|B)$. It is a potential over B, and from Table 5.1 we get the potential $(0.8, 0.7)$. Next, this potential is multiplied by $P(B|A)$ (see Table 5.2).

$B \setminus A$	a_1	a_2
b_1	0.48	0.16
b_2	0.28	0.56

Table 5.2. $P(B|A)\max_C P(C|B)$.

When maximizing Table 5.2 over B, we get the potential $(0.48, 0.56)$ over A. It is multiplied by the prior distribution $(0.4, 0.6)$, and we get $(0.192, 0.336)$. From this, we can conclude that the most-probable configuration has probability 0.336, and the A-component of it must be a_2.

To get the B-component, return to Table 5.2. Since we know that $A = a_2$, we have that the state of maximal value for B is b_2. Actually, when the value 0.56 in Table 5.2 is multiplied by the prior, 0.6, for a_2, we get the maximal value 0.336. In the same way, the C-state is determined from $P(C\,|\,b_2)$ to c_1.

Let $\mathcal{U}$ be the universe for a Bayesian network. The general task of determining the configuration of maximal probability is to determine the X-component for each $X \in \mathcal{U}$. In fact, there may be several configurations of maximal probability. We will leave this problem for a short while and assume that there is exactly one configuration of maximal probability. So the general task is for each variable X to get the distribution resulting from maximizing the remaining variables out. To help in the calculation, we have the following result:

Proposition 5.3 (The distributive law for max).

$$\max_Z f(X,Y)g(Y,Z) = f(X,Y)\max_Z g(Y,Z).$$

So the task is very similar to the task from Chapter 4 except that the operation is "max" instead of "$\sum$." Since the distributive law holds for max too, the propagation methods from Chapter 4 can be applied by substituting "max" for "$\sum$." This is called *max-propagation*, and accordingly we may use the term *sum-propagation* for the methods in Chapter 4. The result of maximizing variables out of a function f is called a *max-marginal* of f.

Theorem 5.1. *Let BN be a Bayesian network representing $P(\mathcal{U})$, and let T be a junction tree corresponding to BN. Let e be the evidence represented by the functions $\{\mathbf{e}_1, \ldots, \mathbf{e}_m\}$, and assume that the evidence functions are attached to appropriate nodes in the junction tree.*

After a full round of (lazy) max-propagation in T we have

i) *for each separator S, $\max_{\mathcal{U} \setminus S} P(\mathcal{U}, e)$ is the product of the two messages in S's mailboxes;*

ii) *for each node V, $\max_{\mathcal{U} \setminus V} P(\mathcal{U}, e)$ is the product of the potential set attached to V and the incoming messages.*

Proof.   Repeat the consideration from Chapter 5 with "max" instead of "$\sum$." Since the potential sets attached to the nodes in the junction tree are never changed, you can always change between max-propagation and sum-propagation.   □

Several Configurations of Maximal Probability

When there is exactly one configuration of maximal probability, then for each variable X we can read the component by taking the state of maximal probability in the max-marginalized distribution for X. However, if there are several configurations of maximal probability, then for some variables $\{Y_1, \ldots, Y_m\}$ there are several states of maximal probability in their max-marginalized distribution. Unfortunately, it does not hold that all combinations of these max-probable states form a configuration of maximal probability. If you request one of them, you can enter a max-probable state as evidence and perform a new max-propagation. If still there are several max-probable states in some of the remaining variables, you can repeat this operation until all variables have only one max-probable configuration.

Working with Subsets of Variables

If the evidence variables $\mathcal{E}$ and the query variables $\mathcal{Q}$ do not constitute all the variables in the Bayesian network, then the above procedure cannot be applied. The problem is that since we are interested only in $\mathcal{Q}$, the remaining variables $\mathcal{U} \setminus (\mathcal{Q}, \mathcal{E})$ should be marginalized out by summation *before* we do the maximization:

$$\max_{\mathcal{Q}} P(\mathcal{Q} \mid e) = \max_{\mathcal{Q}} \sum_{\mathcal{U} \setminus (\mathcal{Q}, \mathcal{E})} P(\mathcal{U} \mid e).$$

The result is the maximum posterior probability (MAP) over the query variables and the associated configuration is called a MAP configuration. Unfortunately, the constraint on the elimination ordering makes it much more difficult to work with MAP problems than MPE problems.

5.4 Axioms for Propagation in Junction Trees

As shown in Section 5.3, the propagation algorithm can be used for other types of tasks than probability updating. Therefore, a general framework and a set of axioms for propagation in junction trees have been established. The framework and the axioms look very much like the properties listed in Section 1.4, and we shall state them in a more general form here.

We have a set ϑ of valuations. Each $v \in \vartheta$ has a set $\mathrm{dom}\,(v) \subseteq \mathcal{U}$ attached. The set $\mathcal{U}$ is called the *universe*. Valuations can be *combined* through a binary operation $\otimes$, and for each $\mathcal{V} \subseteq \mathcal{U}$ there is a *projection operator* $v^{\downarrow V}$.

Axioms

1. $\mathrm{dom}\,(v_1 \otimes v_2) \subseteq \mathrm{dom}\,(v_1) \cup \mathrm{dom}\,(v_2)$,
2. $\mathrm{dom}\,(v^{\downarrow V}) \subseteq V$,
3. Combination is associative: $(v_1 \otimes v_2) \otimes v_3 = v_1 \otimes (v_2 \otimes v_3)$,
4. Combination is commutative: $v_1 \otimes v_2 = v_2 \otimes v_1$,
5. $(v^{\downarrow V})^{\downarrow W} = v^{\downarrow V \cap W}$,
6. The distributive law: If $\mathrm{dom}\,(v_1) \subseteq V$ then $(v_1 \otimes v_2)^{\downarrow V} = v_1 \otimes (v_2)^{\downarrow V}$,
7. $v^{\downarrow \emptyset}$ is a neutral element with respect to combination, and it is denoted by **1**.

The axiom 7 is not needed, but it is customary to include it as an assumption.

With respect to probability updating in Bayesian networks, combination corresponds to multiplication, and projection corresponds to marginalizing out (see Section 1.4). Since the expression $v^{\downarrow V \cap W}$ is symmetric in V and W, axiom 5 includes the property that marginalization is commutative.

In the case of determining the most-probable configuration, projection corresponds to maximizing out.

If you have a valuation framework satisfying the axioms above, you can calculate $(\bigotimes_i v_i)^{\downarrow X}$ for all $X \in \mathcal{U}$ through junction tree propagation. We shall not prove it here, but the interested reader may reread Chapter 4 and check that only the axioms above are used.

5.5 Data Conflict

A Bayesian network represents a closed world with a finite set of variables and causal relations. The causal relations are not universal but reflect relations under certain constraints. Take for example a diagnostic system that on the basis of blood analysis monitors pregnancy. Only diseases and relations relevant for pregnant women are represented in the model. So if the blood originates from a man, the case is not covered by the model. It may happen that findings from male blood are impossible given the model. If so, the inconsistency is easy to detect: the probability of the evidence is 0. However, most often a set of findings is possible in the given model, and the system will not object to it. It will yield posterior probability distributions that may look rather harmless. The same also happens if test results are flawed. In a diagnostic situation, a single flawed test result may turn the investigation in a completely wrong direction (such flawed pieces of information are called *red herrings*.

5.5.1 Insemination

Consider the insemination example from Section 3.1.3, and assume that the farmer also has a scanning test. The model is given in Figure 5.4. To make things easy, assume that all tests have 2% false positives as well as false negatives (the prior for *Pr* is $(0.87, 0.13)$).

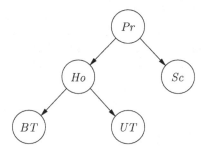

Fig. 5.4. Insemination extended with a scanning test.

Assume that we get the evidence $UT = n$ and $Sc = n$ but $BT = y$. From our knowledge of the network model, we would say that the findings are in conflict. However, a propagation of the evidence does not disclose it. The posterior probabilities for *Pr* are $(0.12, 0.88)$. Since the test results can coexist, we may be facing a rare case, but it may also be the case that the blood test is flawed or that the case is not covered by the model (a bull may have sneaked into the laboratory). We do not really have tools to distinguish between these situations, but it would be good to have a tool that gives a warning, "it seems that the evidence is conflicting."

5.5.2 The Conflict Measure conf

Several approaches for analyzing data for conflicts have been developed. We shall in this section present a measure that requires only two propagations and that gives an indication of a possible conflict. The idea behind the measure is that correct findings originating from a coherent case covered by the model conform to certain expected patterns laid down in the model. In other words, the findings should be positively correlated (see also Section 3.4.3). If $e = \{e_1, \ldots, e_m\}$ is a set of findings, we would expect $P(e)$ to exceed the probability for independent findings: $P(e_1) \cdots P(e_m)$. Hence we define the *conflict measure* as

$$\text{conf}(\{e_1, \ldots, e_m\}) = \log_2 \frac{P(e_1) \cdots P(e_m)}{P(e)}.$$

The reason for the $\log_2$ is sheer convenience; some formulas look nicer. A positive $\text{conf}(e)$ is an indicator of a possible conflict. For the insemination case, the conf-value is 3.1.

To get the required probabilities, you start performing a propagation without evidence entered. From this, you get $P(X)$ for all X in $\mathcal{U}$. If e_i is a finding on X, then $P(e_i)$ can be calculated from $P(X)$ as explained in Section 5.1. To compute $P(e)$, you use Proposition 5.1.

5.5.3 Conflict or Rare Case

It may happen that typical data from a very rare case causes a high conf-value. In the insemination case, a very rare blood type may have the effect of always causing BT to give a positive result.

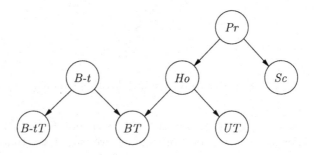

Fig. 5.5. A rare blood type (frequency 0.001) causes BT to always give a positive test result. Here $B\text{-}tT$ is a test for blood-type with 0.1% false positives and negatives.

By extending the model in Figure 5.4 to take the blood type into account, we get the model in Figure 5.5. For this extended model we still get $\text{conf}(UT = n, Sc = n, BT = y) = 3.1$, indicating a possible conflict. The reason is that though the evidence is perfectly coherent for a cow with this particular blood type, it is very rare. Now assume that the blood-type test gives the result y. This resolves the conflict, since conf of the new set of evidence is -1.34.

The problem above calls for a method for pointing out whether a positive conf-value may be explained as a rare case covered by the model.

Let $e = \{e_1, \ldots, e_m\}$ be findings for which $\text{conf}(e) > 0$, and let h be a hypothesis that could explain the findings, $\text{conf}(\{e_1, \ldots, e_m, h\}) \leq 0$.

We have

$$\text{conf}(\{e_1, \ldots, e_m, h\}) = \log_2 \frac{P(e_1) \cdots P(e_m)P(h)}{P(e, h)}$$

$$= \text{conf}(e) + \log_2 \frac{P(h)}{P(h \mid e)}.$$

This means that if

$$\log_2 \frac{P(h \mid e)}{P(h)} \geq \text{conf}(e), \tag{5.1}$$

then h can explain away the conflict. In the blood example the value of the left hand side of (5.1) is 5.4 with $h =$ "$B\text{-}t = y$".

The fraction $\frac{P(h \mid e)}{P(h)}$ is used in various ways, and it is called the *normalized likelihood*. Note that by the fundamental rule

$$\frac{P(h \mid e)}{P(h)} = \frac{P(e \mid h)}{P(e)}.$$

Normalized likelihoods can be monitored automatically for all variables. Therefore, in analyzing for conflict/rare case, it is easy to detect whether a conflict may be due to a particular variable being in a very rare state.

5.5.4 Tracing of Conflicts

After the conflict measure has been found positive, a further task would then be to find out whether a possible conflict is due to flawed findings, and if so, to trace them.

Let us return to the insemination problem with evidence $e = \{e_S =$ "$Sc = n$", $e_U =$ "$UT = n$", $e_B =$ "$BT = y$"$\}$. We have $\text{conf}(e) = 3.1$. We want to trace the origin of the conflict.

The evidence e is in the network communicated to Pr in two sets, $e' = \{e_B, e_U\}$ and $e'' = \{e_S\}$. A further investigation could therefore be to see whether e' contains an internal conflict. To do that we need $P(e')$, which is 0.0196. We get

$$\text{conf}(e') = 3.16,$$

and not surprisingly a conflict is detected in e'.

Another possibility could be that the two sets e' and e'' are conflicting. We define

$$\text{conf}(e', e'') = \log_2 \frac{P(e')P(e'')}{P(e)} = -0.001,$$

which indicates that the two sets of findings are not conflicting, and we conclude that e' is flawed.

To deal with tracing of conflicts, we can use IEJ trees. Using an IEJ tree, we can easily calculate the *local conflict* (see Figure 5.1)

$$\text{conf}(e_V, e_W) = \log_2 \frac{P(e_V)P(e_W)}{P(e)}.$$

The local conflict is a measure of whether the two sets of evidence e_V and e_W are in conflict.

We can also calculate the *partial conflicts* $\text{conf}(e_V)$ and $\text{conf}(e_W)$. The partial conflicts give an indication of possible internal conflicts in the sets e_V and e_W.

For each separator, we can get the probability of the evidence entered to the left and the probability of the evidence entered to the right. We can calculate the local and partial conflicts, and they are used for tracing the origin of the global conflict.

Using the IEJ tree in Figure 5.6, we calculate the following local and internal conflicts: $\mathrm{conf}(e_B, e_U) = 3.16$, $\mathrm{conf}(e_B, e_S) = 2.55$, $\mathrm{conf}(e_S, e_U) = -1.93$, $\mathrm{conf}(\{e_B, e_U\}, e_S) = -0.001$, $\mathrm{conf}(\{e_B, e_S\}, e_U) = 0.615$, $\mathrm{conf}(\{e_S, e_U\}, e_B) = 5.1$.

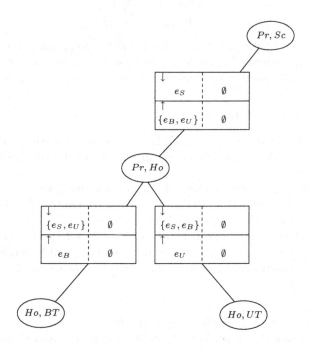

Fig. 5.6. The IEJ tree for the insemination example. The various sets of evidence held are indicated in the mailboxes.

These conflict measures point clearly at e_B as the dubious finding.

To round off this section, we give the following proposition, which relates the three kinds of conflicts.

Proposition 5.4. *The global conflict* $\mathrm{conf}(e)$ *is the sum of a local conflict and partial conflicts.*

$$\mathrm{conf}(e) = \mathrm{conf}(e_V, e_W) + \mathrm{conf}(e_V) + \mathrm{conf}(e_W).$$

Proof.   Exercise 5.8.   □

5.5.5 Other Approaches to Conflict Detection

The conf measure is not the only way of dealing with conflict detection. Another approach to the problem would be to incorporate sources of surprise directly in the model. This can be done by entering variables modeling probabilities for malfunctioning of sensors, and to extend causal variables such as disease variables with the state *other*. This approach, however, has the problem that it is difficult to model malfunctions or *other* unless the types of malfunction and *other* are known. Also, with *other* you can handle only discrepancies that are local in the network.

Another approach is to calculate a so-called *surprise index* for the set of findings. If the findings e are statements on the variables $A, \ldots, B$, the surprise index is the sum of probabilities for all configurations of $(A, \ldots, B)$ with a probability no higher than $P(e)$. If the surprise index is less than 0.1, this should be an indication of a possible conflict. In the insemination case, the surprise index is 0.06. Unfortunately, the calculation of a surprise index is exponential in the number of findings, and it must be considered intractable in general.

5.6 SE Analysis

Evidence e has been entered into a Bayesian network, and some hypotheses $h_1, \ldots, h_n$ are the focus of interest. Sensitivity analysis to evidence will give answers to questions like

– what evidence is in favor of/against/irrelevant for h_i?
– what evidence discriminates h_i from h_j?

We shall call this kind of analysis *SE analysis*.

5.6.1 Example and Definitions

The following example is used for illustration.

> In the morning when Mr Holmes leaves his house, he realizes that his lawn is wet. He wonders whether it has rained during the night or whether he has forgotten to turn off his sprinkler. He looks at the lawn of his neighbors, Dr Watson and Mrs Gibbon. Both lawns are dry, and he concludes that he must have forgotten to turn off his sprinkler.

The network for Holmes' reasoning is shown in Figure 5.7, and the initial probabilities are given in Table 5.3.

The evidence e consists of the three findings e_H, e_W, e_G, and the hypothesis in focus is $h_s :$ "$S = y$". We have $P(h_s) = 0.1$ and $P(h_s \mid e) = 0.9999$.

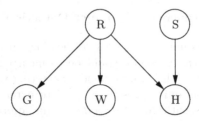

Fig. 5.7. Network for the wet lawn example. Holmes can inspect both Watson's and Mrs Gibbon's lawns.

	$R = y$	$R = n$
$G = y$	0.99	0.1
$G = n$	0.01	0.9
$P(G \mid R) = P(W \mid R)$		

	$R = y$	$R = n$
$S = y$	(1, 0)	(0.9, 0.1)
$S = n$	(0.99, 0.01)	(0, 1)
$P(H \mid R, S)$		

Table 5.3. Tables for the wet lawn example. $P(R) = (0.1, 0.9) = P(S)$.

We have $P(h_s \mid e_H) = 0.51, P(h_s \mid e_W) = 0.1 = P(h_s \mid e_G)$.[1] So neither e_W nor e_G alone has any impact on the hypothesis, but e_H is also not sufficient for the conclusion. Therefore, the immediate conclusion that e_W and e_G are irrelevant for the hypothesis is not correct, and we must conclude that evidence in combination may have a larger impact than the "sum" of the individual impacts.

To investigate further, we must consider the impact of subsets of the evidence. We have

$$P(h_s \mid e_W, e_G) = 0.1, P(h_s \mid e_H, e_G) = 0.988 = P(h_s \mid e_W, e_H).$$

To relate the probabilities above to their impact on the hypothesis, we can divide them by the prior probability $P(h_s)$ to get the *normalized likelihood*.

Other measures can be used, for example *Bayes' factor*

$$\frac{P(e \mid h)}{P(e \mid \neg h)},$$

or the fraction of achieved probability

$$\frac{P(h \mid e')}{P(h \mid e)}.$$

The various normalized likelihoods are given in Table 5.4.

From Table 5.4 we can conclude that no single finding is sufficient for the conclusion. Also, though (e_W, e_G) alone has no impact on h_s, these two

---

[1] A d-separation analysis could yield some of the results. However, this is not the point here.

$W = n$	$G = n$	$H = y$	$\frac{P(h_s \mid e)}{P(h_s)}$
1	1	1	9.999
1	1	0	1
1	0	1	9.88
1	0	0	1
0	1	1	9.88
0	1	0	1
0	0	1	5.1
0	0	0	1

Table 5.4. Normalized likelihoods for the subsets in the example. A "1" in the table indicates that the finding is an element of e'.

findings cannot both be removed. Moreover, we see that the subsets (e_H, e_G) and (e_H, e_W) can account for almost all the change in the probability for h_s.

Definition 5.1. *Let e be evidence and h a hypothesis. Suppose that we want to investigate how sensitive the result $P(h \mid e)$ is to the particular set e. We have that $e' \subseteq e$ is* sufficient *if $P(h \mid e)$ is almost equal to $P(h \mid e')$. We then also say that $e \setminus e'$ is* redundant *evidence.*

The term almost equal *can be made precise by selecting a threshold θ_1 and require that $\left| \frac{P(h \mid e')}{P(h \mid e)} - 1 \right| < \theta_1$. Note that $\frac{P(h \mid e')}{P(h \mid e)}$ is the fraction between the two normalized likelihood ratios.*

- *e' is* minimal sufficient *if it is sufficient, but no proper subset of e' is so.*
- *e' is* crucial evidence *if it is a subset of any sufficient set.*
- *e' is* important evidence *if the probability of h changes too much without it, to be more precise, if $\left| \frac{P(h \mid e \setminus e')}{P(h \mid e)} - 1 \right| > \theta_2$, where θ_2 is some chosen threshold.*

In the example above put $\theta_2 = 0.2, \theta_1 = 0.05$. Then (e_H, c_G) and (e_H, e_W) are minimal sufficient, (e_W, e_G) is important, and e_H is crucial.

In Holmes's universe, there is another possible hypothesis, namely h_r : "$R = y$". To find out which findings discriminate between the two hypotheses, an analysis of h_r can be performed. The probability $P(h_r \mid e')$ is calculated for each subset of e', and the ratio between the two (normalized) likelihoods is used. The ratios are shown in Table 5.5.

Table 5.5 shows that e_W and e_G are good discriminators between the two hypotheses.

As illustrated above, the heart of sensitivity analysis is the calculation of $P(h \mid e')$ for each $e' \subseteq e$. Since the number of subsets grows exponentially with the number of findings, the job may become very heavy, particularly when $P(h \mid e')$ has to be calculated through a propagation in a large network.

$W = n$	$G = n$	$H = y$	$\frac{P(e' \mid h_s)}{P(e' \mid h_r)}$
1	1	1	6622
1	1	0	7300
1	0	1	74
1	0	0	81
0	1	1	74
0	1	0	81
0	0	1	0.92
0	0	0	1

Table 5.5. Likelihood ratios for the hypotheses h_s and h_r.

Note that when $P(h \mid e')$ and $P(h)$ are available, then also Bayes' factors can be calculated:

$$\frac{P(e' \mid h)}{P(e' \mid \neg h)} = \frac{P(h \mid e')P(\neg h)}{P(h)P(\neg h \mid e')} = \frac{P(h \mid e')(1 - P(h))}{P(h)(1 - P(h \mid e'))}.$$

5.6.2 h-Saturated Junction Trees and SE Analysis

A-saturated junction trees – sometimes extended to IEJ trees – can be of great help for SE analysis. If a particular state h of the hypothesis variable H is a focus of interest, another type of junction tree will suffice. Let e be the evidence. After propagating e we insert $H = h$ in an appropriate node R and perform a DISTRIBUTEEVIDENCE from R. The messages from this propagation are stored in the separators too (see Figure 5.8). We call this type of junction tree an h-saturated junction tree.

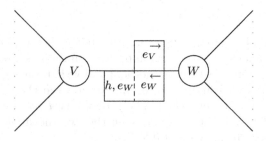

Fig. 5.8. Part of an h-saturated junction tree, where the hypothesis $H = h$ is entered to the right. The evidence handled is indicated.

The specific approach to SE analysis is much dependent on the type of hypothesis, the type and size of the evidence, the topology of the network, etc., and below we shall only give some hints on how the tasks may be approached.

What-If?

Assume that we want to investigate the impact on H if the finding e_X is removed or changed to e'_X.

If you have a single state h in focus, you can use an h-saturated junction tree. Go to the node V, where e_X is placed. Local to V you have messages for all evidence, and substituting e_X with e'_X (e'_X may be empty) will give you $P(e \setminus \{e_X\} \cup \{e'_X\})$. You also have messages involving e together with "$H = h$". Substituting e_X with e'_X will give you $P(e \setminus \{e_X\} \cup \{e'_X\}, h)$. From this you get $P(H = h \,|\, e \setminus \{e_X\} \cup \{e'_X\})$. The same H-saturated junction tree can be used for all findings. What-if? analysis can, for example, sort out redundant findings, and it can also be used to determine the findings acting for or against h.

Note that this technique also allows you to investigate the effect of evidence on a variable for which you have not yet received evidence.

Crucial Findings

Assume that $P(h \,|\, e)$ is high, and we want to determine the set of crucial findings.

Use an h-saturated junction tree. It may happen that some findings are evidence against h, but they are overwritten by the entire set. We assume that findings acting against h have been sorted out (for example through What-if? analysis as above).

For the remaining evidence we assume monotonicity: no insufficient set contains a sufficient subset.

Then e_X is crucial if and only if $e \setminus \{e_X\}$ is not sufficient. Using an h-saturated junction tree, it is easy to determine the crucial findings.

Minimal Sufficient Sets

It will be natural to continue the procedure above and repeatedly remove findings from sufficient sets. However, h-saturated junction trees only allow you to remove findings inserted in the same node in the junction tree. If they are inserted in different nodes, new propagations are required.

An h-saturated IEJ tree can speed up the search using five mailboxes for each separator (see Figure 5.9). An h-saturated junction tree gives you access to $P(h \,|\, e')$ for a large family of subsets $e' \subset e$ (see Table 5.6). From this family you choose the minimal sufficient subsets and continue the search for each of them by establishing a new h-saturated IEJ tree.

As described in Section 5.2.1, the separators can be used to obtain $P(h \,|\, e')$ for the sets e' "communicated" to them. A similar procedure can be used for the nodes in the junction tree. Take for example the node V in Figure 5.9. By selecting appropriate messages from the neighbors, we can handle any union of sets communicated to a separator. This yields a way of calculating for example $P(h \,|\, q, t)$. A full list is given in Table 5.6. Note that some subsets are not in the list, for example $\{t, y\}$.

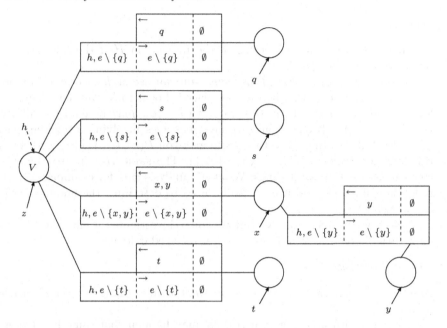

Fig. 5.9. An h-saturated IEJ tree. The evidence "communicated" is indicated in the separators. It is assumed that h is inserted in V. The subsets of the evidence accessed are listed in Table 5.6.

e	$\emptyset$	$\{t\}$	$e \setminus \{t\}$	$\{s\}$	$e \setminus \{s\}$
$\{q\}$	$e \setminus \{q\}$	$\{y\}$	$e \setminus \{y\}$	$\{x\}$	$e \setminus \{x\}$
$\{z\}$	$e \setminus \{z\}$	$\{x,y\}$	$e \setminus \{x,y\}$	$\{z,t\}$	$e \setminus \{z,t\}$
$\{z,s\}$	$e \setminus \{z,s\}$	$\{z,q\}$	$e \setminus \{z,q\}$	$\{t,s\}$	$e \setminus \{t,s\}$
$\{t,q\}$	$e \setminus \{t,q\}$	$\{s,q\}$	$e \setminus \{s,q\}$	$\{x,y\}$	$e \setminus \{x,y\}$
$\{z,x,y\}$	$\{t,s,q\}$	$\{x,y,q\}$	$\{z,t,s\}$	$\{x,y,s\}$	$\{z,t,q\}$
$\{x,y,t\}$	$\{z,s,q\}$				

Table 5.6. A list of sets of evidence e' for which the h-saturated IEJ tree in Figure 5.8 yields $P(h \mid e')$ through a local computation.

5.7 Sensitivity to Parameters

We have a Bayesian network BN with evidence e. Assume that we have a single hypothesis variable H, and let a particular state h of H be in focus of interest. Let $\mathbf{t}$ be a set of parameters for BN (a parameter is an entry in a conditional probability table). We are interested in how $P(h \mid e)$ varies with $\mathbf{t}$.

We must make clear what is meant by "the probabilities are functions of the parameters." Let A be a binary variable, and let π be a configuration of A's parents $pa(A)$. Then, $t = P(A = a \mid \pi)$ is a parameter, but consequently we have $P(A = \neg a \mid \pi) = 1 - t$, and it covaries with t. If A has more than two states, we assume *proportional scaling*: the remaining probabilities are scaled by the same factor. If A has n states, and a_1 is a parameterized state, we assume that $P(A \mid \pi) = (t, (1 - t)x_2, \ldots, (1 - t)x_n)$, where $\sum x_i = 1$.

It is possible to deal with several parameters in the same distribution. If, for example, the first two states are parameterized, we would require $P(A \mid \pi) = (t, s, (1 - t - s)x_3, \ldots, (1 - t - s)x_n)$. Then, s does not scale when t is changed. In the following, we assume proportional scaling, and we also assume that there is at most one parameter per distribution.

Theorem 5.2. *Let BN be a Bayesian network over the universe $\mathcal{U}$. Let t be a parameter and let e be evidence entered in BN. Then, assuming proportional scaling, we have*

$$P(e)(t) = \alpha t + \beta,$$

where α and β are real numbers.

Before proving Theorem 5.2 we need a lemma.

Lemma 5.1. *Let $\phi(V)$ be a potential over the variables $\mathcal{V}$. Let $A \in V$ and let $\mathbf{v}^*$ be a configuration over $V \setminus \{A\}$. Let all entries be real-valued except for $\phi(A, \mathbf{v}^*)$, which has the form $(\alpha_1 t + \beta_1, \ldots, \alpha_k t + \beta_k)$. Then*

$$\sum_V \phi(V) = \alpha t + \beta,$$

where α and β are real numbers.

Proof. Let us first look at the example in Table 5.7. To calculate $\sum_V \phi(A, B, C)$, take first the sum of all numbers in the entries with $B \neq b_2$ and $C \neq c_2$. The result is 56. Finally, add the expressions in the (b_2, c_2)-entry, and you get $4t + 57$.

		B		
		b_1	b_2	b_3
C	c_1	(1, 2, 3)	(2, 4, 7)	(4, 1, 2)
	c_2	(5, 2, 1)	$(t + 1, -2t + 2, 5t - 2)$	(1, 1, 1)
	c_3	(2, 2, 1)	(3, 1, 4)	(2, 2, 2)

Table 5.7. $\phi(A, B, C)$.

In general, let V^* be the set of all configurations in $\mathrm{sp}(\mathcal{V})$ except for the $(A, \mathbf{v}^*)$-configurations. Then

$$\sum_V \phi(V) = \sum_{V^*} \phi(V) + \sum_A \phi(A, \mathbf{v}^*).$$

The first term is a real number β^*, and the second is $(\alpha_1 t + \beta_1) + \cdots + (\alpha_k t + \beta_k)$. Hence

$$\sum_V \phi(V) = \left(\sum_i \alpha_i\right) t + \left(\sum_i \beta_i\right) + \beta^*.$$

$\square$

Proof.   We prove Theorem 5.2. Let $U = \{A\} \cup \{A_1, \ldots, A_n\}$, $\mathrm{fa}(A) = \{A\} \cup \mathrm{pa}(A)$ and let π be a parent configuration for which

$$P(A \mid \pi) = (t, \gamma_2(1 - t), \ldots, \gamma_k(1 - t)).$$

Without loss of generality, assume that the parameter t is attached to the first state of A. Let the evidence potentials be $\mathbf{e}_1, \ldots, \mathbf{e}_m$.

Now

$$P(e) = \sum_U P(U, e) = \sum_U P(A \mid \mathrm{pa}(A)) \prod_i P(A_i \mid pa(A_i)) \prod_j \mathbf{e}_j$$

$$= \sum_{\mathrm{fa}(A)} P(A \mid \mathrm{pa}(A)) \sum_{U \backslash \mathrm{fa}(A)} \prod_i P(A_i \mid pa(A_i)) \prod_j \mathbf{e}_j.$$

The factor $\sum_{U \backslash \mathrm{fa}(A)} \prod_i P(A_i \mid pa(A_i)) \prod_j \mathbf{e}_j$ is a potential, $\phi(\mathrm{fa}(A))$, with only real-numbered values, and we have

$$P(e) = \sum_{\mathrm{fa}(A)} P(A \mid \mathrm{pa}(A)) \phi(\mathrm{fa}(A)).$$

The product $P(A \mid \mathrm{pa}(A)) \phi(\mathrm{fa}(A))$ is a potential satisfying the conditions in Lemma 5.1, and we can conclude that

$$P(e) = \alpha t + \beta.$$

$\square$

Notation: Let $\mathbf{t} = (t_1, \ldots, t_m)$ be a set of parameters, and let $\mathrm{pol}(\mathbf{t})$ be a polynomial over $\mathbf{t}$. The polynomial $\mathrm{pol}(\mathbf{t})$ is said to be *multilinear* if all exponents in the expression are of degree at most 1. If so, it has a term for each subset of $\mathbf{t}$.

Corollary 5.1. *Let BN be a Bayesian network over the universe $\mathcal{U}$. Let $\mathbf{t}$ be a set of parameters for different distributions and let e be evidence entered to BN. Then, assuming proportional scaling, $P(e)(\mathbf{t})$ is a multilinear polynomial over $\mathbf{t}$.*

Proof.   For the sake of notational convenience, let $t = (x, y)$. From Theorem 5.2 we have

$$P(e)(x, y) = \alpha_x(y)x + \beta_x(y) = \alpha_y(x)y + \beta_y(x).$$

Inserting $x = 0$ yields

$$\beta_0(y) = \alpha_y(0)y + \beta_y(0). \tag{5.2}$$

That is, $\beta_x(y)$ is a linear function.
    Inserting $x = 1$ yields

$$\alpha_1(y) + \beta_1(y) = \alpha_y(1)y + \beta_y(1).$$

Using Formula 5.2 we get

$$\begin{aligned}\alpha_x(y) &= \alpha_y(1)y + \beta_y(1) - \alpha_y(0)y - \beta_y(0) \tag{5.3}\\ &= (\alpha_y(1) - \alpha_y(0))y + \beta_y(1) - \beta_y(0).\end{aligned}$$

That is, $\alpha_x(y)$ is a linear function. Combining Formula 5.2 and Formula 5.3 we get

$$P(e)(x, y) = ((\alpha_y(1) - \alpha_y(0))y + \beta_y(1) - \beta_y(0))x + \alpha_y(0)y + \beta_y(0),$$

which is of the form $\alpha xy + \beta x + \gamma y + \delta$.
    If we have more than two parameters, we let $\mathbf{t} = (x, \mathbf{y})$, where $\mathbf{y}$ is a set of parameters. The reasoning above then yields that $\beta_x(\mathbf{y})$ and $\alpha_x(\mathbf{y})$ are multilinear polynomials over $\mathbf{y}$, and we repeat the arguments on $\beta_x(\mathbf{y})$ and $\alpha_x(\mathbf{y})$. □

Corollary 5.2. *Let BN be a Bayesian network over the universe $\mathcal{U}$. Let $\mathbf{t}$ be a set of parameters for different distributions. Let a be a state of $A \in \mathcal{U}$ and let e be evidence. Then $P(a \mid e)(\mathbf{t})$ is a fraction of two multilinear polynomials over $\mathbf{t}$.*

Proof.   Follows from Corollary 5.1 and $P(a \mid e) = \frac{P(a,e)}{P(e)}$.   □

5.7.1 One-Way Sensitivity Analysis

Let e be evidence, h a state of a hypothesis variable H, and s a parameter for the Bayesian network. We wish to establish $P(h \mid e)$ as a function of s. Corollary 5.2 yields that $P(h \mid e)$ has the form

$$P(h \mid e) = \frac{as + \beta}{as + b} = \frac{P(h, e)}{P(e)}.$$

To determine the four constants you can enter two different values, s_0 and s_1, for the parameter and then propagate. You then get $P(h,e)(s_1)$, $P(h,e)(s_0)$, $P(e)(s_1)$, and $P(e)(s_0)$:

$$\alpha s_0 + \beta = P(h,e)(s_0), \qquad \alpha s_1 + \beta = P(h,e)(s_1),$$
$$a s_0 + b = P(e)(s_0), \qquad a s_1 + b = P(e)(s_1),$$

and from the four equations you can determine $P(h \mid e)$ as a function of s.

Another way of determining the coefficients is to establish an h-saturated junction tree. Now, consider a clique with the table for the parameter attached. In this clique (and its neighboring separators) you have all the information needed to calculate $P(h,e)$ and $P(e)$ for different values of s. This method can easily be extended to one-way sensitivity analysis for several variables.

Note that when you have established the h-saturated junction tree, then you can perform one-way sensitivity analysis for any parameter you wish, by looking at a clique and its neighbor separators.

5.7.2 Two-Way Sensitivity Analysis

Let s and u be two parameters. Then $P(e)(s,r) = \alpha sr + \beta s + \gamma r + \delta$, and we wish to determine the coefficients. From the propagation described in Section 5.7.1, we have the value of $P(e)(s_0, r_0)$. By working locally in the clique containing s, we get the values of $P(e)(0, r_0)$ and $P(e)(1, r_0)$, and by working locally in the clique containing r, we get $P(e)(s_0, 0)$ and $P(e)(s_0, 1)$. That is, we have five equations with four unknowns, and we can determine $(\alpha, \beta, \gamma, \delta)$, provided we can pick four equations with an invertible coefficient matrix. Unfortunately, the equations are of rank 3, and we need extra information. By entering a new value s_1 and propagating, we get sufficient information to locally compute all two-way sensitivity analyses involving s.

To calculate three-way sensitivity analysis is much more demanding, and the number of propagations grows exponentially with the number of parameters considered. The complexity of the local computations also increases exponentially. We shall not treat this further.

5.8 Summary

W-Saturated Junction Trees

To calculate the joint probability over a set of variables W, you can either perform a propagation for each configuration of W or you can establish a W-saturated junction tree (the junction tree obtained by performing a full propagation without eliminating the variables in W).

Let T be a W-saturated junction tree with evidence e, and let X be any variable. Then $P(W \mid X, e)$ is calculated through the following procedure:

1. Choose any node V or separator S in T containing X.
2. $P(V \cup W, e)$ is the product of V's set of potentials with the incoming messages ($P(S \cup W, e)$ is the product of the two messages).
3. $P(W, X, e) = \sum_{V \setminus (W \cup \{X\})} P(V \cup W, e)$.
4. $P(X, e) = \sum_W P(W, X, e)$.
5. $P(W \mid X, e) = \frac{P(W, X, e)}{P(X, e)}$.

Finding the Most Probable Explanation (MPE)

The distributive law for max:

$$\max_Z f(X, Y) g(Y, Z) = f(X, Y) \max_Z g(Y, Z).$$

Max propagation corresponds to standard (lazy) junction tree propagation, where marginalizations are performed using the max-operator rather than the $\sum$-operator.

Let BN be a Bayesian network representing $P(\mathcal{U})$, and let T be a junction tree corresponding to BN. Let e be the evidence represented by the functions $\{e_1, \ldots, e_m\}$, and assume that the evidence functions are attached to appropriate nodes in the junction tree.

After a full round of (lazy) max-propagation in T we have

1. for each separator S, $\max_{U \setminus S} P(U, e)$ is the product of the two messages in S's mailboxes;
2. for each node V, $\max_{U \setminus V} P(U, e)$ is the product of the potential set attached to V and the incoming messages.

Axioms for probability updating

1. $\mathrm{dom}\,(v_1 \otimes v_2) \subseteq \mathrm{dom}\,(v_1) \cup \mathrm{dom}\,(v_2)$,
2. $\mathrm{dom}\,(v^{\downarrow V}) \subseteq V$,
3. Combination is associative: $(v_1 \otimes v_2) \otimes v_3 = v_1 \otimes (v_2 \otimes v_3)$,
4. Combination is commutative: $v_1 \otimes v_2 = v_2 \otimes v_1$,
5. $(v^{\downarrow V})^{\downarrow W} = v^{\downarrow V \cap W}$,
6. The distributive law: If $\mathrm{dom}\,(v_1) \subseteq V$ then $(v_1 \otimes v_2)^{\downarrow V} = v_1 \otimes (v_2)^{\downarrow V}$,
7. $v^{\downarrow \emptyset}$ is a neutral element with respect to combination, and it is denoted by **1**.

Data conflict

Conflict measure: To measure how well the evidence fits the model, you can use the conflict measure

$$\mathrm{conf}(\{e_1, \ldots, e_m\}) = \log_2 \frac{P(e_1) \cdots P(e_m)}{P(e)}.$$

Conflict or rare case: If

$$\log_2 \frac{P(h \mid e)}{P(h)} \geq \text{conf}(e),$$

then the hypothesis h can explain the conflict (the conflict is due to e being a rare configuration).

Sensitivity to Evidence

Let e be evidence and h a hypothesis. Suppose that we want to investigate how sensitive the result $P(h \mid e)$ is to the particular set e. We say that $e' \subseteq e$ is *sufficient* if $P(h \mid e)$ is almost equal to $P(h \mid e')$. We then also say that $e \setminus e'$ is *redundant*.

The term *almost equal* can be made precise by selecting a threshold θ_1 and require that $\left| \frac{P(h \mid e')}{P(h \mid e)} - 1 \right| < \theta_1$. Note that $\frac{P(h \mid e')}{P(h \mid e)}$ is the fraction between the two likelihood ratios.

- e' is *minimal sufficient* if it is sufficient, but no proper subset of e' is so.
- e' is *crucial evidence* if it is a subset of any sufficient set.
- e' is *important evidence* if the probability of h changes too much without it, to be more precise, if $\left| \frac{P(h \mid e \setminus e')}{P(h \mid e)} - 1 \right| > \theta_2$, where θ_2 is some chosen threshold.

One can use h-saturated junction trees to find the minimal sufficient sets as well as the crucial findings.

Sensitivity to Parameters

Probability of evidence, $P(e)(t)$: Let BN be a Bayesian network over the universe $\mathcal{U}$. Let t be a parameter and let e be evidence entered in BN. Then, assuming proportional scaling, we have

$$P(e)(t) = \alpha t + \beta,$$

where α and β are real numbers.

Functional expression for $P(X \mid e)(t)$: Let BN be a Bayesian network over the universe $\mathcal{U}$. Let $\mathbf{t}$ be a set of parameters for different distributions. Let a be a state of $A \in \mathcal{U}$ and let e be evidence. Then $P(a \mid e)(\mathbf{t})$ is a fraction of two multilinear polynomials over $\mathbf{t}$.

5.9 Bibliographical Notes

Max-propagation was proposed by Dawid (1992). The axioms for propagation were formulated by Shafer and Shenoy (1990), and Lauritzen and Jensen

(1997) extended them to cover Hugin propagation. A measure for calculating data conflict (surprise index) was first proposed by Habbema (1976). The method presented here is due to Jensen *et al.* (1990a). See also (Laskey, 1991), (Kim and Valtorta, 1995), and (Laskey, 1995). SE analysis is part of *explanation*, which was systematically studied by Suermondt (1992). The presentation here is an extension of (Jensen *et al.*, 1995). Theorem 5.2 establishing the linearity of $P(e)(t)$ was independently proved by Castillo *et al.* (1997) and Coupé and van der Gaag (1998), and the method described here is based on (Kjærulff and van der Gaag, 2000).

5.10 Exercises

Exercise 5.1. Construct the IEJ tree for the Bayesian network from Exercise 4.2 with evidence "$D = y$".

Exercise 5.2. Construct the IEJ tree for the Bayesian network from Exercise 4.3 with the evidence "$C = y$".

Exercise 5.3. Based on the join tree in Figure 4.16, draw the following:

- A junction tree with the evidence $e = \{A = a, F = f\}$.
- An IEJ tree for the evidence $e = \{A = a, F = f\}$.
- An $\{A, F\}$-saturated junction tree. Which messages need to be sent for obtaining this?
- A b-saturated junction tree (b is a state of B) with evidence $e = \{A = a, F = f\}$.

Exercise 5.4. Consider a Bayesian network with two variables A and B, each having two states, and probability distributions defined by $P(A = a_1) = 0.1$, $P(B = b_1 \mid A = a_1) = 0.2$, and $P(B = b_1 \mid A = a_2) = 0.3$. What is the most-probable explanation for $B = b_2$?

Exercise 5.5. [E] Using your implemented model from Exercise 3.14 for the simplified poker game in Sections 3.1.4 and 3.2.3, what is the most-probable explanation for observing $FC = 2$ and $SC = 0$? What is the conflict measure of observing $FC = 0$ and $SC = 2$? What is the conflict measure of observing $FC = 0$, $SC = 2$, and $OH2 = sfl$? Which of the three observations seems to be flawed?

Exercise 5.6. [E] Using your implemented model from Exercise 3.14 for the simplified poker game in Sections 3.1.4 and 3.2.3, let e be the observations $FC = 2$ and $SC = 0$? For hypothesis $OH2 = sfl$ and sensitivity parameters $\theta_1 = \theta_2 = 0.01$, what are the crucial findings? Are the two observations important individually?

Exercise 5.7. [E] This exercise concerns the stud farm from Section 3.2 and the situation in Figure 3.16.

(i) The farmer has to decide on a new mating among the horses Fred, Dorothy, Eric, and Gwenn. Which pair should be chosen to minimize the risk of getting a carrier as offspring?

(ii) What is the most-probable configuration of genotypes of all horses? Does this correspond to the most-probable genotype for each horse?

(iii) The prior frequencies λ_L and λ_K of the a-gene for the outside horses L and K are parameters. Determine intervals for both parameters for which Dorothy as well as Gwenn have a risk above 0.70 of being a carrier.

(iv) Assume that the farmer gets the evidence that Ann is pure, Brian is pure, and Cecily is a carrier. Perform a data conflict analysis.

(v) Assume that a horse is taken out if the probability of it being a carrier is above 0.60. The evidence "$John = aa$" is double checked and considered certain. Perform an SE analysis of the evidence from (iv) for the grandparents of John.

Exercise 5.8. ^{E} This exercise concerns the transmission of symbol strings from Section 3.2.4 and Exercise 3.12 (i).

(i) The sequence $baaca$ is received. What is the most-probable word transmitted?

(ii) Perform a data conflict analysis of the evidence.

(iii) Consider the parameters $t = P(T_4 = a \mid T_3 = a)$, $s = P(R_4 = c \mid T_4 = a)$, and $u = P(R_4 = c \mid T_4 = b)$. Perform an analysis of the sensitivity of the conclusion "the word transmitted is **baaba**." A one-way analysis could, for example, determine the minimal distance to a value where the conclusion changes.

(iv) The parameters s and u are common for all R-variables. Perform a sensitivity analysis as in (iii).

Exercise 5.9. Fill in the intermediate steps of

$$\frac{P(e' \mid h)}{P(e' \mid \neg h)} = \frac{P(h \mid e')P(\neg h)}{P(h)P(\neg h \mid e')} = \frac{P(h \mid e')(1 - P(h))}{P(h)(1 - P(h \mid e'))}.$$

Exercise 5.10. ^{E} Consider the poker model from Exercise 3.13 (ii). Assume that you have seen your opponent change two cards first and then no cards. You have a flush. You know that your opponent sometimes changes no cards in the second round, no matter her hand. Let the frequency of this be t, and let your initial estimate be $t_0 = 0.1$. Analyze the sensitivity of the conclusion with respect to t and determine the value for which you have the best hand with probability 0.67.

Exercise 5.11. Consider the Bayesian network with two variables A and B, each having two states, and probability distributions defined by $P(A = a_1) = t$, $P(B = b_1 \mid A = a_1) = t$, and $P(B = b_1 \mid A = A_2) = 0.3$. For what range of values for t is a_2 the most probable explanation for $B = b_1$?

Exercise 5.12. ^{E} Investigate whether your Bayesian network tool allows for either automated SE analysis or sensitivity analysis of parameters. If so, verify your results from Exercises 5.6 and 5.11.

Exercise 5.13. Prove Proposition 5.4.

6

Parameter Estimation

Assume that you know the structure of a Bayesian network model over the variables $\mathcal{U}$, but you do not have any estimates for the conditional probabilities. On the other hand, you have access to a database of cases, i.e., a set of simultaneous values for some of the variables in $\mathcal{U}$. You can now use these cases to estimate the parameters of the model, namely the conditional probabilities. In this chapter we consider two approaches for handling this problem: First we show how a database of cases can be used to estimate the parameters once and for all (so-called *batch learning*). After that, we shall investigate the situation in which the cases are accumulated sequentially and we wish to *adapt* the model as each new case arrives. The reader is expected to be familiar with Section 1.5.

6.1 Complete Data

Let $M = (S, \boldsymbol{\theta})$ be a Bayesian network with structure S and parameters $\boldsymbol{\theta}$, and let $\mathcal{U}$ be the variables in M. Moreover, let $\mathcal{D}$ be a data set of cases, where each case is a configuration over *all* the variables in $\mathcal{U}$. Such a case is said to be *complete case*. In the learning community, a parameter is typically denoted by θ (rather than t as we have done previously), and in this chapter we shall follow the same convention. Moreover, to ensure that the parameters can be learned independently we shall make the following two assumptions:

- *Global independence* says that the parameters for the various variables are independent. This means that we can modify the tables for the variables independently.
- *Local independence* says that the uncertainties of the parameters for different parent configurations are independent. To be more precise, let (b, c) and (b', c') be different configurations; then the uncertainty on $P(A \mid b, c)$ is independent of the uncertainty on $P(A \mid b', c')$, and the parameters for the two distributions can be modified independently.

6.1.1 Maximum Likelihood Estimation

For each case $\mathbf{d} \in \mathcal{D}$, the probability $P(\mathbf{d}|M)$ is called the *likelihood* of M given $\mathbf{d}$. If we assume that the cases in $\mathcal{D}$ are independent given the model, then the *likelihood of M given $\mathcal{D}$* is

$$L(M \,|\, \mathcal{D}) = \prod_{\mathbf{d} \,\in\, \mathcal{D}} P(\mathbf{d}|M).$$

Often the log is taken, and it is then called the *log-likelihood*:

$$LL(M \,|\, \mathcal{D}) = \sum_{\mathbf{d} \,\in\, \mathcal{D}} \log_2 P(\mathbf{d}|M).$$

If we have to choose among several models for describing the data, then the *principle of maximum likelihood* advises us to choose a model of maximal likelihood given the data. This means that if we want to estimate the conditional probabilities, then our possible models $M_{\boldsymbol{\theta}}$ agree on the structure but differ with respect to the parameters $\boldsymbol{\theta}$. So we choose a parameter estimate $\hat{\boldsymbol{\theta}}$ that maximizes the likelihood:

$$\hat{\boldsymbol{\theta}} = \arg\max_{\boldsymbol{\theta}} L(M_{\boldsymbol{\theta}} \,|\, \mathcal{D}) = \arg\max_{\boldsymbol{\theta}} LL(M_{\boldsymbol{\theta}} \,|\, \mathcal{D}).$$

In what follows we shall use $\hat{\boldsymbol{\theta}}$ to denote a maximum likelihood estimate for the parameters $\boldsymbol{\theta}$.

Example 6.1. We have tossed a thumbtack 100 times. It has landed pin up 80 times, and we look for the best estimate of the probability for *pin up*.

The situation is that we have a family of models, one for each possible value of θ, the probability of *pin up*. Let M_θ denote the model with $P(pin\ up) = \theta$, then by assuming independent tosses, the likelihood of M_θ given the data is

$$P(\mathcal{D} \,|\, M_\theta) = \prod_{d \,\in\, \mathcal{D}} P(d \,|\, M_\theta) = \mu \cdot \theta^{80}(1 - \theta)^{20},$$

where μ is a binomial factor independent of θ. By setting the derivative $\frac{d}{d\theta}P(\mathcal{D} \,|\, M_\theta)$ equal to zero it is easy to see that the likelihood is maximal for $\theta = 0.8$, so $\hat{\theta} = 0.8$.

In general, you get a maximum likelihood estimate as the fraction of positive counts over the total number of counts. This also holds for variables with more than two states. If you want to find a maximum likelihood estimate for the parameters in a Bayesian network model, then this can be done by finding maximum likelihood estimates for each conditional probability distribution in the model. That is, for each conditional probability distribution, e.g., $P(A = a \,|\, B = b, C = c)$, you simply calculate

$$\frac{N(A = a, B = b, C = c)}{N(B = b, C = c)},$$

where $N(A = a, B = b, C = c)$ is the number of cases in the database for which $A = a, B = b, C = c$.

The principle of maximal likelihood therefore supports the intuition of using frequencies as estimates, and to achieve a maximum likelihood estimate you just count. We did so in Section 3.2.4, where Table 3.10 was the result of 10,000 words transmitted.

6.1.2 Bayesian Estimation

When you have a sparse database, maximum likelihood estimation has some drawbacks. Consider Table 6.1, which is the result of collecting 100 transmitted words. If you do maximum likelihood parameter estimation using this table, the outcomes with zero counts would be given zero probability and they are thereby doomed impossible, a rather strong assumption based on only 100 cases.

		Last three letters							
		aaa	aab	aba	abb	baa	bba	bab	bbb
First two letters	aa	2	2	2	2	5	7	5	7
	ab	3	4	4	4	1	2	0	2
	ba	0	1	0	0	3	5	3	5
	bb	5	6	6	6	2	2	2	2

Table 6.1. The table shows the number of five-letter words $(T_1 T_2 T_3 T_4 T_5)$ transmitted over a channel. For example, the word $abaab$ has appeared four times, whereas $bbabb$ has appeared six times.

An alternative to the principle of maximum likelihood is *Bayesian estimation*: start with a prior distribution, and use experience to update the distribution. The approach can be illustrated with a Bayesian network, where each parameter for estimation is made explicit through a node. For the thumbtack experiment, a model for three tosses would be as in Figure 6.1. The conditional probabilities $P(pin\ up|\theta)$ are θ, and the prior distribution $f(\theta)$ is (as always) up to you. If you have no idea at all, a common approach is to use the uniform distribution $f(\theta) = 1, 0 \leq \theta \leq 1$.

Now assume that we have performed one experiment with the result *pin up*. Using Bayes' rule we get

$$f_p(\theta|pin\ up) = \frac{P(pin\ up|\theta)f(\theta)}{P(1\ up)} = \frac{\theta f(\theta)}{P(pin\ up)}$$

for the posterior frequency function f_p. If we let $f(\theta) = 1$, we get

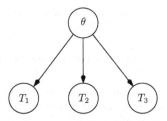

Fig. 6.1. A Bayesian network model for estimating the parameter given the outcome of three tosses.

$$f_p(\theta|pin\ up) = \frac{\theta}{P(pin\ up)}.$$

As usual, $P(pin\ up)$ is calculated as the normalization factor:

$$\int_0^1 \theta\,d\theta = \frac{1}{2},$$

so

$$f_p(\theta|pin\ up) = 2\theta.$$

This yields a distribution of the posterior for θ given *pin up*, and the best single estimate is the mean value of this distribution:

$$\int_0^1 \theta(2\theta)\,d\theta = \frac{2}{3}.$$

Next, assume that we get a toss with pin down. Then we have for the new posterior distribution

$$f_{p_2}(\theta|pin\ down, pin\ up) = \frac{P(pin\ down, pin\ up|\theta)f(\theta)}{P(pin\ up, pin\ down)}$$
$$= \mu P(pin\ down|\theta)P(pin\ up|\theta)f(\theta)$$
$$= \mu P(pin\ down|\theta)\theta \cdot 1 = \mu(1-\theta)\theta,$$

where μ is the normalization constant

$$\frac{1}{\mu} = P(pin\ down, pin\ up) = \int_0^1 (1-\theta)\theta\,d\theta = \frac{1}{6}.$$

The posterior distribution $f_{p2}(\theta|pin\ down, pin\ up)$ can now be written as

$$f_{p_2}(\theta|pin\ down, pin\ up) = 6(1-\theta)\theta,$$

and the single best estimate for θ is

$$\int_0^1 \theta 6(1-\theta)\theta\,d\theta = \frac{1}{2}.$$

Theorem 6.1. *Let X be a binary variable* (yes, no), *and assume that we have performed a number of independent experiments out of which n turned up* yes *and m turned up* no. *Let θ be the probability for* yes. *Then, starting with the even prior distribution for θ, the posterior distribution is*

$$f_p(\theta) = \mu\theta^n(1-\theta)^m,$$

where μ is a normalization constant. The Bayesian estimate for θ is $\frac{n+1}{n+m+2}$.

Parameters estimated through Bayesian estimation are called *maximum a posteriori parameters*.

The theorem can be proved by induction along the lines described above. Moreover, the theorem can be interpreted so that an even prior distribution corresponds to adding two virtual experiments to the data (one for *yes* and one for *no*) and then counting frequencies.

This procedure also holds for distributions over more than two states. To pursue the Bayesian approach, assume for example that you wish to estimate $P(T_2 \mid T_1)$ from Table 6.1. First you marginalize out the other variables to obtain Table 6.2(a).

	T_1	
	a	b
T_2 a	32	17
T_2 b	20	31

(a)

	T_1	
	a	b
T_2 a	33	18
T_2 b	21	32

(b)

Table 6.2. (a) Counts of the first two letters from Table 6.1. (b) The table obtained by adding 1 to all counts in (a).

Next, add 1 to all cells (Table 6.2(b)), and you get the conditional probability table in Table 6.3.

	T_1	
	a	b
T_2 a	$\left(\frac{33}{54}\right)$	$\left(\frac{18}{50}\right)$
T_2 b	$\left(\frac{21}{54}\right)$	$\left(\frac{32}{50}\right)$

Table 6.3. The result of a Bayesian approach for estimating the conditional probability table $P(T_2 \mid T_1)$

6.2 Incomplete Data

In the previous section we saw how the probability parameters in a Bayesian network can be estimated from a complete data set, i.e., a data set in which each case specifies a value for each of the variables. In practice, however, we are often faced with situations in which the data is incomplete. For example, some values may be accidentally missing (for example due to faulty sensor readings), some values may have been intentionally removed, and, in the more extreme case, some variables may simply not be observable (such variables are also called *latent variables* or *hidden variables*). If only some of the cases in the database contain missing values, then you could be tempted to simply throw these cases away and estimate the probability parameters using the remaining (complete) database. This approach, however, may have a serious drawback: Besides the risk of ending up with a very small database, we may unintentionally bias the parameter estimates. For example, assume that we have two binary variables A and B, and we are given a database with 20 cases over A and B. Assume also that the database contains an equal number of cases with $A = a_1$ and $A = a_2$, but when $A = a_2$, then the value of B is missing in 5 of the cases (B is not missing in any of the other cases). Now if we want to find the maximum likelihood estimate for θ, the probability of a_1, using the entire database, then (recall that $P^\#$ is the notation for frequency counts)

$$P^\#(a_1) = \hat{\theta} = \frac{N(a_1)}{N(a_1) + N(a_2)} = \frac{10}{10 + 10} = \frac{1}{2}.$$

However, if we throw away the cases that contain missing values, then the maximum likelihood estimate would be

$$P^\#(a_1) = \hat{\theta}' = \frac{N'(a_1)}{N'(a_1) + N'(a_2)} = \frac{10}{10 + 5} = \frac{2}{3}.$$

The difference in the two estimates is caused by A's influence on B's "missingness." On the other hand, if A does not have an influence on whether the value of B is missing in the database, then we can (if the database is large enough) safely throw away the cases with missing values without affecting the maximum likelihood estimate of A.

The example above illustrates that in order to deal with missing data we need to take into account *how* the data is missing. Consider the incomplete data set as having been produced from a complete data set by a process that hides some of the data.

- If the probability that a particular value is missing depends only on the observed values, then the data is said to be *missing at random* (MAR).
- If this probability is also independent of the observed values, then the data is said to be *missing completely at random* (MCAR).
- If the data is neither MAR nor MCAR, then the process that generated the missing data is said to be *nonignorable*.

In the definitions of MAR and MCAR, the probability that a value is missing is independent of that specific value. In particular, when we have hidden/latent variables, then the data is MCAR, since the variables are unobserved regardless of the values of any of the variables.

To give a few examples. Consider first an exit poll performed during an election, where an extreme right-wing party, ER, is running for parliament. If we expect people who vote for ER to be more likely than others to refuse to answer how they have voted, then the data is neither MCAR nor MAR. This also means that when estimating the parameters, we cannot disregard the underlying process causing the missing data. As another example, assume that we have a database containing the results of two tests. The results of both tests can be either *positive* or *negative*, but whereas the first test is always performed, the second test is performed only as a "backup test" when the result of the first test is *negative*. In this situation the pattern of "missingness" is dependent only on the observed values, hence the data is MAR. Finally, consider a monitoring system equipped with sensors whose values are continuously recorded and stored in a database. The recording system, however, is not completely stable, and sometimes a sensor value is not stored properly (i.e., it will be missing in the database). In this situation, the process causing the data to become missing is independent of all the sensor values, and the data is MCAR.

Today, the majority of the methods used for parameter estimation assume the data to be MAR, and in the remainder of this chapter we shall make the same assumption.

One approach to finding the maximum likelihood parameters could be to simply solve the corresponding likelihood equations. Unfortunately, this approach is not feasible in practice, since an incomplete case may cause the parameters to become dependent. The same holds if we were to consider the *maximum a posteriori parameters* $\boldsymbol{\theta}^*$:

$$\boldsymbol{\theta}^* = \arg\max_{\boldsymbol{\theta}} P(\boldsymbol{\theta} \mid \mathcal{D}). \tag{6.1}$$

Instead, researchers have focused on approximative methods for doing parameter estimation.

6.2.1 Approximate Parameter Estimation: The EM Algorithm

One of the most popular algorithms for doing parameter estimation is the Expectation-Maximization (EM) algorithm. The EM algorithm is a general algorithm for finding maximum likelihood estimates for a set of parameters $\boldsymbol{\theta}$ when one is faced with an incomplete data set. The algorithm basically alternates between a so-called *expectation step* and a *maximization step*: loosely speaking, in the expectation step we "complete" the data set by using the current parameter estimates $\hat{\boldsymbol{\theta}}$ to calculate expectations for the missing values, and in the maximization step we use the "completed" data set to find a new maximum likelihood estimate $\hat{\theta}'$ for the parameters. This estimate is then used to complete the data set in the next iteration of the algorithm. The algorithm continues either for a predetermined number of iterations or until the algorithm has converged.

Example 6.2. Consider the Bayesian network representation M of the simplified insemination problem described in Section 3.1.3 (page 55), and assume that we have the database in Table 6.4.

Cases	Pr	Bt	Ut
1.	?	pos	pos
2.	yes	neg	pos
3.	yes	pos	?
4.	yes	pos	neg
5.	?	neg	?

Table 6.4. A database consisting of five cases covering the variables Pr, Bt, and Ut. The ? indicates that the value of the corresponding variable is missing.

When using the EM algorithm for learning the probability parameters based on this database, we first specify some initial "guesses" for the probability distributions for M, i.e., $P_0(Pr)$, $P_0(Bt|Pr)$ and $P_0(Ut|Pr)$. For the sake of simplicity we let all three probability distributions be even although you would usually start off with random distributions. Now, had the database been complete, then in order to find a new estimate for, say, the distribution $P(Pr = yes)$, we would count the number of cases $N(Pr = yes)$ with $Pr = yes$:

$$P_1^{\#}(Pr = yes) = \frac{N(Pr = yes)}{N}.$$

From the database we see that cases 2, 3, and 4 contain $Pr = yes$, and they therefore contribute with the value 1 to $N(Pr = yes)$. However, for cases 1 and 5 the value for Pr is missing. So to find the contribution from these two cases we use the probability of seeing $Pr = yes$: case 1 therefore contributes with $P_0(Pr = y|Bt = Ut = pos) = 0.5$ and case 5 contributes with $P_0(Pr = y|Bt = neg) = 0.5$. What we are actually calculating here is the expected value for $N(Pr = yes)$, denoted by $\mathbb{E}[N(Pr = yes)]$:

$$\mathbb{E}[N(Pr = y)] = P_0(Pr = y|Bt = Ut = pos) + 1 + 1 + 1$$
$$+ P_0(Pr = y|Bt = neg) = \frac{1}{2} + 1 + 1 + 1 + \frac{1}{2} = 4;$$
$$\mathbb{E}[N(Pr = n)] = P_0(Pr = n|Bt = Ut = pos) + 0 + 0 + 0$$
$$+ P_0(Pr = n|Bt = neg) = \frac{1}{2} + 0 + 0 + 0 + \frac{1}{2} = 1.$$

In general, the expected value of $N(Pr = yes)$ is given by

$$\mathbb{E}[N(Pr = yes)] = \sum_{i=1}^{N} P_0(Pr = yes \mid \mathbf{d}_i).$$

We can now use the expected counts to calculate a new estimate for $P(Pr)$, but before we come that far we should also calculate the counts necessary for finding new estimates for the remaining probabilities. To estimate, say, $P(Ut = pos \mid Pr = yes)$, we need estimates for $P(Ut = pos, Pr = yes)$ and $P(Pr = yes)$:

$$P_1^{\#}(Ut = pos \mid Pr = yes) = \frac{P^{\#}(Ut = pos, Pr = yes)}{P^{\#}(Pr = yes)}$$

$$= \frac{\left[\dfrac{N(Ut = pos, Pr = yes)}{N}\right]}{\left[\dfrac{N(Pr = yes)}{N}\right]}$$

$$= \frac{N(Ut = pos, Pr = yes)}{N(Pr = yes)}.$$

Here $N(Ut = pos, Pr = yes)$ denotes the number of cases containing both $Ut = pos$ and $Pr = yes$. However, as for Pr, we cannot find $N(Ut = pos, Pr = yes)$ when there are missing values, so again we use the expected value/count

$$\mathbb{E}[N(Ut = pos, Pr = yes)] = \sum_{i=1}^{N} P(Ut = pos, Pr = yes \mid \mathbf{d}_i).$$

For the database above we get

$$\begin{aligned}
\mathbb{E}[N(Ut = pos, Pr = yes)] &= P(Ut = pos, Pr = yes \mid Bt = pos, Ut = pos) + 1 \\
&\quad + P(Ut = pos, Pr = yes \mid Bt = pos, Pr = yes) \\
&\quad + 0 + P(Ut = pos, Pr = yes \mid Bl = neg) \\
&= \frac{1}{2} + 1 + \frac{1}{2} + 0 + \frac{1}{4} = 2.25.
\end{aligned}$$

These counts are sufficient for finding new estimates for the probability parameters in the network (see Section 6.1). For example,

$$P_1^{\#}(Pr = yes) = \frac{\mathbb{E}[N(Pr = yes)]}{N} = \frac{4}{5} = 0.8,$$

$$P_1^{\#}(Ut = pos \mid Pr = yes) = \frac{\mathbb{E}[N(Ut = pos, Pr = yes)]}{\mathbb{E}[N(Pr = yes)]} = \frac{2.25}{4} = 0.5625.$$

When a new estimate has been found for all the probabilities, the procedure starts over again, but this time you should use the newly found probability estimates when calculating the expected counts. The procedure continues until the probabilities no longer change or until another termination criterion is met. In the special case that the database is complete, the algorithm converges after one iteration and returns the maximum likelihood estimates for the parameters.

Calculation of Family Counts

In the example above, we saw that in order to find a new estimate for a conditional probability distribution $P(X \mid \mathrm{pa}(X))$ we should calculate the expected counts for the family $\{X\} \cup \mathrm{pa}(X)$ of variables. That is, for a specific configuration of the family we calculate the expected number of cases that contain this configuration. Intuitively, we can consider the following three situations:

1. If a case is inconsistent with the configuration (i.e., the case and the configuration disagrees on at least one value), then it counts as 0.
2. If a case contains the entire configuration, then it counts as 1.
3. If the value for a variable is missing in a case, then it contributes with a fractional count corresponding to the conditional probability of seeing the configuration.

The situations 1 and 2 are in fact special cases of situation 3.

From a computational point of view, the calculation of the expected counts is the main difficulty of the EM-algorithm: when a case does not contain a value for all the variables in question, then we need to calculate the conditional probability distribution for these variables given that particular case. We shall consider two situations: First, assume that we are interested in a specific configuration $\mathrm{fa}(A) = \mathbf{a}$ for a family of variables, and let $\mathbf{d}$ be a case with a missing value for exactly one variable, X, in $\mathrm{fa}(A)$. If $\mathbf{a}$ specifies $X = x$, then the probability for $\mathbf{a}$ given $\mathbf{d}$ is equal to the probability $P(X = x \mid \mathbf{d})$, which in turn can be calculated by a single propagation in the Bayesian network. Second, and more generally, assume that $\mathbf{d}$ contains missing values for a set of variables $\mathcal{X} \subseteq \mathrm{fa}(A)$ in the family. In this situation the probability for $\mathbf{a}$ can be read directly from the joint probability $P(\mathcal{X} \mid \mathbf{d})$, but this is not immediately provided by the Bayesian network. Fortunately, in order to calculate this probability we can exploit the junction tree architecture (see Section 4.4). In particular, the construction of the underlying junction tree ensures that each family of variables is contained in at least one clique, say V, having variables $\mathcal{V}$. Hence, after a single propagation of the evidence corresponding to case $\mathbf{d}$, all the required probabilities can be read directly from the potentials associated with V and its neighboring separators. Specifically, from Theorem 4.5 we see that if V is a clique with the set of potentials Φ_V and with k neighboring separators containing the V-directed sets of potentials $\Phi_1, \ldots, \Phi_k$, then

$$P(\mathcal{V}, \mathbf{d}) = \prod_{\phi_V \in \Phi_V} \phi_V \prod_{\phi_1 \in \Phi_1} \phi_1 \cdots \prod_{\phi_k \in \Phi_k} \phi_k.$$

From this joint probability we can find the required probability $P(\mathcal{X}, \mathbf{d})$ by marginalizing out the irrelevant variables:

$$P(\mathcal{X}, \mathbf{d}) = \sum_{\mathcal{V} \setminus \mathcal{X}} P(\mathcal{V}, \mathbf{e}).$$

We return to our previous example. In order to calculate all the expected counts, we use the junction tree structure shown in Figure 6.2.

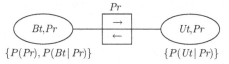

Fig. 6.2. A junction tree representation of the simplified insemination problem.

In particular, when calculating the contribution from case 5, we perform a full propagation with the evidence $Bt = neg$, and we get the annotated junction tree in Figure 6.3.

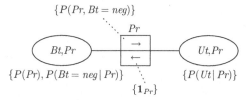

Fig. 6.3. A junction tree representation of the simplified insemination problem after inserting and propagating the evidence $Bt = neg$.

The required probability, e.g., $P(Ut, Pr \mid Bt = neg)$, can now be calculated directly from the potential in the clique containing Ut and the potential in the separator directed toward that clique:

$$P(Ut, Pr, Bt = neg) = P(Ut \mid Pr)P(Pr, Bt = neg),$$

$$P(Bt = neg) = \sum_{Ut, Pr} P(Ut, Pr, Bt = neg),$$

$$P(Ut, Pr \mid Bt = neg) = \frac{P(Ut, Pr, Bt = neg)}{P(Bt = neg)}.$$

Similarly, if we use the junction tree to calculate the contribution from case 5 to the expected counts for the family $\{Bt, Pr\}$, then we need $P(Bt, Pr \mid Bt = neg) = P(Pr \mid Bt = neg)$. This probability can be found using the same method as above:

$$P(Pr, Bt = neg) = P(Bt = neg \mid Pr)P(Pr)\mathbf{1}_{Pr},$$

$$P(Bt = neg) = \sum_{Pr} P(Pr, Bt = neg),$$

$$P(Pr \mid Bt = neg) = \frac{P(Pr, Bt = neg)}{P(Bt = neg)}.$$

The EM-Algorithm for Bayesian Networks

We describe the algorithm more formally. Assume that we have a model structure B over the variables $\mathcal{U} = \{X_1, \dots, X_n\}$, and let θ_{ijk} denote the parameter corresponding to the conditional probability $P(X_i = k \mid \mathrm{pa}(X_i) = j)$, i.e., the conditional probability for variable X_i being in its kth state given the jth configuration of the parents of X_i. Using this notation we can find a maximum likelihood estimate, $\hat{\theta}_{ijk}$, for the parameters θ_{ijk} given a data set $\mathcal{D} = \{\mathbf{d}_1, \dots, \mathbf{d}_m\}$ with m cases as follows:

Algorithm 6.1 [The EM algorithm]

1. *Choose an $\epsilon > 0$ to regulate the stopping criterion.*
2. *Let $\boldsymbol{\theta}^0 = \{\theta_{ijk}\}$, where $1 \le i \le n$, $1 \le k \le |\mathrm{sp}(X_i)| - 1$, and $1 \le j \le |\mathrm{sp}(\mathrm{pa}(X_i))|$, be some initial estimates of the parameters (chosen arbitrarily).*
3. *Set $t := 0$.*
4. *Repeat:*
   E-step: For each $1 \le i \le n$ calculate the table of expected counts:

$$\mathop{\mathbb{E}}_{\boldsymbol{\theta}^t}[N(X_i, \mathrm{pa}(X_i)) \mid \mathcal{D}] = \sum_{\mathbf{d} \in \mathcal{D}} P(X_i, \mathrm{pa}(X_i) \mid \mathbf{d}, \boldsymbol{\theta}^t).$$

   M-step: Use the expected counts as if they were actual counts to calculate a new maximum likelihood estimate for all θ_{ijk}:

$$\hat{\theta}_{ijk} = \frac{\mathbb{E}_{\boldsymbol{\theta}^t}[N(X_i = k, \mathrm{pa}(X_i) = j) \mid \mathcal{D}]}{\sum_{h=1}^{|\mathrm{sp}(X_i)|} \mathbb{E}_{\boldsymbol{\theta}^t}[N(X_i = h, \mathrm{pa}(X_i) = j) \mid \mathcal{D}]}.$$

   Set $\boldsymbol{\theta}^{t+1} := \hat{\boldsymbol{\theta}}$ and $t := t + 1$.
   Until $|\log_2 P(\mathcal{D} \mid \boldsymbol{\theta}_t) - \log_2 P(\mathcal{D} \mid \boldsymbol{\theta}_{t-1})| \le \epsilon$.

□

The EM-algorithm has been generalized for estimating the *maximum a posteriori* parameters (or penalized likelihood) instead of the maximum likelihood parameters. In this approach, virtual counts are added to both the denominator and numerator in the M-step, hence the method follows the idea of the Bayesian estimation method for complete data (see Section 6.1.2). As before, the virtual values can be interpreted as counts from a virtual database.

6.2.2 *Why We Cannot Perform Exact Parameter Estimation

When we have access to a complete database we can find the exact maximum likelihood parameters by simply counting frequencies in the database, or we can express the posterior probability distribution of the parameters in closed form. However, we are not that lucky when working with incomplete data. For example, assume that we have a probability distribution $P(\mathcal{U} \,|\, \boldsymbol{\theta})$ and that we get a single case $\mathbf{d}$ that specifies a configuration $\mathbf{x}$ over $\mathcal{X} \subset \mathcal{U}$; the variables $\mathcal{Y} = \mathcal{U} \setminus \mathcal{X}$ are therefore not observed. In order to find an estimate for the maximum likelihood parameters we should maximize the following expression with respect to $\boldsymbol{\theta}$:

$$P(\mathbf{x} \,|\, \boldsymbol{\theta}) = \sum_{\mathcal{Y}} P(\mathbf{x} \,|\, \mathcal{Y}, \boldsymbol{\theta}) P(\mathcal{Y} \,|\, \boldsymbol{\theta}).$$

That is, we maximize a sum having one term for each configuration of the unobserved variables. When performing the maximization we cannot consider the terms independently, since $P(\mathcal{Y} \,|\, \boldsymbol{\theta})$ will, in general, depend on all the parameters involved. Moreover, we have such a weighted sum for each case in the database; hence the number of terms may become intractably large.

6.3 Adaptation

When constructing a Bayesian network, you will almost always be uncertain of the correctness of the conditional probabilities specified, whether they are specified manually or learned from data. Usually you would allow each probability to range within an interval, and a number in this interval is then chosen. This type of uncertainty is called *second-order uncertainty*.

Second-order uncertainty raises two questions:

- Does the second-order uncertainty have an impact on the conclusions from the model?
- Are there systematic ways of reducing the second-order uncertainty?

The first question was discussed in Section 3.4 and was addressed in Section 5.6. In this section, we address the second question. We will look at a situation in which certain parameters are open for modification.

When a system is at work, you repeatedly get new cases, and you would like to learn from these cases. The situation may be that you are fairly certain of the structure of the network. However, the conditional probabilities are dependent on a context that varies from place to place, and you want to build a system that automatically adapts to the particular context in which it is placed.

In Figure 6.4(a), the variable A is directly influenced by B and C, and the strength is modeled by $P(A \,|\, B, C)$. The uncertainty in $P(A \,|\, B, C)$ may be modeled explicitly by introducing an extra parent, T, for A (Figure 6.4(b)). The variable T can be considered as a type variable. To reflect the frequencies of the context types, a prior distribution $P(T)$ is given.

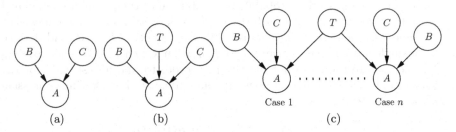

Fig. 6.4. Adaptation through a type variable T. The distribution of T is updated by *Case 1* and used in the next case.

When a case, $\mathbf{e}$, is entered into the network, the propagation will yield a new distribution $P^*(T) = P(T \mid \mathbf{e})$, and we may say that the change of the distribution for T reflects what has been learned from the case. Now $P^*(T)$ can be used as a new prior distribution when we get the next case. All variables whose tables are dependent on the context will be children of T. The way $P(T)$ is updated can also be made explicit in the network structure as shown in Figure 6.4(c). The network contains a copy of the variables for each case that will be considered, and when the ith case arrives, the corresponding variables are instantiated and $P^*(T) = P(T \mid \mathbf{e}_1, \ldots, \mathbf{e}_{i-1})$ is updated to $P(T \mid \mathbf{e}_1, \ldots, \mathbf{e}_i)$.

Example 6.3. Consider again the milk test problem described in Section 3.2.1, and assume that the farmer is not always as careful as he ought to be when performing the test. When this is the case, the risk of getting a false positive or a false negative is ten times as high as it otherwise would have been. Let us initially assume that there is an 80% chance that the farmer performs the test carefully.

One way of modeling this situation is to introduce a type variable *Type* (with states *careful* and *careless*) representing how the farmer performs the test (see Figure 6.5).

The probability $P(Inf)$ is as before, and the conditional probability distribution $P(Test \mid Inf, Type = careful)$ is as specified in Section 3.2.1. The probability distributions $P(Test \mid Inf, Type = careless)$ and $P(Type)$ can be derived from the description above, i.e., $P(Type) = (0.8, 0.2)$ and $P(Test \mid Inf, Type = careless)$ are as specified in Table 6.5.

Now assume that a test is performed and the result is negative. When updating the probabilities with this piece of evidence you get $P^*(Type) = P(Type \mid Test = neg) = (0.815, 0.185)$. This probability distribution represents our updated belief in how the farmer performs the test. That is, the next time

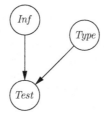

Fig. 6.5. The type variable *Type* models whether the farmer performs the milk test properly.

	$Inf = yes$	$Inf = no$
$Test = pos$	0.9	0.1
$Test = neg$	0.1	0.9

Table 6.5. The table shows the conditional probability distribution $P(Test \mid Inf, Type = careless)$.

you get new evidence you should use this conditional probability distribution (i.e., $P^*(Type)$) as the prior distribution for the variable *Type*.

Finally, it should be noted that you have to be a bit careful when working with several type variables. To illustrate the problem, assume that we get the case $A = a$ for the Bayesian network shown in Figure 6.6. When inserting this piece of evidence, we see, from d-separation, that S and T become dependent. Hence we cannot use their updated marginal distributions as prior distributions for the next case (by doing so we would have to assume that they are independent, which we have just seen is not the case). That is, in order for the above procedure to work correctly with several type variables, the evidence from a case should d-separate the type variables.

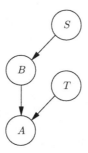

Fig. 6.6. A Bayesian network augmented with two type variables, S and T.

6.3.1 Fractional Updating

If the uncertainty of the conditional probabilities cannot be modeled explicitly through type variables, statistical methods can be used. The statistical task is first to specify a prior probability distribution over the parameters, and then iteratively update this distribution as new cases are entered. The correct approach for updating this distribution is basically the same as the task of learning exact parameter estimates from a database, but as we also saw in Section 6.2.2, this is infeasible in practice when we have missing values. Instead, approximative techniques are usually applied.

Consider $P(A \mid B, C)$, and let all variables be ternary. Under the assumptions of global and local independence, we may now think of $P(A \mid b_i, c_j) = (x_1, x_2, x_3)$ as a distribution established through a number of past cases in which (B, C) was in state (b_i, c_j). We can then express our certainty of the distribution by a fictitious sample size s. The larger the sample size, the smaller the second-order uncertainty, so we work with a *sample size* s, a set of counts (n_1, n_2, n_3) such that $s = n_1 + n_2 + n_3$, and

$$P(A \mid b_i, c_j) = \left(\frac{n_1}{s}, \frac{n_2}{s}, \frac{n_3}{s}\right).$$

That is, s represents the number of cases with (b_i, c_j), and n_1 is the number of these cases that also include a_1.

Let us first consider a couple of simple cases before we take the general case.

1. We get a new case $\mathbf{e}$ with $B = b_i$ and $C = c_j$ and with $A = a_1$. Then $n_1 := n_1 + 1$ and $s := s + 1$, and the probabilities are updated as follows:

$$x_1 := \frac{(n_1 + 1)}{(s + 1)}; \qquad x_2 := \frac{n_2}{(s + 1)}; \qquad x_3 := \frac{n_3}{(s + 1)}.$$

2. We get a new case $\mathbf{e}$ with $B = b_i$ and $C = c_j$, but for A we have only a distribution $P(A \mid \mathbf{e}) = P(A \mid b_i, c_j, \mathbf{e}) = (y_1, y_2, y_3)$. Then we cannot work with integer counts, and we update $n_k := n_k + y_k$ and $s := s + 1$. Accordingly, we get

$$x_1 := \frac{(n_1 + y_1)}{(s + 1)}; \qquad x_2 := \frac{(n_2 + y_2)}{(s + 1)}; \qquad x_3 := \frac{(n_3 + y_3)}{(s + 1)}.$$

3. We get a new case $\mathbf{e}$ with $A = a_1$, but for B and C we have only $P(B = b_i, C = c_j \mid \mathbf{e}) = z$. As before, we cannot work with integer counts, so instead we update with a fractional count:

$$x_1 := \frac{(n_1 + z)}{(s + z)}; \qquad x_2 := \frac{n_2}{(s + z)}; \qquad x_3 := \frac{n_3}{(s + z)}.$$

In general, we may get a case with $P(b_i, c_j \mid \mathbf{e}) = z$ and $P(A \mid b_i, c_j, \mathbf{e}) = (y_1, y_2, y_3)$. To update the counts, we use these distributions; because the sample size is increased only with z we take $n_k := n_k + zy_k$, and we get

$$x_k := \frac{(n_k + zy_k)}{(s + z)} = \frac{n_k + P(a_k, b_i, c_j \mid \mathbf{e})}{s + P(b_i, c_j \mid \mathbf{e})}.$$

This scheme is known as *fractional updating*. Unfortunately, the scheme has a serious drawback, namely that it tends to overestimate the count of s, thereby overestimating our certainty of the distribution. Assume for example that $e = \{B = b_i, C = c_j\}$. Then the case tells us nothing about $P(A \mid b_i, c_j)$, but nevertheless fractional updating will add a count of 1 to s and take it as a confirmation of the present distribution:

$$x_k := \frac{n_k + P(a_k \mid b_i, c_j)}{s + 1} = \frac{n_k + \frac{n_k}{s}}{s + 1} = \frac{n_k}{s}.$$

In Section 6.3.6 we shall return to this issue and consider another approximative updating method, which does not have the same drawback as mentioned above.

6.3.2 Fading

It is often a problem for fractional updating that the initial counts are kept when the system is trying to adapt to the environment. Particularly, when the conditional probabilities in the environment change over time, the accumulated counts will prevent the system from following the changes. Also, because fractional updating has a tendency to overestimate counts, vacuous counts will build up and make the parameters too resistant to change. Therefore, to keep the flexibility of parameters, it may be a good policy to prevent the sample size from growing unbounded.

An idea for solving this problem is the following: For example, let a ternary variable X have sample size s and counts (n_1, n_2, n_3), and assume that we get a count of 1 for x_1. Now, instead of increasing n_1 by one, we first multiply the counts by a fading factor, $q \in (0, 1)$. Hence, we get

$$s := sq + 1; n_1 := n_1 q + 1; n_2 := n_2 q; n_3 := n_3 q.$$

If we assume that all counts are of value 1, the influence from the past will fade away exponentially. In the limit where $s \to \infty$, we get a sample size s^*, where

$$s^* = \frac{1}{(1 - q)}.$$

The number s^* is called the *effective sample size*, and it represents a steady-state situation. If $s = s^*$ and we get a new count, we have

$$s := s^* q + 1 = \frac{q}{(1 - q)} + 1 = \frac{1}{(1 - q)} = s^*.$$

Instead of declaring a fading factor, you may declare an effective sample size s^*, and the fading factor is then

$$q^* = \frac{(s^* - 1)}{s^*}.$$

This idea can be used for each distribution $P(X \mid \mathrm{pa}(X) = \pi)$ that we wish to adapt to the evidence. The effective sample size need not be the same for all distributions. The effective sample size to declare is dependent on how resistant to change you wish the distribution to be. The higher the resistance, the higher the effective sample size.

Fading can be implemented such that the effective sample size is preserved. In other words, if the sample size for a distribution is equal to the declared effective sample size s^*, then it will not be changed in adapting to a new case.

Let $P(X \mid \pi)$ be declared with an effective sample size s^*, and assume we have $P(\pi \mid \mathbf{e}) = y$ for a case. Then fractional updating yields a new count of y. To preserve the sample size in the steady-state situation we have to adapt the fading factor q to the count y:

$$s^* q + y = s^*.$$

Hence

$$q = \frac{(s^* - y)}{s^*}.$$

Note that if $P(\pi \mid \mathbf{e}) = 1$, then $q = q^*$, and if $P(\pi \mid \mathbf{e}) = 0$, then $q = 1$.

6.3.3 *Specification of an Initial Sample Size

Frequently, the uncertainty of a parameter is expressed as an interval $[x, y]$. To exploit the technique for adaptation, the second-order uncertainty expressed by this interval will be translated to an initial sample size and a set of counts. The specification of the interval $[x, y]$ for $t = P(A = a)$ can be interpreted as, "I expect the value of t to be somewhere in the middle of the interval, and I am 90% sure that the value is in the interval." In other words, you have a distribution of t with mean close to $\frac{1}{2}(x + y)$ and with 90% of the density mass inside $[x, y]$.

As an example, take the interval $[0.3, 0.4]$ for the state a of the binary variable A. We interpret the interval as before, and assume that the distribution is the result of s samples out of which n were in state a. The distribution for t is a beta distribution, $\mathrm{Beta}(n_1, n_2)$, with mean $\mu = \frac{n_1}{s}$ and with variance $\sigma^2 = \frac{\mu(1-\mu)}{(s+1)}$, where $s = n_1 + n_2$ (see Figure 6.7 for examples). It holds that at least 90% of the probability mass lies in the interval $[\mu - 3\sigma, \mu + 3\sigma]$, so we seek values for s and n such that $\mu \approx 0.35$ and $\sigma \approx 0.0167$, and we get $n_1 = 285.16$ and $s = 814.73$.

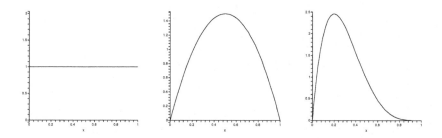

Fig. 6.7. The figure shows the density functions for the three beta distributions Beta(1, 1), Beta(2, 2), and Beta(2, 5).

6.3.4 Example: Strings of Symbols

Consider the transmission of symbols example from Section 3.2.4 with the model from Figure 3.18. Assume that every tenth word is sent through an error-correcting code, so that you know for certain the word transmitted. You wish to adapt the parameters of the model to the words actually transmitted and received.

First, you can use the coded words to adapt the distribution of the error rates: $P(R_i | T_i)$. Choose the effective sample size 100 for all parameters. This gives the fading factor 0.99. Also, let the initial sample be 100. The counts are given in Table 6.6.

	$T = a$	$T = b$
$R = a$	80	15
$R = b$	10	80
$R = c$	10	5

Table 6.6. Initial counts for $P(R | T)$.

Whenever a coded word is received, you have five cases (excluding the redundancy bits in the code). Assume that **baaba** was sent but **baaca** received. This means that the distribution $P(R | a)$ is modified three times and the distribution $P(R | b)$ is changed twice. For $P(R | b)$ we get the faded counts $((80 \cdot 0.99 + 1) \cdot 0.99 + 1) \cdot 0.99 + 1, 10 \cdot 0.99^3, 10 \cdot 0.99^3) = (80.6, 9.7, 9.7)$, and for $P(R | b)$ we get the faded counts $(15 \cdot 0.99^2, (80 \cdot 0.99 + 1) \cdot 0.99, 5 \cdot 0.99^2 + 1) = (14.7, 79.4, 5.9)$.

The noncoded words cannot be used for adaptation of $P(R | T)$, but they can be used for modifying $P(T_1)$ as well as $P(T_{i+1} | T_i)$. Assume that we receive the word $e = baaca$. Let us concentrate on modifying $P(T_2 | T_1)$. Let the initial sample size be 50 for $T_1 = a$ and 150 for $T_1 = b$. From Table 3.11, we infer the count table as given in Table 6.7.

	$T_1 = a$	$T_1 = b$
$T_2 = a$	30	60
$T_2 = b$	20	90

Table 6.7. Initial counts for $P(T_2 \,|\, T_1)$.

The model from Exercise 3.13 yields $P(T_1 \,|\, \mathbf{e}) = (0.13, 0.87)$, $P(T_2 \,|\, T_1 = a, \mathbf{e}) = (0.81, 0.19)$, and $P(T_2 \,|\, T_1 = b, \mathbf{e}) = (0.66, 0.34)$. The fading factors are $(100 - 0.13)/100 = 0.9987$ and $(100 - 0.87)/100 = 0.9913$. We get for $P(T_2 \,|\, a)$ the counts $(30 \cdot 0.9987 + 0.13 \cdot 0.81, 20 \cdot 0.9987 + 0.13 \cdot 0.19) = (30.07, 20.00)$ and for $P(T_2 \,|\, b)$ we get $(60 \cdot 0.9913 + 0.87 \cdot 0.66, 90 \cdot 0.9913 + 0.87 \cdot 0.34) = (60.05, 89.5)$.

Note that the sample size increases for the part with initial sample size smaller than the effective sample size and decreases for the part with initial sample size larger than the effective sample size.

6.3.5 Adaptation to Structure

As for the parameters in a model, it may happen that the structure of the model does not fit the cases you meet. If you use incremental adaptation of parameters, you will often experience that the changes in parameter values to a large degree will compensate for a slightly incorrect structure. Anyway, the structural inaccuracy may be so substantial that parameter adjustments cannot compensate. Unfortunately, no handy method for incremental adaptation of structure has been constructed. The reason is that structural changes are performed in jumps, and the justification for a jump is based on accumulated experience rather than a single case.

Basically, there are two ways out: you can accumulate the cases and run a batch learning algorithm (see Chapter 7) now and then, or you can work concurrently with several models. The second way is similar to the "expert disagreement approach."

Assume that you have three alternative models M_1, M_2, M_3 with initial normalized weights w_1, w_2, w_3; these weights can be interpreted as the probabilities for the models, $P(M_1)$, $P(M_2)$, and $P(M_3)$. A case with evidence $\mathbf{e}$ is entered into all models, and propagation yields $P(A \,|\, M_i, \mathbf{e})$ as well as $P(\mathbf{e} \,|\, M_i)$, where A is any variable. Then we can calculate new weights for the models

$$w_i := P(M_i \,|\, \mathbf{e}) = \frac{P(\mathbf{e} \,|\, M_i) P(M_i)}{P(\mathbf{e})} = \frac{P(\mathbf{e} \,|\, M_i) w_i}{\sum_j w_j P(\mathbf{e} \,|\, M_j)},$$

as well as the probability for the variable A:

$$P(A \,|\, \mathbf{e}) = w_1 P(A \,|\, M_1, \mathbf{e}) + w_2 P_2(A \,|\, M_2, \mathbf{e}) + w_3 P_3(A \,|\, M_3, \mathbf{e}).$$

6.3.6 *Fractional Updating as an Approximation

As we saw in Section 6.3.1, fractional updating has a serious drawback, namely that it tends to overestimate the sample size. To overcome this problem an alternative updating method (called *incremental updating*) has been proposed. Both fractional updating and incremental updating have their origins in the same problem: exact updating of the probability parameters is intractable, since it requires us to keep track of a mixture of Dirichlet distributions, where the number of mixture components may grow exponentially in the number of cases. More specifically, given evidence $\mathbf{e}$, both updating methods look for an approximation of the posterior distribution $P(\boldsymbol{\theta}|\mathbf{e})$, which determines the conditional probability distributions in the network.

In order to illustrate the updating method, we will first revisit the initial problem and show some of the derivations that underlie both fractional updating and incremental updating. Based on this, we will consider where the two updating methods differ.

Consider again the conditional probability distribution $P(A \,|\, B, C)$, where all variables are ternary. We set $P(A = a_k \,|\, B = b_i, C = c_j, \boldsymbol{\theta}) = \theta_{ijk}$ (see Figure 6.8 for a graphical representation) such that $\boldsymbol{\theta} = \{\theta_{ijk}\}$ and $1 \leq i \leq 3$, $1 \leq j \leq 3$, and $2 \leq k \leq 3$; the parameter θ_{ij1} is given by $1 - (\theta_{ij2} + \theta_{ij3})$. We will sometimes use the shorthand notation $\boldsymbol{\theta}_{ij} = \{\theta_{ij1}, \theta_{ij2}, \theta_{ij3}\}$, and we also assume that the prior distribution for $\boldsymbol{\theta}_{ij}$ follows a Dirichlet distribution with hyperparameters (n_1, n_2, n_3), denoted by $\mathrm{Dir}[\boldsymbol{\theta}_{ij}|n_1, n_2, n_3]$.

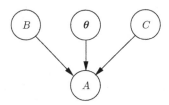

Fig. 6.8. An explicit representation of the parameter $\boldsymbol{\theta}$, which determines the conditional probability distribution $P(A = a_k \,|\, B = b_i, C = c_j)$.

Now assume that we have the simple case with evidence $\mathbf{e} = \{A = a_2, B = b_i, C = c_j\} \cup \mathbf{e}'$. Since A, B, and C constitute the Markov blanket for $\boldsymbol{\theta}$, we can disregard $\mathbf{e}'$ when updating the distribution for the parameters $\boldsymbol{\theta}_{ij}$, i.e., $f(\boldsymbol{\theta}_{ij}|a_2, b_i, c_j, \mathbf{e}') = f(\boldsymbol{\theta}_{ij}|a_2, b_i, c_j)$. Moreover, due to the choice of prior distribution for $\boldsymbol{\theta}$, we have

$$f(\boldsymbol{\theta}_{ij}|a_2, b_i, c_j) = \mathrm{Dir}[\boldsymbol{\theta}_{ij}|n_1, n_2 + 1, n_3].$$

As we did in the thumbtack problem (Section 6.1.2), we can similarly find a single point estimate for $P(A = a_k \,|\, B = b_i, C = c_j)$ by calculating the expectation of θ_{ijk} given $\mathbf{e} = \{A = a_2, B = b_i, C = c_j\} \cup \mathbf{e}'$:

$$P'(A = a_k \mid B = b_i, C = c_j) = \int_{\boldsymbol{\theta}_{ij}} \theta_{ijk} \mathrm{Dir}[\boldsymbol{\theta}_{ij} \mid n_1, n_2 + 1, n_3] d\boldsymbol{\theta}_{ij}$$

$$= \begin{cases} \frac{n_k + 1}{n_1 + n_2 + n_3 + 1} & \text{for } k = 2, \\ \frac{n_k}{n_1 + n_2 + n_3 + 1} & \text{otherwise.} \end{cases}$$

These updating rules are identical to those for fractional updating.

Consider now the more general situation in which the evidence does not necessarily include A, B, and C. In this case, we first express the posterior distribution $f(\boldsymbol{\theta}_{ij} \mid e)$ as follows (recall that $f(\boldsymbol{\theta} \mid A, B, C) = f(\boldsymbol{\theta} \mid A, B, C, e)$):

$$f(\boldsymbol{\theta} \mid e) = \sum_A \sum_B \sum_C f(\boldsymbol{\theta} \mid A, B, C) P(A, B, C \mid e).$$

From the assumption of local parameter independence (see Section 6.3.1) we can derive that $f(\boldsymbol{\theta}_{ij}) = f(\boldsymbol{\theta}_{ij} \mid A, B = b_{i'}, C = c_{j'})$, for $i' \neq i$ or $j' \neq j$. This allows us to decompose the above expression into two parts, one with $j' = j$ and $i' = i$ and the other with $j' \neq j$ and $i' \neq i$:

$$f(\boldsymbol{\theta}_{ij} \mid e) = \sum_A f(\boldsymbol{\theta}_{ij} \mid A, B = b_i, C = c_j) P(A, B = b_i, C = c_j \mid e)$$

$$+ \sum_{j' \neq j} \sum_{i' \neq i} \sum_A f(\boldsymbol{\theta}_{ij}) P(A, B = b_{i'}, C = c_{j'} \mid e)$$

$$= \sum_A f(\boldsymbol{\theta}_{ij} \mid A, B = b_i, C = c_j) P(A, B = b_i, C = c_j \mid e)$$

$$+ f(\boldsymbol{\theta}_{ij})(1 - P(B = b_i, C = c_j \mid e)).$$

As we also used above, we have that, for example, $f(\boldsymbol{\theta}_{ij} \mid A = a_2, B = b_i, C = c_j) = \mathrm{Dir}[\boldsymbol{\theta}_{ij} \mid n_1, n_2 + 1, n_3]$; hence the above expression can be rewritten as

$$\begin{aligned} f(\boldsymbol{\theta}_{ij} \mid e) = {}& \mathrm{Dir}[\boldsymbol{\theta}_{ij} \mid n_1 + 1, n_2, n_3] P(A = a_1, B = b_i, C = c_j \mid e) \\ &+ \mathrm{Dir}[\boldsymbol{\theta}_{ij} \mid n_1, n_2 + 1, n_3] P(A = a_2, B = b_i, C = c_j \mid e) \\ &+ \mathrm{Dir}[\boldsymbol{\theta}_{ij} \mid n_1, n_2, n_3 + 1] P(A = a_3, B = b_i, C = c_j \mid e) \\ &+ \mathrm{Dir}[\boldsymbol{\theta}_{ij} \mid n_1, n_2, n_3](1 - P(B = b_i, C = c_j \mid e)). \end{aligned} \quad (6.2)$$

Note that the last term models the situation in which the specified parent configuration is not observed, and if it is observed then the term contributes with zero.

This equation readily generalizes to a variable A with r states and parent configuration π:

$$\begin{aligned} f(\boldsymbol{\theta}_\pi \mid e) = {}& \sum_{k=1}^r \mathrm{Dir}[\boldsymbol{\theta}_\pi \mid n_1, \dots, n_k + 1, \dots, n_r] P(A = a_k, \mathrm{pa}(A) = \pi \mid e) \\ &+ \mathrm{Dir}[\boldsymbol{\theta}_\pi \mid n_1, \dots, n_r](1 - P(\mathrm{pa}(A) = \pi \mid e)). \end{aligned} \quad (6.3)$$

Unfortunately, there is a computational problem with this expression, namely that the number of mixture components may grow exponentially in the number of cases that we process. This problem has led to the development of approximate updating methods such as fractional updating and incremental updating. Both of these methods approximate the mixture above using a single Dirichlet distribution, but there is a difference in how they estimate the parameters.

Fractional Updating Revisited

In fractional updating, equation (6.3) is approximated with a single Dirichlet distribution. The hyperparameters for this approximate distribution are formed by taking the linear combination (as defined by the mixture) of the corresponding hyperparameters in the mixture (disregarding the last term). For example, for the first hyperparameter n'_1 in equation (6.2) we get

$$
\begin{aligned}
n'_1 &= (n_1 + 1)P(A = a_1, B = b_i, C = c_j | e) \\
&\quad + n_1 P(A = a_2, B = b_i, C = c_j | e) \\
&\quad + n_1 P(A = a_3, B = b_i, C = c_j | e) \\
&= n_1 + P(A = a_1, B = b_i, C = c_j | e).
\end{aligned}
$$

That is, the mixture in equation (6.2) is approximated by

$$
\begin{aligned}
f'(\boldsymbol{\theta}_{ij} | e) = \mathrm{Dir}[&n_1 + P(A = a_1, B = b_i, C = c_j | e), \\
&n_2 + P(A = a_2, B = b_i, C = c_j | e), \\
&n_3 + P(A = a_3, B = b_i, C = c_j | e)],
\end{aligned}
$$

and the new estimate for $P(A = a_k |, B = b_i, C = c_j)$ is then given by the mean value of θ_{ijk}:

$$
\begin{aligned}
P'(A = a_k |, B = b_i, C = c_j) = \int_{ij} \theta_{ijk} \mathrm{Dir}[&n_1 + P(A = a_1,, B = h_i, C = c_j | e), \\
&n_2 + P(A = a_2,, B = b_i, C = c_j | e), \\
&n_3 + P(A = a_3,, B = b_i, C = c_j | e)]d\boldsymbol{\theta}_{ij}.
\end{aligned}
$$

Hence

$$
\begin{aligned}
P'(A = a_k |, B = b_i, C = c_j) &= \frac{n_k + P(A = a_k,, B = b_i, C = c_j | e)}{n_1 + n_2 + n_3 + P(B = b_i, C = c_j | e)} \\
&= \frac{n_k + P(A = a_k,, B = b_i, C = c_j | e)}{s + P(B = b_i, C = c_j | e)},
\end{aligned}
$$

and by comparing this result with the updating rule presented in Section 6.3.1 we see that they are identical. Thus, the intuitive appeal of fractional updating that we saw in Section 6.3.1 rests on a mathematical foundation.

The Incremental Updating Rule

Analogously to fractional updating, when doing incremental updating we also estimate the mixture of Dirichlet distributions in equation (6.3) with a single Dirichlet distribution. However, the hyperparameters for the approximate Dirichlet distribution are determined by equating the means and average variance of the mixture to the means and the average variance of the approximating distribution. To be more specific, let θ^*_{ik} denote the mean of $\theta_{..k}$ in the ith component in equation (6.3). The mean for the kth parameter in the mixture is then (θ^*_{0k} denotes the mean of $\theta_{..k}$ in the last term)

$$\theta^*_k = \sum_{i=1}^{r} \theta^*_{ik} P(A = a_i, \pi \,|\, \mathbf{e}) + \theta^*_{0k}(1 - P(B = b_j, C = c_j | \mathbf{e})).$$

Thus estimating the mixture with a single Dirichlet distribution having the parameters $(s\theta^*_1, \ldots, s\theta^*_r)$ will provide the correct means for the parameters θ_k. The value for s is found by setting the average variance of the approximating distribution

$$\tilde{v} = \sum_{i=1}^{r} \theta^*_i \frac{\theta^*_i(1 - \theta^*_i)}{s+1},$$

equal to the average-mean-weighted variance of the mixture. This gives the following updating value for s:

$$s = \frac{\sum_{k=1}^{r} \theta^{*2}_k(1 - \theta^*_k)}{\sum_{k=1}^{r} \theta^*_k v_k} - 1,$$

where v_k is the variance of θ^*_k in the mixture. Although this updating rule does not have the same intuitive appeal as fractional updating, it has the property that the sample size will not increase when no relevant evidence is entered. In fact, it is actually possible for the sample size to decrease if the evidence does not reflect an event with high prior probability.

6.4 Tuning

We have a Bayesian network BN. For this network, we have some evidence $\mathbf{e}$, and for a particular variable A we have $\mathbf{x} = P(A \,|\, \mathbf{e}) = (x_1, \ldots, x_n)$. We may have a prior request $\mathbf{y} = (y_1, \ldots, y_n)$ for $P(A \,|\, \mathbf{e})$, so we want to tune the network such that $P(A \,|\, \mathbf{e}) = \mathbf{y}$. Assume that the structure of BN is fixed, but for the conditional probabilities we have some freedom described by a set of modifiable parameters $\mathbf{t} = (t_1, \ldots, t_m)$ with an initial set of values $\mathbf{t}_0$; to emphasize that we consider a subset of the parameters we use t_i to represent a parameter rather than θ_{ijk} as we previously have used. We want to set the parameters so that $P(A \,|\, \mathbf{e})$ is sufficiently close to $\mathbf{y}$. One way to measure how close the two distributions are would be to use the Euclidean distance:

Definition 6.1. *Let* $\mathbf{x} = (x_1, \ldots, x_n)$ *and* $\mathbf{y} = (y_1, \ldots, y_n)$ *be two probability distributions. Then the* Euclidean distance *between* $\mathbf{x}$ *and* $\mathbf{y}$ *is (although we do not take the square root):*

$$\text{dist}\,(\mathbf{x}, \mathbf{y}) = \sum_{i=1}^{n}(x_i - y_i)^2.$$

The Euclidean distance measure is a *metric*, meaning that:

1. $\text{dist}(\mathbf{x}, \mathbf{y}) = 0$ if and only if $\mathbf{x} = \mathbf{y}$.
2. $\text{dist}(\mathbf{x}, \mathbf{y}) \leq \text{dist}(\mathbf{x}, \mathbf{z}) + \text{dist}(\mathbf{z}, \mathbf{y})$.
3. $\text{dist}(\mathbf{x}, \mathbf{y}) = \text{dist}(\mathbf{y}, \mathbf{x})$.

Another distance measure frequently used is the Kullback-Leibler divergence:

Definition 6.2. *Let* $\mathbf{x} = (x_1, \ldots, x_n)$ *and* $\mathbf{y} = (y_1, \ldots, y_n)$ *be two probability distributions. Then the* Kullback-Leibler divergence *between* $\mathbf{x}$ *and* $\mathbf{y}$ *is:*

$$\text{KL}(\mathbf{x}, \mathbf{y}) = \sum_{i=1}^{n} x_i \log_2\left(\frac{x_i}{y_i}\right),$$

where $0 \log_2(0/y_i) = 0$ *and* $x_i \log_2(x_i/0) = \infty$.

Note that the Kullback-Leibler divergence does not satisfy property 3 above so it is not a metric. In the remainder of this section we shall consider only the Euclidean distance.

If $(t_1, \ldots, t_n)$ are parameters in the Bayesian network BN (parameters are entries in conditional probability tables, see also Section 5.7) over the universe $\mathcal{U}$, then $P(\mathcal{U})$ is a function of $(t_1, \ldots, t_n)$, as are also $P(A \mid \mathbf{e})$ and $P(\mathbf{e})$. In the following, we assume proportional scaling, and we also assume that there is at most one parameter per distribution.

The task is to set the parameters such that the distance is as small as possible. If the parameters cannot be set in such a way that the distance is close to zero, then it is an indication of an incorrect structure.

If it is possible to determine $\text{dist}(\mathbf{x}, \mathbf{y})$ as a function of $\mathbf{t}$, you might be so fortunate that the problem can be solved directly. However, usually the problem cannot be solved directly even when the function is known, and a *gradient descent* method can be used:

1. Calculate **grad** $\text{dist}(\mathbf{x}, \mathbf{y})$ with respect to the parameters $\mathbf{t}$.
2. Give $\mathbf{t}_0$ a displacement $\triangle \mathbf{t}$ in the direction opposite to the direction of the gradient **grad** $\text{dist}\,(\mathbf{x}, \mathbf{y})\,(\mathbf{t}_0)$; that is, choose a step size $\alpha > 0$ and let $\triangle \mathbf{t} = -\alpha\,\textbf{grad}\,\text{dist}\,(\mathbf{x}, \mathbf{y})\,(\mathbf{t}_0)$.
3. Iterate this procedure until the gradient is close to $\mathbf{0}$.

From the definition of the Euclidean distance measure, we see that

$$\frac{\partial}{\partial t} \operatorname{dist}(\mathbf{x}, \mathbf{y}) = \sum_i 2(x_i - y_i)\frac{\partial x_i}{\partial t}.$$

The y_i's are known, and the x_i's are available through updating in BN, so what we need are $\mathbf{grad}\, x_i(\mathbf{t})$ for all i. If the variable A is binary, we have $\mathbf{x} = (x, 1 - x)$, $\mathbf{y} = (y, 1 - y)$, and

$$\operatorname{dist}(\mathbf{x}, \mathbf{y}) = 2(x - y)^2$$

and

$$\mathbf{grad}\operatorname{dist}(\mathbf{x}, \mathbf{y}) = 4(x - y)\,\mathbf{grad}\, x,$$

From these formulas, we see that the gradient is $\mathbf{0}$ if and only if either x is independent of all the parameters or $x = y$.

6.4.1 Example

Let BN be the Bayesian network in Figure 6.9 with initial probabilities from Table 6.8. Let C be the information variable and A the variable of interest. Assume also that the parameters are $t = P(\neg a)$ and $s = P(\neg c \mid \neg b)$. Initially, we have $\mathbf{t}_0 = (0.5, 0.4)$.

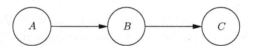

Fig. 6.9. A small Bayesian network for illustration.

$B \setminus A$	a	$\neg a$
b	1	0.3
$\neg b$	0	0.7

$C \setminus B$	b	$\neg b$
c	1	0.6
$\neg c$	0	0.4

Table 6.8. Parameters for the network in Figure 6.9, $P(A) = (0.5, 0.5)$.

Assume that we require $P(A \mid c) = (0.4, 0.6) = (y, 1 - y)$. Through updating, we get $x = P(a \mid c) = 0.58$. We calculate $P(a \mid c)$ as a function of $\mathbf{t}$:

$$P(A, c) = \sum_B P(A)P(B \mid A)P(c \mid B) = (1 - t, t - 0.7ts),$$

$$P(a \mid c) = \frac{P(a, c)}{\sum_A P(A, c)}.$$

We get

$$P(a \mid c) = x(t, s) = \frac{(1 - t)}{(1 - 0.7ts)}.$$

The request is

$$\frac{1 - t}{1 - 0.7ts} = 0.4,$$

which yields

$$s = \frac{t - 0.6}{0.28t} = \frac{25}{7} - \frac{15}{7t}.$$

The set of parameters t meeting the request is shown in Figure 6.10.

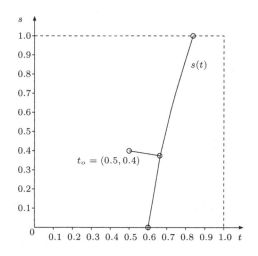

Fig. 6.10. The graph of $s(t)$ consists of the parameter pairs (t, s) meeting the request $P(a \mid c) = 0.4$.

Out of the infinite number of parameter pairs $(t, s(t))$, we choose one. If we do not wish to choose either of the extremes $(0.6, 0)$ and $(\frac{5}{6}, 1)$, it would be natural to choose the point closest to $\mathbf{t_0} = (0.5, 0.4)$. This point is characterized by the property that the normal contains $\mathbf{t_0}$ (see Figure 6.10). Through standard calculations, we get the following equation in t:

$$t^4 - \frac{1}{2}t^3 + \frac{666}{98}t - \frac{225}{49} = 0.$$

A root is $t = 0.668$, and we get $s = 0.364$. For this very simple example, it was possible to calculate the closest parameter setting meeting the request.

The situation need not be much more complex before a direct calculation becomes intractable.

The gradient descent method will in this example go as follows:

$$\mathbf{grad}\, x(t) = \frac{1}{(1 - 0.7ts)^2}(0.7s - 1, (1 - t)0.7t),$$

$$\mathbf{grad}\, x(t_0) = (-0.97, 0.24).$$

Formula (6.4) yields

$$\mathbf{grad}\, \text{dist}\, (\mathbf{x}, \mathbf{y}) = 4(0.58 - 0.4)(-0.97, 0.24) = (-0.70, 0.18).$$

Using a step size of 0.2, we get

$$\triangle \mathbf{t} = (0.14, -0.036)$$

and

$$\mathbf{t}_1 = (0.640, 0.364); \quad P^1(a\,|\,c) = 0.43.$$

The process is repeated:

$$\mathbf{grad}\, x(t_1) = (-1.06, 0.23),$$
$$\mathbf{grad}\, \text{dist}\, (x, y) = (-0.13, 0.03),$$
$$\mathbf{t}_2 = (0.686, 0.358); \quad P^2(a\,|\,c) = 0.380.$$

Repeating once more yields

$$\mathbf{t}_3 = (0.672, 0.361); \quad P^3(a\,|\,c) = 0.395.$$

6.4.2 Determining grad dist(x, y) as a Function of t

The gradient descent method seems to require that we be able to calculate $\mathbf{x}$ and $\mathbf{grad}\,\mathbf{x}$ as a function of the parameters $\mathbf{t}$. It was possible for the preceding small example, but the method used will in general be intractable.

Instead, the results form Section 5.7 can be used. By using proportional scaling we have

$$x = \frac{\alpha t + \beta}{at + b}.$$

This yields

$$\frac{\partial x}{\partial t} = \frac{\alpha(at + b) - a(\alpha t + \beta)}{(at + b)^2} = \frac{\alpha b - a\beta}{(at + b)^2},$$

where the constants can be found as described in Section 5.7.

6.5 Summary

Maximum Likelihood Estimation

For each case $\mathbf{d} \in \mathcal{D}$, the probability $P(\mathbf{d}|M)$ is called the *likelihood* of M given $\mathbf{d}$. If we assume that the cases in $\mathcal{D}$ are independent given the model, then the *likelihood of M given $\mathcal{D}$* is

$$L(M \mid \mathcal{D}) = \prod_{\mathbf{d} \in \mathcal{D}} P(\mathbf{d}|M).$$

The parameters $\boldsymbol{\theta}$ maximizing the likelihood are called the maximum likelihood parameters (and denoted by $\hat{\boldsymbol{\theta}}$):

$$\hat{\boldsymbol{\theta}} = \arg \max_{\boldsymbol{\theta}} L(M_{\boldsymbol{\theta}} \mid \mathcal{D}) = \arg \max_{\boldsymbol{\theta}} LL(M_{\boldsymbol{\theta}} \mid \mathcal{D}),$$

where

$$LL(M \mid \mathcal{D}) = \sum_{\mathbf{d} \in \mathcal{D}} \log_2 P(\mathbf{d}|M).$$

If the database does not contain missing values, then the likelihood of a Bayesian network is maximized by the (local) maximum likelihood estimates for the conditional probability tables, say $P(A \mid \mathrm{pa}(A))$, in the network:

$$\frac{N(A, \mathrm{pa}(A))}{N(\mathrm{pa}(A))}.$$

Bayesian Estimation

Let X be a binary variable (*yes, no*), and assume that we have performed a number of independent experiments out of which n turned up *yes* and m turned up *no*. Let θ be the probability for *yes*. Then, starting with the even prior distribution for θ, the posterior distribution is

$$f_p(\theta) = \mu \theta^n (1 - \theta)^m,$$

where μ is a normalization constant. The Bayesian estimate for θ is $\frac{n+1}{n+m+2}$.

This result can be interpreted so that an even prior distribution corresponds to adding two virtual experiments to the data (one for *yes* and one for *no*) and then counting frequencies. The procedure generalizes to distributions over variables with more than two states.

Incomplete Data

- If the probability that a particular value is missing depends only on the observed values, then the data is said to be *missing at random* (MAR).
- If this probability is also independent of the observed values, then the data is said to be *missing completely at random* (MCAR).
- If the data is neither MAR nor MCAR, then the process that generated the missing data is said to be *nonignorable*.

The EM algorithm

To find an estimate for the maximum likelihood parameters when the data is incomplete, you may run the EM algorithm; note that you are guaranteed only to find a local maximum likelihood estimate.

1. Choose an $\epsilon > 0$ to regulate the stopping criterion.
2. Let $\boldsymbol{\theta}^0 = \{\theta_{ijk}\}$, where $1 \leq i \leq n$, $1 \leq k \leq |\mathrm{sp}(X_i)| - 1$, and $1 \leq j \leq |\mathrm{sp}(\mathrm{pa}(X_i))|$, be some initial estimates of the parameters (chosen arbitrarily).
3. Set $t := 0$.
4. Repeat:
   E-step: For each $1 \leq i \leq n$ calculate the table of expected counts:

$$\mathbb{E}_{\boldsymbol{\theta}^t} [N(X_i, \mathrm{pa}(X_i)) \,|\, \mathcal{D}] = \sum_{\mathbf{d} \in \mathcal{D}} P(X_i, \mathrm{pa}(X_i) \,|\, \mathbf{d}, \boldsymbol{\theta}^t).$$

M-step: Use the expected counts as if they were actual counts to calculate a new maximum likelihood estimate for all θ_{ijk}:

$$\hat{\theta}_{ijk} = \frac{\mathbb{E}_{\boldsymbol{\theta}^t} [N(X_i = k, \mathrm{pa}(X_i) = j) \,|\, \mathcal{D}]}{\sum_{h=1}^{|\mathrm{sp}(X_i)|} \mathbb{E}_{\boldsymbol{\theta}^t} [N(X_i = h, \mathrm{pa}(X_i) = j) \,|\, \mathcal{D}]}.$$

Set $\boldsymbol{\theta}^{t+1} := \hat{\boldsymbol{\theta}}$ and $t := t + 1$.
Until $|\log_2 P(\mathcal{D} \,|\, \boldsymbol{\theta}_t) - \log_2 P(\mathcal{D} \,|\, \boldsymbol{\theta}_{t-1})| \leq \epsilon$.

The probabilities required in the E-step are easily calculated using junction tree propagation.

Adaptation

Adaptation through type variables: The second-order uncertainty can be characterized as uncertainty about which table out of $\mathbf{t}_1, \ldots, \mathbf{t}_m$ is the correct one for $P(A \,|\, pa(A))$.

Add a type variable T with states $t_1, \ldots, t_m$ and with A as child. The prior probability $P(t_1, \ldots, t_m)$ reflects your belief in the various tables. Put $P(A \,|\, pa(A), t_i) = \mathbf{t}_i$.

Whenever a case e has been processed, the probability $P(t_1, \ldots, t_n \,|\, e)$ is used as the new prior for the next case.

Fractional updating: Assume that the second-order uncertainty obeys both the global and local independence requirements. For each parent configuration π, choose a fictitious sample size n expressing the present certainty of $P(A \,|\, \pi)$. This yields a fictitious sample size $n_a = nP(a \,|\, \pi)$ for the configuration (a, π).

When a case has been processed, it yields $P(a, \pi \,|\, e)$. Add $P(a, \pi \,|\, e)$ to n_a. Thereby the sample is increased by $P(\pi \,|\, e)$.

Warning: fractional updating reduces the second-order uncertainty too quickly.

Fading: Instead of counting up with n_a, first multiply the counts for π by a fading factor. A fading factor q can be established from an effective sample size s^*

$$q = \frac{(s^* - P(\pi \mid e))}{s^*}.$$

The alternative model approach: If there is explicit uncertainty in the model – that is, if there are alternative models $M_1, \ldots, M_m$ – they can be weighted initially and run in parallel. After each case, the weights are modified.

Tuning

The set of parameters $\mathbf{t}$ open for modification; $\mathbf{x}(t)$ the current distribution in the model; $\mathbf{y}$ the target distribution.

1. Calculate **grad** dist$(\mathbf{x}, \mathbf{y})$ with respect to the parameters $\mathbf{t}$.
2. Give $\mathbf{t}_0$ a displacement $\triangle \mathbf{t}$ in the direction opposite to the direction of the gradient **grad** dist $(\mathbf{x}, \mathbf{y})(\mathbf{t}_0)$; that is, choose a step size $\alpha > 0$ and let $\triangle \mathbf{t} = -\alpha\,\textbf{grad}\,\text{dist}\,(\mathbf{x}, \mathbf{y})\,(\mathbf{t}_0)$.
3. Iterate this procedure until the gradient is close to $\mathbf{0}$.

We have

$$\frac{\partial}{\partial t}\,\text{dist}\,(\mathbf{x}, \mathbf{y}) = \sum_i 2(x_i - y_i)\frac{\partial x_i}{\partial t}.$$

Because $P(e)(t) = \alpha t + \beta$, we know that $x_i(t)$ is the ratio of two linear functions, and the partial derivatives can be calculated for all parameters through two propagations (Chapter 4).

6.6 Bibliographical Notes

The characterization of the different ways in which data may be missing/incomplete was suggested by Rubin (1976). With outset in incomplete data, the EM algorithm was proposed by Dempster *et al.* (1977) for learning maximum likelihood parameter estimates. Green (1990) described how the EM-algorithm can be used to find penalized maximum likelihood estimates, and Lauritzen (1995) showed how the junction tree architecture can be exploited in calculating the expected counts in the E-step of the algorithm.

When data arrives sequentially, the probability parameters can be adapted using fractional updating (Titterington, 1976). In some cases, however, fractional updating may overestimate the sample size and an improved version of the algorithm (known as incremental updating) was proposed by Spiegelhalter and Lauritzen (1990). Later this algorithm was extended by Olesen *et al.* (1992) to also allow for fading.

The tuning method was proposed by Jensen (1999), based on work by Russell *et al.* (1995) and Castillo *et al.* (1996).

6.7 Exercises

Exercise 6.1. Consider Example 6.1. Prove that the maximum likelihood estimate for the model given the data is $\theta = 0.8$.

Exercise 6.2. In the thumbtack experiment, let the nonnormalized prior distribution for θ be

$$f(\theta) = \begin{cases} \theta \text{ if } \theta \leq 1/2 \\ (1-\theta) \text{ if } 1/2 \leq \theta \leq 1 \end{cases}$$

(i) What is the normalization constant?

We have performed one experiment resulting in *up*.

(ii) What is the functional part of f_p, the posterior distribution for θ?
(iii) What is normalization constant for f_p?
(iv) What is the posterior Bayesian estimate?

Exercise 6.3. Consider the data in Table 6.1 and a Bayesian network consisting of two nodes T_1 and T_2, with T_1 being a parent of T_2. What are the maximum likelihood parameter estimates for the model given the data? What are the Bayesian parameter estimates for the model given the data?

Exercise 6.4. Prove the distribution part of Theorem 6.1.

Exercise 6.5. Establish a Bayesian estimate of the conditional probability $P(a|b)$ from the counts in Table 6.1.

Exercise 6.6. Characterize the type (MAR, MCAR, or nonignorable) of missingness that underlies the database for the variables A and B described in the beginning of Section 6.2.

Exercise 6.7. Without taking the size of the database into account, when would it be safe to throw away cases with missing values, i.e., should the data be MAR, MCAR, or neither of the two?

Exercise 6.8. [E]

(i) Update the remaining probabilities in Example 6.2.
(ii) Use the updated probabilities to perform another iteration of the EM-algorithm.

Exercise 6.9. Refer back to the example of EM parameter estimation in Example 6.2. What are the estimated parameters after a full iteration? And after two full iterations? What are the maximum likelihood parameter estimates using only the complete cases? What are the Bayesian parameter estimates using only the complete cases?

Exercise 6.10. ^{E} Consider the model in Exercise 3.28.

(i) What happens when you adapt to the following sequence of (A, B) states: $\langle(n, y)(n, y)(y, n)\rangle$?
(ii) Process a sequence of cases with $A = y$ in which the states of B are

$$(n, y, n, y, y, n, n, y, y, y).$$

What are your beliefs in the experts now, and what is $P(B \mid A)$?

Exercise 6.11. You have the same model as in Exercise 6.10, but $P(B \mid A)$ is the one in Table 6.9.

$B \backslash A$	y	n
y	0.75	0.4
n	0.25	0.6

Table 6.9. Table for Exercise 6.11.

For $P(B \mid A = y)$, you have an initial sample size of 12.

(i) Perform fractional updating from the sequence in Exercise 6.10 (iii).
(ii) Perform fractional updating on the same sequence but with fading factor 0.9.

Exercise 6.12. The network from Example 6.4.1 in its initial state has sample sizes $s_t = 25$, $s_s = 10$, and $s_u = 25$ for the three parameters. It now receives 20 cases with $C = c$ out of which 10 have $A = a$ (the rest have $A = \neg a$). For the cases with $A = a$, all cases have $B = b$, and in the rest, 4 had $B = \neg b$.

1. Adapt the network without fading.
2. Adapt the network with effective sample sizes 25, 10, and 25 for $t, s,$ and u, respectively.
3. Adapt the network to the same cases but without the information on B.

Exercise 6.13. Perform the calculations of Example 6.4.1 by use of a direct representation of the parameters t, s.

Exercise 6.14. Assume that in Example 6.4.1 we require $P(A \mid c) = (0.5, 0.5)$, and assume that $t = 0.6$ is fixed. Use the technique from Example 6.4.1 to tune the parameters s and u.

Exercise 6.15. Let D be a child of C, and let C have parents A and B, all variables being binary. $P(A)$ and $P(B)$ have even distributions; $P(D \mid c) = (0.1, 0.9)$, $P(D \mid \neg c) = (0.6, 0.4)$, and $P(c \mid A, B)$ are as specified in Table 6.10. Tune the parameters t, s to the prescribed behavior $P(a \mid d) = 0.8$.

$B \setminus A$	a	$\neg a$
b	$1 - ts$	$1 - s$
$\neg b$	$1 - t$	0

Table 6.10. The conditional probability table $P(C = c \mid A, B)$ for Exercise 6.15.

7

Learning the Structure of Bayesian Networks

Consider the following situation. Some agent produces samples of cases from a Bayesian network N over the universe $\mathcal{U}$. The cases are handed over to you, and you are asked to reconstruct the Bayesian network from the cases. This is the general setting for structural learning of Bayesian networks. In the real world you cannot be sure that the cases are actually sampled from a "true" network, but this we will assume. We will also assume that the sample is fair. That is, the set $\mathcal{D}$ of cases reflects the distribution $P_N(\mathcal{U})$ determined by N. In other words, the distribution $P_{\mathcal{D}}^{\#}(\mathcal{U})$ of the cases is very close to $P_N(\mathcal{U})$. Furthermore, we assume that all links in N are *essential*, i.e., if you remove a link, then the resulting network cannot represent $P(\mathcal{U})$. Mathematically, it can be expressed as follows: if pa(A) are the parents of A, and B is any of them, then there are two states b_1 and b_2 of B and a configuration $\mathbf{c}$ of the other parents such that $P(A|b_1, \mathbf{c}) \neq P(A|b_2, \mathbf{c})$.

The task is now to find a Bayesian network, M, close to N. In principle this can be done by performing parameter learning for all possible structures, and then selecting as candidates those models for which $P_M(\mathcal{U})$ is close to $P_{\mathcal{D}}^{\#}(\mathcal{U})$. However, by following this very simple approach we are faced with three problems, which are fundamental for learning Bayesian networks. First of all, the space of all Bayesian network structures is extremely large. In fact, it has been shown that the number of different structures, $f(n)$, grows more than exponentially in the number n of nodes (some example calculations can be found in Table 7.1):

$$f(n) = \sum_{i=1}^{n}(-1)^{i+1}\frac{n!}{(n-i)!n!}2^{i(n-i)}f(n-1). \qquad (7.1)$$

Secondly, when searching through the network structures, we may end up with several equally good candidate structures. Since a Bayesian network over a complete graph can represent any distribution over its universe, we know that we will always have several candidates, but a Bayesian network

Nodes	Number of DAGs	Nodes	Number of DAGs
1	1	13	$1.9 \cdot 10^{31}$
2	3	14	$1.4 \cdot 10^{36}$
3	25	15	$2.4 \cdot 10^{41}$
4	543	16	$8.4 \cdot 10^{46}$
5	29281	17	$6.3 \cdot 10^{52}$
6	$3.8 \cdot 10^{6}$	18	$9.9 \cdot 10^{58}$
7	$1.1 \cdot 10^{9}$	19	$3.3 \cdot 10^{65}$
8	$7.8 \cdot 10^{11}$	20	$2.35 \cdot 10^{72}$
9	$1.2 \cdot 10^{15}$	21	$3.5 \cdot 10^{79}$
10	$4.2 \cdot 10^{18}$	22	$1.1 \cdot 10^{87}$
11	$3.2 \cdot 10^{22}$	23	$7.0 \cdot 10^{94}$
12	$5.2 \cdot 10^{26}$	24	$9.4 \cdot 10^{102}$

Table 7.1. The table shows the number of different DAGs that can be generated for a given number of nodes. For example, there exist $4.2 \cdot 10^{18}$ different DAGs with 10 nodes.

over a complete graph will hardly be the correct answer. If so, it is a very disappointing answer.

Thirdly, we have the problem of *overfitting*: the selected model is so close to $P_{\mathcal{D}}^{\#}(\mathcal{U})$ that it also covers the smallest deviances from $P_N(\mathcal{U})$. Again, a complete graph can represent $P_{\mathcal{D}}^{\#}(\mathcal{U})$ exactly, but $\mathcal{D}$ may have been sampled from a sparse network.

There are basically two methods used for learning the structure of Bayesian networks; *constraint-based* and *score-based*. The constraint based methods establish a set of conditional independence statements holding for the data, and use this set to build a network with d-separation properties corresponding to the conditional independence properties determined. The score-based methods produce a series of candidate Bayesian networks, calculate a score for each candidate, and return a candidate of highest score.

To emphasize the focus on structural learning we shall use the following convention: A Bayesian network $M = (S, \boldsymbol{\theta}_S)$ consists of a network structure, S, and a set of parameters, $\boldsymbol{\theta}_S$, where the parameters determine the conditional probabilities of the model. The structure S consists of an acyclic directed graph, $G = (\mathcal{U}, \mathcal{E})$, together with a specification of the state space for each node/variable in the graph.

7.1 Constraint-Based Learning Methods

We shall first consider the following problem: we have to determine the structure of a Bayesian network, and the only source of information is an oracle that correctly answers queries of the type, "is the variable A d-separated from the variable B given the set $\mathcal{X}$?", later we shall replace the oracle with a database for answering the queries. We let $I(A, B, \mathcal{X})$ denote that A is d-separated from B given $\mathcal{X}$. We use $I(A, B)$ as shorthand for $I(A, B, \emptyset)$, and if $\mathcal{X}$ consists of only one element C, we write $I(A, B, C)$.

The method consists in first determining the skeleton of the network, and afterward directing the links.

Definition 7.1. *The* skeleton *of a Bayesian network N is the undirected graph obtained by removing directions from all arcs in N.*

The skeleton can quite easily be established through a series of questions to the oracle: if there is a link between A and B, then they cannot be d-separated. That is, the link $A - B$ is part of the skeleton if and only if $\neg I(A, B, \mathcal{X})$ for all $\mathcal{X}$ not containing A or B. As a starting point, let us assume that we have the skeleton.

7.1.1 From Skeleton to DAG

Consider the skeleton in Figure 7.1(a), and assume that the only conditional independence found is $I(A, B)$. This means that A and B are not independent given C, and therefore Figure 7.1(b) is the only possible directed graph with d-separation properties corresponding to the conditional independences found.

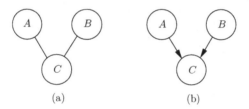

(a)   (b)

Fig. 7.1. (a) A skeleton for the set $\{I(A, B)\}$. (b) the corresponding DAG.

This observation can be generalized as illustrated in Figure 7.2, where C must be a child of A and B if $I(A, B)$ or $I(A, B, D)$.

Rule 1 [introduction of v-structures]: If you have three nodes, A, B, C, such that $A - C$ and $B - C$, but not $A - B$, then introduce the v-structure $A \to C \leftarrow B$ if there exists an $\mathcal{X}$ (possibly empty) such that $I(A, B, \mathcal{X})$ and $C \notin \mathcal{X}$.

As an example, consider the skeleton in Figure 7.3(a) with independences $I(A, B)$, $I(B, C)$, $I(A, B, C)$, $I(B, C, A)$, $I(C, D, A)$, $I(B, C, \{D, A\})$, $I(C, D, \{A, B\})$, $I(B, E, \{C, D\})$, $I(A, E, \{C, D\})$, $I(B, C, \{A, D, E\})$, $I(A, E, \{B, C, D\})$, $I(B, E, \{A, C, D\})$. Consider the chain $C - E - D$. Since E is not a member of any conditioning set yielding C and D independent, we introduce the v-structure $C \to E \leftarrow D$. In the same way we introduce the v-structure

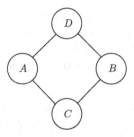

Fig. 7.2. $\{A, C, B\}$ are connected in an undirected chain, but there is another path between A and B. If also $I(A, B)$ or $I(A, B, D)$, then C must be a child of A and B.

$A \rightarrow D \leftarrow B$ (see Figure 7.3(b)) With these two v-structures, there cannot be more of them. This is also confirmed by the conditional independences, and since they give no clue as to the remaining link, $A - C$, it may be oriented in any direction (see Figure 7.3(c)).

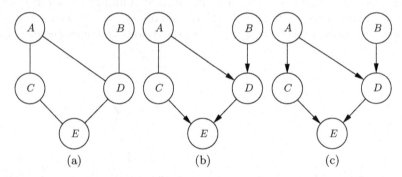

Fig. 7.3. (a) A skeleton. (b) Two v-structures introduced through rule 1. (c) A full DAG

After the v-structures have been introduced, we give a direction to the remaining links using the following rules:

Rule 2 [Avoid new v-structures]: When Rule 1 has been exhausted, and you have $A \rightarrow C - B$ (and no link between A and B), then direct $C \rightarrow B$.

Rule 3 [Avoid cycles]: If $A \rightarrow B$ introduces a directed cycle in the graph, then do $A \leftarrow B$.

Rule 4 [Choose randomly]: If none of the rules 1–3 can be applied anywhere in the graph, choose an undirected link and give it an arbitrary direction.

For example, having found the v-structures in Figure 7.3 (b), we can choose any direction for $A - C$ (Figure 7.3 (c)).

Example 7.1. Consider the graph in Figure 7.4(a). The only v-structure found is $C \to F \leftarrow D$. Rule 2 yields the direction $F \to G$ (Figure 7.4(b)). None of the Rules 1–3 can be applied, and we choose the direction $D \leftarrow E$ (Figure 7.4(c)). Now Rule 2 yields $D \to A$ and $D \to B$ (Figure 7.4(d)), and in turn, Rule 2 yields $A \to C$ (Figure 7.4(e)). Now none of the Rules 1–3 can be applied. We use rule 4 and choose $A \to B$ (Figure 7.4(f)).

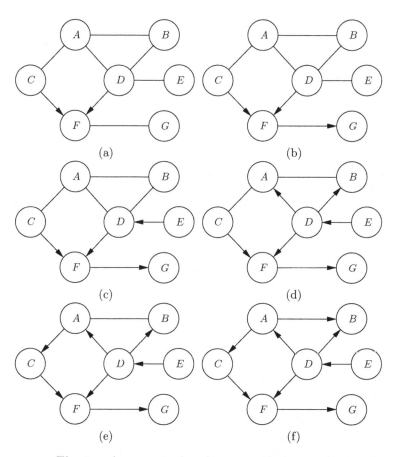

Fig. 7.4. A sequence of applications of Rules 2 and 4.

The application of Rules 2–4 raises various questions. For example, Rule 4 opens up for several solutions. If in the example above we had chosen $D \to E$ rather than $D \leftarrow E$, the solution could have been the DAG in Figure 7.5.

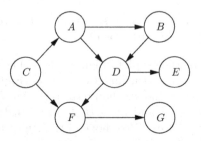

Fig. 7.5. A resulting DAG if we had chosen $D \to E$ rather than $D \leftarrow E$.

A solution represents a family of probability distributions over the universe $\mathcal{U}$; a distribution for each setting of the parameters. From a statistical point of view, the various solutions are equivalent; they have the same d-separation properties. Equivalently, they represent the same family of distributions. Furthermore, a maximal likelihood setting of the parameters in one graph will have a corresponding parameter setting in any other graph, and this parameter setting is also of maximal likelihood.

A more fundamental problem is whether there in fact is a solution. If the oracle is reliable, then the skeleton and the v-structures are correct, and therefore there must be a way to direct the remaining links so that the generative model is established. Moreover, any other solution will also be valid (see Section 7.3.2). Finally, you might fear that the choice in Rule 4 may lead us into a blind alley. However, it has been proven that this will not happen.

7.1.2 From Independence Tests to Skeleton

Since consulting the oracle has a price, we wish to reduce the number of questions. We use the answers from the oracle when establishing the skeleton and when introducing v-structures. The following theorem helps to reduce the number of questions.

Theorem 7.1. *The nodes A and B are not linked in N if and only if $I(A, B, \mathrm{pa}(A))$ or $I(A, B, \mathrm{pa}(B))$.*

Proof.   Clearly, if $I(A, B, \mathcal{X})$ for any $\mathcal{X}$, then A and B are not linked.

Assume now that A and B are not linked in N, and construct the ancestral graph for $\{A, B\}$ (see Section 2.2.1). If there is a path in this graph from B to A not passing through $\mathrm{pa}(B)$, then B is an ancestor of A, and all paths from B to A must pass through $\mathrm{pa}(A)$.

$\square$

The theorem ensures that it is sufficient to ask questions of the form $I(A, B, \mathcal{X})$, where $\mathcal{X}$ is a subset of A's or B's neighbors. It is used in the *PC algorithm* to focus on local independence questions.

Algorithm 7.1 [The PC algorithm: test sequence]

1. *Start with the complete graph;*
2. *i := 0;*
3. **while** *a node has at least $i + 1$ neighbors*
   - **for all** *nodes A with at least $i + 1$ neighbors*
     - **for all** *neighbors B of A*
       - **for all** *neighbor sets $\mathcal{X}$ such that $|\mathcal{X}| = i$ and $\mathcal{X} \subseteq (\text{nb}(A)\backslash\{B\})$*
         - **if** $I(A, B, \mathcal{X})$ **then** *remove the link $A - B$ and store "$I(A, B, \mathcal{X})$"*
   - $i := i + 1$

□

7.1.3 Example

Assume that the cases are a faithful sample of the Bayesian network in Figure 7.6(a). We start with the complete graph in Figure 7.6(b) and ask the questions $I(A, B)?$, $I(A, C)?$, $I(A, D)?$, $I(A, E)?$, $I(B, C)?$, $I(B, D)?$, $I(B, E)?$, $I(C, D)?$, $I(C, E)?$, $I(D, E)?$.

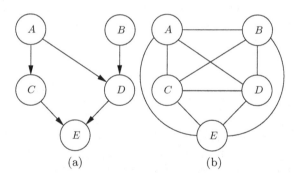

(a)   (b)

Fig. 7.6. (a) The Bayesian network from which the cases have been sampled.(b) The starting graph for the PC algorithm.

We get "yes" for $I(A, B)?$ and $I(B, C)?$; the links $A - B$ and $B - C$ are removed (see Figure 7.7(a)), and i is set to 1.

We now ask $I(A, C, E)?$, $I(B, C, D)?$, $I(B, C, E)?$, $I(B, D, C)?$, $I(B, D, E)?$, $I(B, E, C)?$, $I(B, E, D)?$, $I(C, B, A)?$, $I(C, D, B)?$, $I(C, D, A)?$. The last question has the answer "yes"; we remove the link $C - D$ and continue; $I(C, E, A)?$, $I(C, E, B)?$, $I(D, B, E)?$, $I(D, E, B)?$, $I(E, A, B)?$, $I(E, A, D)?$, $I(E, B, A)?$, $I(E, C, B)?$, $I(E, C, D)?$, $I(E, D, A)?$, $I(E, D, C)?$.

Next, for $i = 2$ we ask questions like $I(A, C, \{D, E\})?$, and we get affirmative answers for $I(B, E, \{C, D\})?$ and $I(A, E, \{C, D\})?$. The result is shown

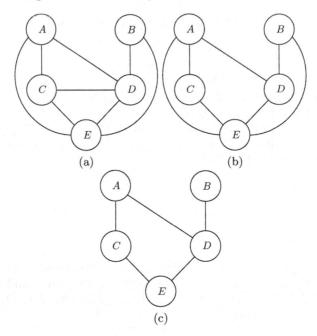

Fig. 7.7. (a) The result after testing all unconditional independences ($i = 0$). (b) After testing with a single conditioning variable. (c) After testing with two conditioning variables.

in Figure 7.7(c). Setting $i = 3$ we realize that no node has four neighbors, and the algorithm terminates.

To sum up, we get the skeleton in Figure 7.7(c) together with the conditional independences $I(A,B)$, $I(B,C)$, $I(C,D,A)$, $I(A,E,\{C,D\})$, and $I(B,E,\{C,D\})$. They are sufficient for applying Rules 1–4.

The PC-algorithm has the following property, which is easily seen from the construction and Theorem 7.1.

Property 1: If the case set is a faithful sample from a Bayesian network, N, then the graph resulting from the PC-algorithm is the skeleton of N.

We also have the following property, which allows us to establish the direction of the arcs.

Property 2: The conditional independences found by the PC-algorithm are sufficient for determining the v-structures.

Let namely $A - C - B$ be a chain, and assume that the PC-algorithm found $I(A,B,\mathcal{X})$. We know that the two links are part of the skeleton, and if $C \notin \mathcal{X}$ then the only way to direct the links will be to introduce the v-structure $A \to C \leftarrow B$. On the other hand, if $C \in \mathcal{X}$ we cannot have a v-structure.

The Necessary Path Condition

The number of queries to the oracle can be further reduced. Consider the situation in Figure 7.8, where the links $A - D$ and $A - C$ have been removed. Then we need not ask for $I(A, B, D)$ (or $I(A, B, C)$), since no path between A and B passes D (similar for C). This is called the *necessary path condition*: only ask $I(A, B, \mathcal{X})$ for sets $\mathcal{X}$, where all members of $\mathcal{X}$ occur on a path between A and B.

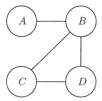

Fig. 7.8. D cannot block any path between A and B.

7.1.4 Constraint-Based Learning on Data Sets

When learning structure, you do not have an oracle for queries of the type $I(A, B, \mathcal{X})$. Instead, you have a data set $\mathcal{D}$, which you may analyze for conditional independences. We shall use the notation $I_\mathcal{D}(A, B, \mathcal{X})$ for conditional independence in the distribution determined by $\mathcal{D}$. We shall assume that $\mathcal{D}$ is sampled from a Bayesian network N.

Definition 7.2. $\mathcal{D}$ *is a* faithful sample *from N if the following holds: A and B are d-separated in N given $\mathcal{X}$ if and only if $I_\mathcal{D}(A, B, \mathcal{X})$.*

If $\mathcal{D}$ is faithful to N, we can use a test for independence in $\mathcal{D}$ as oracle. For this, *conditional mutual information* can be used.

$$\mathrm{CMI}(A, B|\mathcal{X}) = \sum_\mathcal{X} P^\#(\mathcal{X}) \sum_{A,B} P^\#(A, B|\mathcal{X}) \log_2 \frac{P^\#(A, B|\mathcal{X})}{P^\#(A|\mathcal{X})P^\#(B|\mathcal{X})}. \quad (7.2)$$

It holds (Exercise 7.5) that

$$I_\mathcal{D}(A, B, \mathcal{X}) \Leftrightarrow \mathrm{CMI}(A, B|\mathcal{X}) = 0.$$

Based on the data set, the oracle will calculate an estimate of $\mathrm{CMI}(A, B|\mathcal{X})$, and then it performs a χ^2-test on the hypothesis $\mathrm{CMI}(A, B|\mathcal{X}) = 0$, and the user decides on a significance level. A high significance level means that fewer

links are removed. Because any test has false positives as well as false negatives, there is a risk that links that should have been removed are not removed,
and vice versa. The error rate is closely related to the sample size. The smaller
the sample size, the more independences will be accepted and the fewer links
inserted.

In a real-world learning situation, you may, for example, get into the situation illustrated in Figure 7.9: you have $\neg I(A,B), \neg I(A,C), \neg I(B,C)$, but
$I(A,B,C), I(A,C,B), I(B,C,A)$; hence you cannot direct the links without
violating the independences found by the test. Then a solution may be to
remove one link and direct the remaining as a chain.

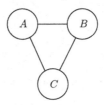

Fig. 7.9. The tests yield all pairs dependent, but all pairs independent given the
third variable.

This is called an *uncertain region*: removal of a link is dependent on how
you treat the other links. Note that for this example, the PC algorithm will
stop after $I(A,B,C)$ and $I(A,C,B)$ and removal of the links $A-B$ and $A-C$.
If the necessary path condition is used, the process will stop after $I(A,B,C)$
and removal of the link $A-B$.

There may be other reasons why it is not possible to direct links without
violating some of the independences returned by the tests. Assume you have
the four variables A,B,C,D, and you get the independences $I(A,C), I(A,D)$,
and $I(B,D)$ for $i=0$. Then the PC algorithm extended with the necessary
path condition will stop, and you have the skeleton in Figure 7.10. Now there
is no proper way of directing the links.

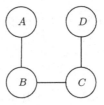

Fig. 7.10. A skeleton that cannot be directed.

Rule 1 grants the introduction of two v-structures, $A \rightarrow B \leftarrow C$ and $B \rightarrow C \leftarrow D$; but then the link $B - C$ receives two directions. For this particular case, the inconsistency need not be due to the test, but it can be caused by a hidden variable as illustrated in Figure 7.11.

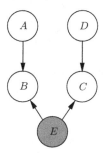

Fig. 7.11. The problem illustrated in Figure 7.10 may be caused by a hidden variable (E).

You cannot always assume that a problem related to directing the skeleton is due to erroneous tests or hidden variables. It may happen that the cases have not been sampled from a Bayesian network. Anyhow, you have to enforce directions inconsistent with the test results. Beware that violating dependence results makes it impossible to represent the joint probability distribution of the case set.

It is tempting to conclude that the PC algorithm discovers causality from observed data. for a century it has been a commonly accepted view that causality can be discovered only through controlled experiments, where an outside agent fixes some variables to certain states. The new algorithms for learning Bayesian network structures have questioned this view. The PC algorithm (and other preceding constraint-based algorithms) works on observed nonmanipulated data, and it allows you to introduce v-structures. However, you can conclude that you have discovered a causal relation only if you can be sure that there are no hidden variables obscuring the picture.

For example, consider the structure in Figure 7.12 with D and E hidden. The PC algorithm will yield $I(A,C)$, $\neg I(A,B)$, $\neg I(B,C)$ and stop. However, A and C are not causes of B. We shall not go deeper into this very lively and interesting discussion.

Finally, it shall be mentioned that even a completely correct statistical test for independence may not provide the correct d-separation properties (even if you have a very large database); the conditional probabilities in the network may hide dependencies. Take for example two switches A and B for the light C. The light is on if and only if A and B are in the same position. The prior probabilities for A and B are even. Although both links in this example

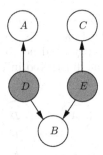

Fig. 7.12. A structure with *confounding variables*:: D and E are hidden and obscure the learning of causality.

are essential, then for a fair sample $\mathcal{D}$ we have $I_{\mathcal{D}}(A, C)$ and $I_{\mathcal{D}}(B, C)$; the problem is that the faithfulness assumption is violated.

7.2 Ockham's Razor

When learning structure from experiments, there is a general principle of inductive learning, called *Ockham's razor* (after William of Ockham, 1285–1349). It recommends that one choose the simplest hypothesis consistent with the observations.

In the case of learning Bayesian networks, this principle has a justification of its own. The complexity of a Bayesian network can be measured in number of links or in number of independent parameters.

Proposition 7.1. *Let M be a Bayesian network over the variables $\mathcal{U}$, and assume that the parameters $\boldsymbol{\theta}_M$ are both locally and globally independent (see Section 6.3.1). Then the number of independent parameters (or the* size *of M) is given by:*

$$\text{size}(M) = \sum_{X \in \mathcal{U}} |\text{pa}(X)| \cdot (|\text{sp}(X)| - 1). \qquad (7.3)$$

For example, assuming that all variables are binary, then the size of the model in Figure 7.12 is $1+2+1+4+2 = 10$. On the other hand, when the assumption about either local and global parameter independence is violated, then the number of independent parameters is usually lower.

Proposition 7.2. *Let N be a Bayesian network over $\mathcal{U}$ with only essential links. Then no other Bayesian network M representing $P_N(\mathcal{U})$ can have fewer links or a smaller size than N.*

Proof. Let M represent $P(\mathcal{U})$. Since all links are essential, it must hold that whenever A and B are linked in N they are also in M. If there is a chance for M to have smaller size than N, then it must be because some links in M have the direction opposite to that of the corresponding links in N.

Let L be a link from A to C, which is reversed. For simplicity we assume that C has only one parent more, and that A has a single parent (See Figure 7.13(a)). Figure 7.13(b) depicts the situation in which the link has been reversed.

In Figure 7.13(b) we have that C and D are independent, and A and B are independent given C. To compensate for this, M must have extra links. The cheapest will be to add a link from C to D and from B to A (See Figure 7.13(c)). Elementary arithmetic (see Exercise 7.6) now yields that the size of the Bayesian network in Figure 7.13(c) is larger than for the Bayesian network in Figure 7.13(a). Equality is only possible if A has no parents, and A is the only parent of C.   □

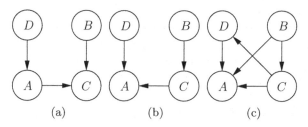

Fig. 7.13. In (a), C and D are dependent. If the link between A and C is reversed, then C and D become independent (b), and to compensate for this you can insert extra links (c).

Note that the proposition does not hold if we count probability parameters rather than size. Note also that we use conditional independence properties rather than probabilities in the proof.

The proposition justifies a search for minimal models: If the real world is a Bayesian network (with all links essential), and if the sample set is faithful, then among all the models representing the distribution, the true one is minimal with respect to links as well as size.

7.3 Score-Based Learning

When doing structural learning, we look for a Bayesian network structure that on the one hand can represent our database sufficiently well (when augmented with a set of probabilities) and on the other hand is not overly complex. In Section 7.1 we saw how to perform structural learning based on independence tests, and in this section we shall focus on another type of learning, called *score-based learning*. Score-based learning assigns a number (a score) to each

Bayesian network structure. The score reflects the "usefulness" of a structure, where the term "usefulness" can for example cover how likely it is that the structure could have been used to generate the database at hand.

If we have a score function that takes a Bayesian network structure as argument and returns a value, then the task of score-based learning can be considered a search problem: we simply look for the model structure with the highest score. This also means that a score based learning algorithm can in principle be completely described by specifying two components, (1) a score function, and (2) a search procedure.

7.3.1 Score Functions

When specifying a score for a network structure S with respect to a database $\mathcal{D}$, your first attempt might be to consider the Euclidean distance (see Definition 6.1) between the probability distribution, $P_D^{\#}(\mathcal{U})$, represented by the database $\mathcal{D}$ and your "best shot" at the probability distribution that can be encoded in S over the same set of variables. By "best shot" we mean the conditional probabilities for S that bring $P_S(\mathcal{U})$ closest to $P_D^{\#}(\mathcal{U})$. An immediate attempt might be to use the maximum likelihood estimates $\hat{\boldsymbol{\theta}}_S$ (see Section 6.1), in which case the distance measure can be specified as

$$\text{dist}\left(P_D(\mathcal{U}), P_S(\mathcal{U} \mid \hat{\boldsymbol{\theta}})\right) = \sum_{\mathbf{x} \in \text{sp}(\mathcal{U})} \left(P_D(\mathbf{x}) - P_S(\mathbf{x} \mid \hat{\boldsymbol{\theta}}_S)\right)^2.$$

Unfortunately, there are (at least) two rather severe problems in using this distance as the score of a structure. First of all, since a complete network structure can encode any probability distribution, we know that in order to minimize the distance above, we should simply select any complete network structure; this is obviously not satisfactory. To avoid this problem you could augment the score with a term penalizing model complexity (see Proposition 7.2). This means that the score of a structure should be defined as a trade-off between how good it is at representing the distribution encoded by the database and the complexity/size of the structure. A possible suggestion for such a score could then be

$$\text{score}\left(P_D(\mathcal{U}), P_S(\mathcal{U} \mid \hat{\boldsymbol{\theta}})\right) = \text{dist}\left(P_D(\mathcal{U}), P_S(\mathcal{U} \mid \hat{\boldsymbol{\theta}})\right) + c \cdot \text{size}(S),$$

where c is a (user specified) constant used to control the trade-off between model accuracy and model complexity.

However, even though we may have found a suggestion for a score function that reliably reflects the usefulness of a structure, we still have another problem to address: from a computational perspective, the Euclidean distance can be extremely difficult to work with, since it is a function of $P_D^{\#}(\mathcal{U})$. That is, it basically requires us to deal with a table over the joint state space of all the variables, and we are therefore faced with the same combinatorial explosion, which we again and again try to avoid.

To summarize the discussion above, we look for a score function that should (at least) have the following properties:

- It should balance the accuracy of a structure with the complexity of the structure.
- It should be computationally tractable to evaluate.

The Bayesian Information Criterion

An example of a score function satisfying the above two properties is the *Bayesian information criterion* (BIC), which contains a term measuring how well the data fits the model as well as a term that accounts for model complexity:[1]

$$\text{BIC}(S \mid \mathcal{D}) = \log_2 P(\mathcal{D} \mid \hat{\theta}_S, S) - \frac{\text{size}(S)}{2} \log_2(N), \tag{7.4}$$

where $\hat{\theta}$ is an estimate of the maximum likelihood parameters for the structure S. If we furthermore assume that the cases are independent given the model, then

$$\text{BIC}(S \mid \mathcal{D}) = \sum_{i=1}^{N} \log_2 P(\mathbf{d}_i \mid \hat{\theta}_S, S) - \frac{\text{size}(S)}{2} \log_2(N).$$

In order to score a model using BIC, you start off by estimating the maximum likelihood parameters for the model. If the database is complete, then this is just a matter of frequency counting, but if some of the cases contain missing values you may run the EM algorithm. Based on these estimates you calculate the probability for each case in the database. This can be done by simply inserting the case as evidence in the Bayesian network and performing a propagation; the probability of the case is then the probability of the evidence. If all the cases are complete, then this task is even simpler, you just multiply the appropriate entries in the conditional probability tables, which, in turn, are frequency counts derived from the database! This also means that the calculation of the BIC score has been reduced to a counting problem: let r_i denote the number of states for variable X_i, and let $q_i = \prod_{X_l \in \Pi_i} r_l$ denote the number of configurations over the *parents* for X_i in S (if X_i does not have any parents then we let $q_i = 1$). With this notation we now have (the derivation is left as an exercise)

$$\text{BIC}(S \mid \mathcal{D}) = \sum_{i=1}^{n} \sum_{j=1}^{q_i} \sum_{k=1}^{r_i} N_{ijk} \log_2 \left(\frac{N_{ijk}}{N_{ij}} \right) - \frac{\log_2 N}{2} \sum_{i=1}^{n} q_i(r_i - 1), \tag{7.5}$$

where N_{ijk} denotes the number of cases in the database with X_i in its kth configuration and $\text{pa}(X_i)$ in the jth configuration.

---

[1] The exact form of the BIC score can be derived from a Taylor expansion of $P(\mathcal{D} \mid S)$.

Example 7.2. Consider the two Bayesian network structures over the two binary variables X_1 and X_2 shown in Figure 7.14 (we shall refer to them as
B_a and B_b, respectively), and assume that we have the database shown in
Table 7.2.

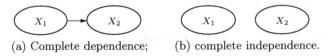

(a) Complete dependence;   (b) complete independence.

Fig. 7.14. Two BN model structures for the domain $\mathbf{X} = (X_1, X_2)$.

Case	X_1	X_2
1.	yes	positive
2.	yes	positive
3.	yes	positive
4.	yes	positive
5.	yes	positive
6.	yes	positive
7.	yes	negative
8.	yes	negative
9.	no	negative
10.	no	negative

	yes	no
X_1	8	2

		X_1	
		yes	no
X_2	pos	6	0
	neg	2	2

Table 7.2. A database for the two binary variables X_1 and X_2 as well as the counts
N_{11k} and N_{2jk} derived from the database.

In order to calculate the BIC score for B_a we first calculate the counts
(the states *yes* and *positive* correspond to state number 1) shown in Table 7.2.
By substituting these values into equation (7.5), we get

$$\text{BIC}(S \mid \mathcal{D})$$

$$= \left[8 \cdot \log \left(\frac{8}{8+2} \right) + 2 \cdot \log \left(\frac{2}{8+2} \right) + 6 \cdot \log \left(\frac{6}{6+2} \right) + 2 \cdot \log \left(\frac{2}{6+2} \right) \right.$$

$$\left. + 0 \cdot \log \left(\frac{0}{0+2} \right) + 2 \cdot \log \left(\frac{2}{0+2} \right) \right] - \frac{1+2}{2} \log (10)$$

$$= -18.69.$$

For the network BN_b we calculate the following counts, which can be read
and derived from the counts in Table 7.2: $N'_{111} = 8$, $N'_{112} = 2$, $N'_{211} = N_{211} + N_{221} = 6$ and $N'_{212} = N_{212} + N_{222} = 4$. This gives us

$\mathrm{BIC}(S \mid \mathcal{D})$

$$= \left[8 \cdot \log \left(\frac{8}{8+2} \right) + 2 \cdot \log \left(\frac{2}{8+2} \right) + 6 \cdot \log \left(\frac{6}{6+4} \right) + 6 \cdot \log \left(\frac{4}{6+4} \right) \right]$$
$$- \frac{1+1}{2} \log(10)$$
$$= -20.25.$$

That is, according to the BIC score we should choose B_a rather than B_b.

7.3.2 Search Procedures

Given a score function, the task is to find the highest-scoring Bayesian network structure among the set of all possible network structures. That is, the task of structural learning has been reduced to a search problem. The challenging part of this problem is that the size of the space of all structures is super-exponential in the number of nodes (see equation (7.1)) so an exhaustive enumeration of all the structures is not possible.

Instead, researchers have considered heuristic search strategies that move around in the search space by iteratively performing small changes to the current structure. Most commonly, these search methods work directly on the space of Bayesian network structures; hence each point in such a *search space* corresponds to a particular DAG; in the remainder of this section we shall use the terms structure and DAG interchangeably, since the state spaces of the variables in the structure is fixed.

The definition of the search space determines the definition of the *search operators* used to move from one structure to another. In turn, these operators determine the *neighborhood* of a DAG, namely the DAGs that can be reached in one step from the current DAG. Typically, the operators consist of:

- *arc addition*: insert a single arc between two nonadjacent nodes.
- *arc deletion*: remove a single arc between two nodes.
- *arc reversal*: reverse the direction of a single arc.

In what follows we let $op(S, A)$ represent the result of performing the arc operation A on the structure S, i.e., $op(S, A)$ is a DAG that differs from S with respect to one arc.

One important property of these operators is that they result only in local changes to the current structure; for example, if an arc is inserted from node X_i to X_j, then only the family of node X_j is changed, and similarly if an arc is deleted; if an arc is reversed, then the families of both X_i and X_j are changed. This property can be exploited when we have a so-called decomposable score function.

Definition 7.3. *A score function is said to be* decomposable *if it can be expressed as a sum of local scores, one for each family of nodes in the structure:*

$$\text{score}(\mathcal{D}, S) = \sum_{i=1}^{n} \text{score}(X_i, \text{pa}(X_i), \mathcal{D}).$$

The BIC score is an example of a decomposable score function for complete data, since it can be written as

$$\text{BIC}(S \mid \mathcal{D}) = \sum_{i=1}^{n} \left[\sum_{j=1}^{q_i} \sum_{k=1}^{r_i} N_{ijk} \log \left(\frac{N_{ijk}}{N_{ij}} \right) - \frac{1}{2} q_i (r_i - 1) \log N \right].$$

This decomposition property can be used when we evaluate the benefit of making an arc change. For example, if we insert an arc from X_i to X_j, then only the local score for X_j will change, i.e., when evaluating whether such a move is beneficial we need to evaluate only the score difference (or gain)

$$\Delta(X_i \to X_j) = \text{score}(X_j, \text{pa}(X_j) \cup \{X_i\}, \mathcal{D}) - \text{score}(X_j, \text{pa}(X_j), \mathcal{D}). \quad (7.6)$$

Greedy Search

A simple heuristic search procedure is greedy search: choose some initial structure (usually the empty structure, a randomly chosen structure, or a prior structure specified by the user) and calculate the gain for each legal arc operation; by legal we mean that the resulting graph should be acyclic. Next, perform the arc operation A with highest gain (if positive) and use the resulting model as your current model. More formally:

Algorithm 7.2 [Greedy search]

1. *Let S be an initial structure.*
2. *Repeat*
   a) *Calculate $\Delta(A)$ for each legal arc operation A*
      - *Let $\Delta^* = \max_A \Delta(A)$ and $A^* = \arg\max_A \Delta(A)$.*
   b) *If $\Delta^* > 0$, then*
      - *Set $S = op(S, A^*)$.*
3. *Until $\Delta^* \leq 0$.*

□

It should be noted that in the greedy algorithm above you can further exploit the decomposition property of the score function: If the parents sets of two nodes, say X_i and X_j, do not change from one iteration to another, then the gain (equation (7.6)) of any arc operation involving X_i and X_j will remain unchanged. This gain can therefore be cached for subsequent iterations so that the calculations can be reused.

Obviously, when we work with heuristic search algorithms we are not guaranteed to find a global optimal structure but only a local optimal structure. Several methods have been proposed to escape local maxima. An example of this is greedy search with multiple restarts: after a local maximum is found the search is reinitialized with a random structure. This reinitialization is then repeated for a fixed number of iterations, and the best structure found throughout the entire process is selected.

Prior Information

A way of reducing the search space (and thereby also the risk of ending up in a local maximum) is to incorporate prior information, thus constraining the models under investigation.

There are various standard ways of constraining the models to consider. First, causality can be exploited. If possible, the nodes are clustered in a causal hierarchy. You may, for example, consider a medical domain in which you have disease nodes **D**, symptom nodes **S**, risk factor nodes **R**, and treatment nodes **T**. Then, you need not consider links from a node in **S** to a node in **T**. The full hierarchy is shown in Figure 7.15.

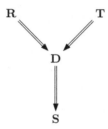

Fig. 7.15. A causal hierarchy for clusters of nodes. Directed links are allowed only inside a cluster or downward in the hierarchy.

If **R** and **T** have two nodes, and **D** and **S** have three nodes, then the hierarchy allows approximately 10^{15} different DAGs. This is a considerable reduction compared to $4.2 \cdot 10^{18}$, but still it is extremely many. This prior knowledge could then be included directly in the search algorithm by considering an arc operation as being legal only if it adheres to the causal hierarchy.

A more general approach could be to specify a partial ordering, $\preceq$, over the variables, such that we allow an arc from X_i to X_j only if $X_i \preceq X_j$. In the special case that we have a linear ordering, then the i'th node can have at most $i - 1$ parents producing 2^{i-1} different parent sets. The number of structures consistent with the ordering is therefore

$$\prod_{i=1}^{n} 2^{n-1} = 2^{\sum_{i=1}^{n-1} i} = 2^{n(n-1)/2}.$$

Although the number of structures is still exponential, specifying a linear ordering provides a substantial reduction. For example, with 10 nodes we have

$3.5 \cdot 10^{13}$ different structures as opposed to $4.2 \cdot 10^{18}$ in the unrestricted case. Whether it is reasonable to specify such an ordering is heavily dependent on the domain in question. However, you could imagine different rules of thumb. For instance, if the variables represent events that manifest themselves at different points in time, then you may be able to order the variables according to these time points. An example of this could be variables representing components in a physical production process, where there is a time delay for an item to move from one component to another.

Finally, you could also use more specific expert statements when reducing the model space. All positive as well as negative statements on the presence of links reduce the number by a factor between 2 and 3. Consider again the medical domain above. If, for example, the expert states that the nodes in **D** are independent given **R** and **T** (three links missing), then the model space is reduced by a factor of 25.

*Equivalence Class Search

It can sometimes be advantageous to define the search space using a more abstract representation than DAGs. An example is a procedure called *greedy equivalence search*.

The search is based on the observation that data alone cannot be used to discriminate among structures with the same d-separation properties (see also Section 7.1.1).

Definition 7.4. *Two network structures B_1 and B_2 are said to be* equivalent *if they have the same d-separation properties.*

The equivalence relation is reflexive, symmetric, and transitive; hence the relation defines a collection of *equivalence classes*.

A score function assigning the same score to equivalent structures is said to be *score equivalent*; the BIC score is an example of a score-equivalent function (see Figure 7.16). This also means that if we have identified a particular structure using a score equivalent function, then we could just as well pick any other structure equivalent to the one identified. A way of making this observation explicit is to define the search space such that each point in the search space corresponds to an equivalence class.

In order to move around in the space of equivalence classes, we should also specify a set of search operators, but due to the nature of the search space, these operators are a bit more complex than the ones used in DAG spaces. Instead we shall only define the *neighborhood* for an equivalence class: the set of structures reachable by a single change to the current structure or one of its equivalents. Since the equivalence classes are defined in terms of independence statements, we define the neighborhood of an equivalence class in this way. We have an *upper neighborhood* consisting of equivalence classes with fewer dependence statements, and a *lower neighborhood* with more dependence statements. The two neighborhoods are defined as the equivalence

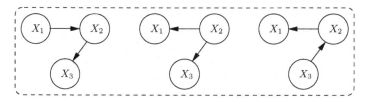

Fig. 7.16. The three DAGs constitute an equivalence class, and any score equivalent scoring function will assign the same score to all three structures.

classes that can be obtained by either adding or deleting a single arc from a DAG in the current equivalence class. Figure 7.17 illustrates the different equivalence classes for three variables; the arcs attached to an equivalence class identify the upper and lower neighborhoods.

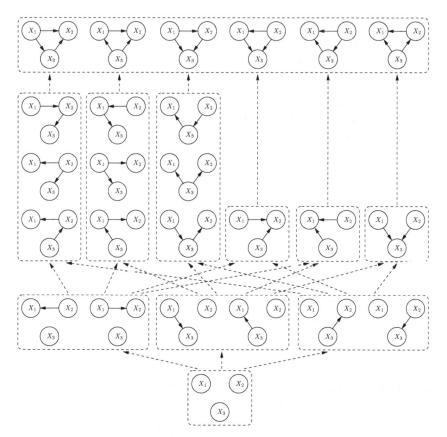

Fig. 7.17. The equivalence class hierarchy for the possible structures over three variables.

Based on this specification of the search space, the greedy equivalence search algorithm consists of two steps: First, start with the equivalence class representing no dependencies among the variables (the bottom equivalence class in Figure 7.17), and perform a greedy search upward until a local maximum is reached. Next, starting from the equivalence class just identified, perform a greedy search downward until a local maximum is reached. It has been proved that if the database is sufficiently large, then the resulting equivalence class is guaranteed to include the Bayesian network from which the data was generated.

Finally, it should be emphasized that even though we have made another specification of the search space, we have unfortunately not solved the general complexity problem that we faced in DAG spaces: the number of equivalence classes also grows super-exponentially in the number of variables.

7.3.3 Chow–Liu Trees

The BIC score function incorporates a penalty term to control model complexity. Another way of dealing with this issue is to put restrictions on the allowable network structures so that overly complex structures are not considered. A particular simple class of Bayesian network structures is the set of tree-shaped structures, where each node is allowed at most one parent (see Figure 7.18). Not only is probability updating very easy in these networks, but Chow and Liu also showed that a network of maximal likelihood can be learned efficiently from a database; due to this result, these tree structures are also called *Chow–Liu trees*.

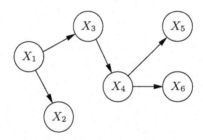

Fig. 7.18. An example of a Chow–Liu tree structure.

Theorem 7.2 (Learning of Chow-Liu trees). *Let $\mathcal{D}$ be a data set over the variables $\{X_1, \ldots, X_n\}$. A Chow-Liu tree of maximal likelihood can be constructed as follows:*

1. *Calculate the mutual information $\mathrm{MI}(X_i, X_j)$ for each pair (X_i, X_j).*
2. *Consider the complete MI-weighted graph: the complete undirected graph over $\{X, \ldots, X_n\}$, where the links (X_i, X_j) have the weight $\mathrm{MI}(X_i, X_j)$.*

3. Build a maximal-weight spanning tree for the complete MI-weighted graph.
4. Direct the resulting tree by choosing any variable as a root and setting the directions of the links to be outward from it.
5. Learn the parameters.

Notes:

- The likelihood of a Bayesian network B given a data set $\mathcal{D}$ is the same as described in Section 7.3.1: $P(\mathcal{D}|B)$.
- The formula for mutual information is

$$\mathrm{MI}(X, Y) = \sum_{X,Y} P(X, Y) \log_2 \left(\frac{P(X, Y)}{P(X)P(Y)} \right).$$

- A maximal-weight spanning tree can be constructed through Kruskal's algorithm: choose repeatedly a link of maximal weight not producing a cycle.
- Calculating the mutual information for a pair of variables requires one sweep through the data. If the database consists of N cases, then this can be done in time $O(N)$, and since we need to perform this calculation for all pairs of variables, the overall time complexity of the Chow–Liu algorithm becomes $O(n^2 \cdot N)$.

Example 7.3. Consider the *Cold or Angina* problem described in Section 3.1.2 and assume that we have a database of cases from this domain. For simplicity we assume that the cases are a faithful sample from the model in Figure 3.6, with the probabilities specified as in Section 3.2.5, Table 3.15 (Page 76) and Table 3.20 (Page 96).

In order to learn a Chow–Liu tree for this domain, we start by calculating the mutual information between each pair of variables (the following calculations are based on the specified model). For the variables *Cold* and *SoreThroat? (Sore)* we get

$$\mathrm{MI}(\mathit{Cold}, \mathit{Sore}) = \sum_{\mathit{Cold}, \mathit{Sore}} P(\mathit{Cold}, \mathit{Sore}) \log_2 \left(\frac{P(\mathit{Cold}, \mathit{Sore})}{P(\mathit{Cold})P(\mathit{Sore})} \right)$$
$$= 0.02101216.$$

The mutual information for all pairs of variables is given in Table 7.3.

Based on these calculations we can construct a maximal-weight spanning tree by starting from the empty graph, and iteratively adding an edge with maximum weight as long as no cycle is created. The resulting structure is shown in Figure 7.19(a), and by picking *Fever?* as a root and directing the edges away from *Fever?* we obtain the Chow–Liu tree in Figure 7.19(b).

Since the learned model has a tree structure, the model may specify (conditional) independences that are not reflected in the data. On the other hand,

MI($Cold, Angina$) = 0	MI($Fever?, Angina$) = 0.015076
MI($SoreThroat?, Angina$) = 0.018016	MI($SeeSpots?, Angina$) = 0.0180588
MI($Cold, Fever?$) = 0.014392	MI($Cold, SoreThroat?$) = 0.0210122
MI($Cold, SeeSpots?$) = 0	MI($SoreThroat, Fever?$) = 0.0015214
MI($Fever?, SeeSpots?$) = 0.0017066	MI($SeeSpots?, SoreThroat?$) = 0.0070697

Table 7.3. The mutual information for each pairs of variables.

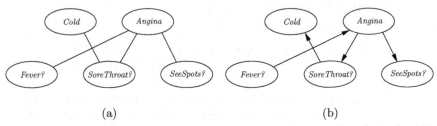

(a)   (b)

Fig. 7.19. Figure (a) shows a maximal weight spanning tree based on the MI-calculations in Table 7.3. Figure (b) shows the Chow–Liu tree obtained by selecting the variable *Fever?* as root and directing the edges away from *Fever?*.

even though the independence properties are inaccurate, it has turned out that the model may still provide a good approximation. We shall return to this issue in Chapter 8, where we will use Chow–Liu trees in a classification context.

Proof. [Learning of Chow–Liu trees, Theorem 7.2]

The proof involves some pencil pushing. First we rewrite the log-likelihood of a Bayesian network B given the data $\mathcal{D} = (\mathbf{d}_1, \ldots, \mathbf{d}_N)$:

$$\log_2 P(\mathcal{D} \mid B) = \log_2 \prod_{j=1}^{N} P(\mathbf{d}_j \mid B) = \sum_{j=1}^{N} \log_2 P(\mathbf{d}_j \mid B)$$

$$= \sum_{j=1}^{N} \sum_{i=1}^{n} \log_2 P(X_i = \mathbf{d}_j \mid \mathrm{pa}(X_i) = \mathbf{d}_j, B).$$

The number of cases in $\mathcal{D}$ that agree on a particular configuration of X_i and $\mathrm{pa}(X_i)$ is given by $N(X_i, \mathrm{pa}(X_i)) = N \cdot P^{\#}(X_i, \mathrm{pa}(X_i))$. Hence instead of summing over all the cases, we can write

$$\log_2 P(\mathcal{D} \mid B) = N \cdot \sum_{i=1}^{n} \sum_{X_i, \mathrm{pa}(X_i)} P^{\#}(X_i, \mathrm{pa}(X_i)) \cdot \log_2 P(X_i \mid \mathrm{pa}(X_i), B).$$

Since we are looking for a Bayesian network of maximal likelihood, we can assume that the parameters of B are maximum likelihood parameters (see Section 6.1.1), i.e., $P(X \mid \mathrm{pa}(X), B) = P^{\#}(X \mid \mathrm{pa}(X))$, and therefore

$$\log_2 P(\mathcal{D} \mid B) = N \cdot \sum_{i=1}^{n} \sum_{X_i, \mathrm{pa}(X_i)} P^{\#}(X_i, \mathrm{pa}(X_i)) \cdot \log_2 P^{\#}(X_i \mid \mathrm{pa}(X_i)).$$

This equation can be rewritten as

$$\log_2 P(\mathcal{D} \mid B) = N \cdot \sum_{i=1}^{n} \sum_{X_i, \mathrm{pa}(X_i)} P^{\#}(X_i, \mathrm{pa}(X_i)) \cdot \left(\log_2 \frac{P^{\#}(X_i, \mathrm{pa}(X_i))}{P^{\#}(X_i) P^{\#}(\mathrm{pa}(X_i))} \right.$$
$$\left. + \log_2 P^{\#}(X_i) \right),$$

and since the parent sets contain at most one variable, we get

$$\log_2 P(\mathcal{D} \mid B) = N \cdot \sum_{i=1}^{n} \mathrm{MI}(X_i, \mathrm{pa}(X_i)) + \sum_{i=1}^{n} \sum_{X_i} P^{\#}(X_i) \cdot \log_2 P^{\#}(X_i).$$

This expression is maximized by choosing parents such that the sum of the MI terms is maximized. Since B should be a tree and each parent set contains at most one variable, step three in the theorem is guaranteed to maximize the log-likelihood. Finally, by choosing an arbitrary root and directing the arcs away from the root, we ensure that each node will get at most one parent, and from the d-separation properties we also see that we get the same independence properties regardless of the choice of root.   □

7.3.4 *Bayesian Score Functions

The BIC score is an example of a score function combining a maximum likelihood term with a term measuring complexity. Another approach for measuring the fitness of a Bayesian network model structure, S, is to calculate the posterior probability that the data was generated by a distribution with the same independence properties as S. If we abuse the notation slightly, and also use S to denote the hypothesis that the data is sampled from a distribution with the same independence properties as S, then we have:

$$P(S \mid \mathcal{D}) = \frac{P(\mathcal{D}, S)}{P(\mathcal{D})} = \frac{P(S) P(\mathcal{D} \mid S)}{P(\mathcal{D})} = \mu P(S) P(\mathcal{D} \mid S), \qquad (7.7)$$

where $\mu = P(\mathcal{D})$ is the normalization constant. This constant does not depend on S, and it is therefore not necessary to calculate it when we compare two network structures. Actually, if you were to calculate $P(\mathcal{D})$ you would be faced with a computational problem, because the calculation of this constant involves summing over all possible model structures, i.e., $P(\mathcal{D}) = \sum_B P(B) P(\mathcal{D} \mid B)$.

From equation (7.7) we see that in order to score a structure based on its posterior probability given the data, we only need two terms, namely the prior probability of the structure $(P(S))$ and the *marginal likelihood* of the structure given the data $(P(\mathcal{D}\,|\,S))$. Typically you would choose a prior probability distribution for the structures that is relatively easy to calculate, and the main computational problem is therefore the calculation of the marginal likelihood, where we will have to deal with the parameters of the model $\boldsymbol{\theta}_S$ (we shall return to the specification of structure priors in Section 7.3.4):

$$P(\mathcal{D}|S) = \int_{\boldsymbol{\theta}_S} P(\mathcal{D}|S,\boldsymbol{\theta}_S)f(\boldsymbol{\theta}_S|S)d\boldsymbol{\theta}_S, \tag{7.8}$$

where $f(\boldsymbol{\theta}_S\,|\,S)$ is a prior probability distribution over the parameters (conditional probabilities) for S. The integral in the above equation is over all parameters, and, in effect, over all possible Bayesian networks with the same structure but with different conditional probability distributions. Intuitively, the marginal likelihood can therefore be interpreted as the probability that we could generate the database $\mathcal{D}$ if we were to randomly select the parameters for S according to the parameter prior $f(\boldsymbol{\theta}_S\,|\,S)$.

As hinted above, the hard part in the calculation of $P(S\,|\,\mathcal{D})$ is the evaluation of the integral in equation (7.8). Fortunately, it has been shown that the evaluation of this integral can be reduced to a simple counting problem based on the following six assumptions:

1. the database $\mathcal{D}$ is a faithful sample from some Bayesian network;
2. the cases in the database $\mathcal{D}$ are independent given the BN model;
3. the database is complete;
4. the prior distribution of the parameters in every Bayesian network is uniform;
5. [local independence] for any two configurations over the parents for a variable X_i, the parameters for the conditional probability distributions associated with X_i are independent; and
6. [global independence] the densities of the parameters for the conditional probability distributions for X_i and X_j are independent for $i \neq j$.

Now let us again use N_{ijk} to denote the number of cases in the database that include the configuration $(X_i = k, \mathrm{pa}(X_i) = j)$. Based on the assumptions above, the following theorem has been proved.

Theorem 7.3. *Let $\mathcal{D}$ be a database over the variables $X_1, X_2, \ldots, X_n$, and consider the Bayesian network structure B_s over the same set of variables. Given the six assumptions above, it holds that*

$$P(\mathcal{D}\,|\,S) = \prod_{i=1}^{n} \prod_{j=1}^{q_i} \frac{(r_i - 1)!}{(N_{ij} + r_1 - 1)!} \prod_{k=1}^{r_i} (N_{ijk})! \;\;, \tag{7.9}$$

where $N_{ij} = \sum_{k=1}^{r_i} N_{ijk}$.

This means that the evaluation of the integral in equation (7.8) is reduced to a counting problem, which can be carried out in polynomial time.

Example 7.4. Consider again the two Bayesian network structures from Fig. 7.14, and assume that we have the database from Table 7.2 for the two binary variables X_1 and X_2.

Let us also assume that we have a priori the same belief in the two network structures, $P(BN_a) = P(BN_b)$. In order to select between B_a and B_b, the task then reduces to the calculation of

$$P(\mathcal{D}|S) = \prod_{i=1}^{n}\prod_{j=1}^{q_i} \frac{(r_i - 1)!}{(N_{ij} + r_1 - 1)!} \prod_{k=1}^{r_i}(N_{ijk})!$$

for both networks. To start off, consider network B_a ($X_1 \to X_2$). As for the calculation of the BIC score (see Example 7.2), we also need the following counts, which can be found from the database: $N_{111} = 8$, $N_{112} = 2$, $N_{211} = 6$, $N_{212} = 2$, $N_{221} = 0$, and $N_{222} = 2$. By using these values we get

$$P(\mathcal{D}|BN_a) = \frac{(2-1)!8!2!(2-1)!6!2!(2-1)!2!0!}{(10+2-1)!(8+2-1)!(2+2-1)!} = 2.67 \cdot 10^{-6}.$$

For the network BN_b we have $N_{111} = 8$, $N_{112} = 2$, $N_{211} = 6$, and $N_{212} = 4$. This gives us

$$P(\mathcal{D}|BN_b) = \frac{(2-1)!8!2!(2-1)!6!4!}{(10+2-1)!(10+2-1)!} = 8.75 \cdot 10^{-7}.$$

So with a uniform prior distribution over both structure and parameters we should prefer B_a over B_b.

Although the metric provides a simple expression for calculating the likelihood of a structure, it also rests on assumptions that may not always be appropriate. Most notably, it requires that the database be complete. When these assumptions are not fulfilled you will have to resort to other methods, such as the BIC score or constraint-based algorithms.

Prior Distribution over Structures

In order to score a network structure using the metric above, you have to specify a prior distribution $P(S)$ over the network structures. The specification of this prior can be used to guide the subsequent structure search, although the contribution from the prior distribution is usually dominated by the likelihood term $P(\mathcal{D}\,|\,S)$ when the database gets large ($P(\mathcal{D}\,|\,S)$ decreases exponentially fast as cases are added to the database). One exception, however, occurs when some of the network structures are given zero probability a priori, in which case the data cannot change that belief.

Common to most (if not all) prior distributions over structures currently used is that they can be expressed as a product (or sum) with one term for each family of nodes in the network:

$$P(S) = c \cdot \prod_{i=1}^{n} \rho(X_i, \mathrm{pa}(X_i)),$$

where c is a normalization constant that does not depend on S. These types of prior probability distributions are *decomposable*, which means that equation (7.9) is also decomposable (see Definition 7.3).

The simplest prior distribution is the one that encodes complete ignorance, i.e., we use an even distribution over the possible network structures:

$$\rho(X_i, \mathrm{pa}(X_i)) = 1.$$

A more informative prior that has been suggested is based on the difference between the families in the current network S and the families in a user-specified prior network B_p. Specifically, let δ_i denote the number of parents that B_P and S disagree on for X_i:

$$\delta_i = |(\mathrm{pa}(X_i)_S \cup \mathrm{pa}(X_i)_{B_P}) \setminus (\mathrm{pa}(X_i)_S \cap \mathrm{pa}(X_i)_{B_P})|. \qquad (7.10)$$

Then we can give a low prior probability to structures that are far away from the prior network structure, B_P, by setting

$$\rho(X_i, \mathrm{pa}(X_i)) = \kappa^{\sum_{i=1}^{n} \delta_i},$$

where $0 < \kappa \leq 1$ is a user-specified constant.

Example 7.5. Consider the four Bayesian network structures depicted in Fig. 7.20, and assume that Figure 7.20(B_P) is a prior network specified by the user.

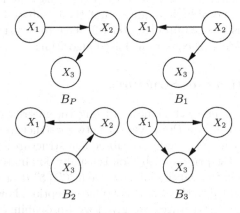

Fig. 7.20. The candidate structures B_1, B_2, and B_3 are assigned a prior distribution based on the number of arcs they have in common with the prior network B_P.

When using equation (7.10) to calculate the unnormalized prior probability for the candidate structure B_1, we first calculate the differences in the node families:

$$\delta_1^{B_1} = |(\{X_2\} \cup \emptyset) \setminus (\{X_2\} \cap \emptyset)| = 1;$$
$$\delta_2^{B_1} = |(\emptyset \cup \{X_1\}) \setminus (\emptyset \cap \{X_1\})| = 1;$$
$$\delta_3^{B_1} = |(\{X_2\} \cup \{X_2\}) \setminus (\{X_2\} \cap \{X_2\})| = 0.$$

Hence, the total difference between the two structures is measured as $\delta^{B_1} = \delta_1^{B_1} + \delta_1^{B_2} + \delta_1^{B_3} = 2$, which gives the prior probability

$$P(B_1) = c \cdot \kappa^\delta = c \cdot \kappa^2. \tag{7.11}$$

For B_2 we get $\delta_1^{B_2} = 1$, $\delta_2^{B_2} = 2$, and $\delta_3^{B_2} = 1$, and therefore $\delta = 4$ and $P(B_2) = c \cdot \kappa^4$. Finally, for B_3 we have $\delta_1^{B_3} = 0$, $\delta_2^{B_3} = 1$ and $\delta_3^{B_3} = 1$, and therefore $P(B_3) = c \cdot \kappa^2$. That is, a priori, B_p would be given the highest probability, then B_1 and B_3, and finally, B_2 would be given the lowest probability; observe that the normalization constant is of no importance when comparing structures.

Finally, it should be emphasized that although we can easily come up with elaborate prior distributions, there is also a caveat: the prior distribution does not necessarily assign the same score to equivalent network structures (as in Example 7.5). When this is the case, then if it is used to define, say, the score function in equation (7.9), the resulting score function is not score equivalent. As an example, consider equation (7.10), and use any prior network structure that is different from the empty graph.

Regulating Model Complexity

An attractive property of the BD score (and likelihood based scoring functions in general) is that it has an intrinsic property that no extra term is needed for penalizing complexity.

The intuition why the BD score is less likely to pick out an overfitted network structure is closely related to the Bayesian version of Ockham's razor: A complex structure with few conditional independences can generate many possible data sets, so it is unlikely that it has generated this particular data set at hand (see Figure 7.21 for an illustration). Obviously, models that are too simple are also unlikely to have generated the data.

To provide a specific example, consider again the Bayesian network structures depicted in Figure 7.14. From the model in Figure 7.14(a) you can sample a database, and then use it to score the model structure S in Figure 7.14(b), where X_1 and X_2 are independent. Specifically, let the databases be generated according to the following probability distributions: $P(X_1) = (0.5, 0.5)$,

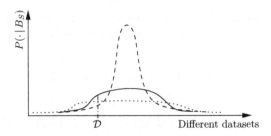

Fig. 7.21. The figure illustrates the marginal likelihood for three different structures; the dotted line represents a structure that is too complex, the dashed line represents a structure that is too simple, and the solid line represents an appropriate structure.

$$P(X_2 = 1 | X_1 = 0) = 0.5(1 - \epsilon),$$
$$P(X_2 = 1 | X_1 = 1) = 0.5(1 + \epsilon),$$

where the parameter ϵ varies between 0 and 1 and is used to control the strength of the dependency between X_1 and X_2; the larger the value of ϵ, the stronger the dependency. A plot of $P(S | \mathcal{D})$ for four different database sizes (generated for various values of ϵ) is depicted in Figure 7.22. In particular, we can see that when the database is not too large, S is acceptable even for relatively large values of ϵ. For example, with $\epsilon = 0.2$ we have that $P(S | \mathcal{D}) \approx 0.6$ for a database with 100 cases.

7.4 Summary

Constraint-Based Methods

The structure of a Bayesian network can be learned from independence statements of the form, "A independent of B given C" (denoted by $I(A, B, C)$):

1. Find the skeleton of the Bayesian network: the link $A - B$ is part of the skeleton if and only if $\neg I(A, B, \mathcal{X})$ for all $\mathcal{X}$ not containing A or B.
2. Direct the links:
   Introduction of v-structures: If you have three nodes, A, B, C, such that $A - C$ and $B - C$, but not $A - B$, then introduce the v-structure $A \rightarrow C \leftarrow B$ if there exists an $\mathcal{X}$ (possibly empty) such that $I(A, B, \mathcal{X})$ and $C \notin \mathcal{X}$.
   Avoid new v-structures: When Rule 1 has been exhausted, and you have $A \rightarrow C - B$ (and no link between A and B), then direct $C \rightarrow B$.
   Avoid cycles: If $A \rightarrow B$ introduces a directed cycle in the graph, then do $A \leftarrow B$.
   Choose randomly: If none of the rules 1–3 can be applied anywhere in the graph, choose an undirected link and give it an arbitrary direction.

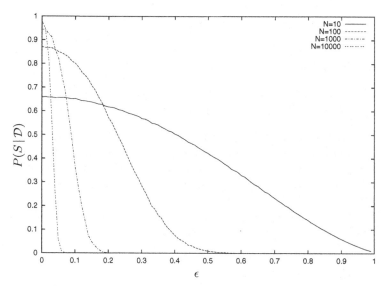

Fig. 7.22. A plot of $P(\text{complete independence} \mid \mathcal{D})$ for four different database sizes. The databases were sampled from a model with two variables, where the level of dependency between the variables was controlled by the value of ϵ (a high value implies a high dependency).

The independence statements can be established through statistical tests on a database.

The PC algorithm: The initial skeleton can be found using the PC-algorithm:

1. Start with the complete graph;
2. $i := 0$;
3. **while** a node has at least $i + 1$ neighbors
   - **for all** nodes A with at least $i + 1$ neighbors
     - **for all** neighbors B of A
       - **for all** neighbor sets $\mathcal{X}$ such that $|\mathcal{X}| = i$ and $\mathcal{X} \subseteq (\text{nb}(A) \setminus \{B\})$
         - **if** $I(A, B, \mathcal{X})$ **then** remove the link $A - B$ and store "$I(A, B, \mathcal{X})$"
   - $i := i + 1$

Score-Based Methods

A Bayesian network can be learned from a database by performing a search in the space of all DAGs and selecting the one with the highest score.

The BIC Score:

$$\mathrm{BIC}(S \mid \mathcal{D}) = \sum_{i=1}^{N} \log_2 P(\mathbf{d} \mid \hat{\theta}_S, S) - \frac{\mathrm{size}(S)}{2} \log_2(N).$$

Chow–Liu trees

A tree-shaped Bayesian network of maximal likelihood can be learned in polynomial time using the Chow–Liu algorithm:

1. Calculate the mutual information $\mathrm{MI}(X_i, X_j)$ for each pair (X_i, X_j).
2. Consider the complete MI-weighted graph: the complete undirected graph over $\{X, \ldots, X_n\}$, where the links (X_i, X_j) have the weight $\mathrm{MI}(X_i, X_j)$.
3. Build a maximal-weight spanning tree for the complete MI-weighted graph.
4. Direct the resulting tree by choosing any variable as a root and setting the directions of the links to be outward from it.
5. Learn the parameters.

Mutual information:

$$\mathrm{MI}(X, Y) = \sum_{X,Y} P(X, Y) \log_2 \left(\frac{P(X, Y)}{P(X)P(Y)} \right).$$

7.5 Bibliographical Notes

The first method for automated learning of Bayesian networks was the method of Chow and Liu (1968), which learned tree-structured models. A statistical approach for learning Bayesian networks through manually selected independence tests was given by Edwards and Havranek (1985). The PC algorithm was developed by Spirtes *et al.* (1993); see also (Spirtes *et al.*, 2000). It is an extension of work by Wermuth and Lauritzen (1990) and Verma and Pearl (1991). The necessary path condition and uncertain areas are due to Steck (2001). Improved algorithms, which theoretically should be more robust in face of flawed independence tests, are given by Margaritis and Thrun (1999) and Cheng *et al.* (2002). A discussion on observing causality can be found in (Pearl, 2000).

The dimensionality of models with hidden variables has been explored by Geiger *et al.* (1996) in the context of model selection. Here the BIC score (Schwarz, 1978) was extended to Bayesian networks with hidden variables. The BIC score of a model is an asymptotic approximation to the marginal likelihood of that model, and it is equivalent to the minimum description length proposed by Rissanen (1987), and adopted to a decomposable consistent score for Bayesian networks by Lam and Bacchus (1994) and Friedman and Goldszmidt (1998). A Bayesian metric for scoring models was proposed by Cooper and Herskovits (1991) and generalized in (Cooper and Herskovits, 1992). Cooper and Herskovits (1992) also proposed a search algorithm (known as the K2 algorithm) that performs a greedy search conditioned on a linear ordering of the variables (for literature on search in general, see (Michalewicz and Fogel, 2000)). Heckerman *et al.* (1995b) considered the specification of prior information such that equivalent network structures (Chickering, 1995) are given the same score. In the context of equivalent structures, greedy search procedures have been proposed by Chickering (2002); Chickering and Meek (2002) that are guaranteed to identify the correct structure when the amount of data grows large. (Chickering *et al.*, 2004) is one of the latest in a line of results that show that the task of learning Bayesian network structures is NP-hard, and Cowell (2001) has shown that, under often-quoted assumptions, constraint-based learning and score based learning are equivalent. In the context of missing data, Friedman (1998) has proposed a structural learning method that follows the intuition of the EM-algorithm. Finally, Friedman and Koller (2003) provide a method for calculating the posterior probability of absence or presence of individual arcs in the generating net given the data. Other sources of literature that can be recommended for further reading include (Buntine, 1996), (Heckerman, 1998), (Jordan, 1998), and (Cowell *et al.*, 1999).

7.6 Exercises

Exercise 7.1. Apply the PC algorithm to learn a skeleton over the six variables A, B, C, D, E, and F (use the network structure in Figure 7.23 as an oracle). Using rules 1–4, exploit the identified independence statements to set directions on the links in the skeleton.

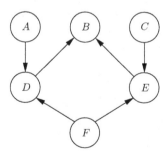

Fig. 7.23. Use the Bayesian network structure as an oracle for Exercise 7.1.

Exercise 7.2. Use rules 1–4 to set the directions on the remaining links in the structure in Figure 7.24.

Exercise 7.3. Assume that the PC-algorithm is run on five variables A, B, C, D, and E. During its running time, the algorithm gets positive

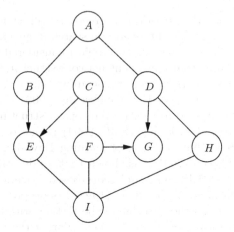

Fig. 7.24. A partial Bayesian network structure found by the PC algorithm and rule 1.

replies only to the following oracle queries: $I(A, B, E)$, $I(A, C, E)$, $I(A, D, E)$, $I(B, E, \{C, D\})$, and $I(C, D)$. The result of the run is a Bayesian network M. What does the skeleton of M look like? Which graphs can M be?

Exercise 7.4.

(i) Find a (tight) upper bound on the number of independence tests performed by the PC algorithm.
(ii) Discuss an implementation strategy for the PC algorithm with focus on the time used to perform the independence tests required by the algorithm (that is, calculating the conditional mutual independence expression, equation (7.2)).

Exercise 7.5. Prove that

$$I(A, B, \mathcal{X}) \Leftrightarrow \mathrm{CMI}(A, B | \mathcal{X}) = 0.$$

Exercise 7.6. Show that the size of the BN in Figure 7.13(c) is larger than the size of the BN in Figure 7.13(a).

Exercise 7.7. What is the size (see Proposition 7.1) of the BN shown in Figure 7.3(c) assuming that all variables are ternary.

Exercise 7.8. What is the BIC score, based on the data in Table 7.4, for the structure in Figure 7.25? What is the score for the structure in Figure 7.26?

Exercise 7.9. [E] Calculate the BIC score for the model of the simplified insemination problem described in Section 3.1.3, based on the (incomplete) database in Example 6.2.

C	B	A	C	B	A
1	1	1	1	1	2
1	1	1	1	1	2
1	1	2	2	2	1
2	1	2	1	1	2
1	1	1	2	1	2
1	1	2	1	1	1
2	1	2	1	1	1
1	1	1	2	1	2
1	1	1	1	1	2
1	1	2	1	1	1
1	1	1	1	1	1
1	1	1	2	1	2
1	1	2	1	1	2
2	2	1	1	1	2
1	1	2	2	2	1
2	1	2	1	1	2

Table 7.4. A number of configurations over binary variables A, B, and C.

Fig. 7.25. A Bayesian network for Exercise 7.8.

Fig. 7.26. A Bayesian network Exercise 7.8.

Exercise 7.10. What is the result of running greedy search based on the BIC score and the data in Table 7.4 starting from the empty graph?

Exercise 7.11. Show that the two expressions (in equation (7.4) and equation (7.5)) for the BIC score are identical.

Exercise 7.12. Tabu search is a general search technique based on greedy search. The technique tries to avoid getting stuck in local minima by prohibiting moves that involve aspects that were changed by a recent move. How could Algorithm 7.2 be modified to incorporate this behavior?

Exercise 7.13. Simulated annealing is a general search technique based on greedy search. The technique tries to explore more parts of the search space by making totally random moves at first, ignoring the score of the parts of the search space it moves to. Gradually it starts letting the scores influence the search, and finally ends up moving only if the score improves (as greedy search always behaves). How could Algorithm 7.2 be modified into a simulated annealing search algorithm? (Hint: Use a counter i that is decreased at each iteration, and an error term like e^{-i} to modify scores.)

Exercise 7.14. What network structures are equivalent to the network in Figure 7.27?

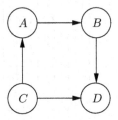

Fig. 7.27. A Bayesian network for Exercise 7.14.

Exercise 7.15. Learn a Chow–Liu tree from the data in Table 7.4.

Exercise 7.16. Complete Example 7.4 by calculating the BD score for the Bayesian network structure shown in Figure 7.28 based on the database in Table 7.2. As in Example 7.4 we assume that all network structures are a priori equally probable and that we have a uniform prior over all the possible parameters.

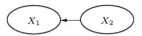

Fig. 7.28. Together with the network structures shown in Figure 7.14, this BN structure defines the space of model structures covering two variables.

(i) Calculate $P(\mathcal{D})$, the prior probability of the data.
(ii) Calculate the conditional probabilities for the three network structures given the database.
(iii) What should the prior probability for the empty graph (at least) have been for it to be picked by the BD score? Give an intuitive reason.

Exercise 7.17. Consider the database in Table 7.2 and a prior network structure consisting of an arc from X_2 to X_1. What is the result of learning with a greedy search and the BD score introduced in (7.7)?

Exercise 7.18. Show that when using a nonempty prior network structure together with equation (7.10), the resulting prior distribution cannot be score equivalent.

8

Bayesian Networks as Classifiers

You receive an email and wish to determine whether it is spam; you see a bird and wish to determine its species; you examine a patient and wish to diagnose him. These are only a few examples of the very common human task of classification.

Formally, you have a set of variables, $\{F_1, \ldots, F_n\}$, called features (or attributes) and a *class variable*, C, where the states of C correspond to the possible classes. For the bird example above, the feature variables would encode various characteristics of the bird, and the class variable would represent the possible species. Since it often happens that some feature values are not known, feature variables are often extended with state "?" for unknown (or "missing value"). A case is said to be *complete* if there are no missing values. A case set is said to be *consistent* if two complete cases with the same values on the features are of the same class.

A *classifier* is a function from $F_1 \times \cdots \times F_n$ to C. We shall deal only with classification tasks over a finite set of classes and with discrete features.

If you have a Bayesian network model, it can be used for classification. In fact, if there is only one hypothesis variable, the network is a model for classification. In the pregnancy model (Section 3.1.3), for example, test information is used to classify the state of the cow, the class being the state of highest probability.

In this chapter we consider learning of classifiers. Let $\mathcal{D}$ be a data set of cases over features $\{F_1, \ldots, F_m\}$ and class variable C; we do not require the data set to be consistent. We wish to use the data set for constructing a classifier. If the space of feature configurations is small and the amount of data is relatively large, you may use the data set to establish a look-up table: given a complete case $\mathbf{f}$ of features, look up $\mathbf{f}$ in the data set. If there are cases in $\mathcal{D}$ with feature values $\mathbf{f}$, then return the majority class value. If $\mathbf{f}$ is not present in $\mathcal{D}$, then return the most frequent class value in $\mathcal{D}$. However, this method is tractable only for small configuration sets; even with a moderate number of feature variables you will need a more compact representation of the classification function. Any other method for learning classifiers should predominantly produce better classifiers.

8.1 Naive Bayes Classifiers

Consider the poker game model introduced in Section 3.2.3, and extend the model with a variable for my hand (*MH*) and for best hand (*BH*) (see Exercise 3.14). A Bayesian network model would be like the one in Figure 8.1.

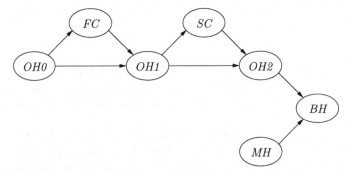

Fig. 8.1. A Bayesian network for the poker game extended with a node for my hand and best hand.

Assume that you have a set of cases over the observable features *MH*, opponent's change of cards, *FC*, *SC*), and the class (*BH*). Exploiting structural learning will most likely result in the model in Figure 8.2. The reader may test this by a manual run of the PC algorithm on Figure 8.1 with the variables (*OHi*) hidden.

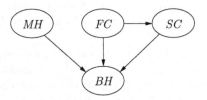

Fig. 8.2. A Bayesian network learned from a case set of poker games.

The model in Figure 8.2 does not provide a compact representation of the classification function, since the class variable has all features as parents, and therefore the conditional probability table for the class variable is as large as a look-up table for the classification problem. Unfortunately, it is often seen in connection to Bayesian network classifiers that in the correct model, the class variable has (almost) all feature variables as parents, and the network therefore becomes intractably large. Instead, you can insist on working with a class of simpler structures and search for the model that best approximates the correct structure.

One such class of models could be naive Bayes networks (see Section 3.1.5), and for the poker game, the structure will be the one in Figure 8.3.

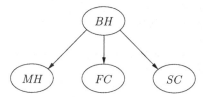

Fig. 8.3. Naive Bayes structure for the poker game.

In general, in a *naive Bayes classifier* (NBC) each feature variable has the class variable as its only parent. This means that the structure is fixed, and the only task involved in learning is to estimate the parameters.

The parameters for an NBC are easily determined by the methods presented in Chapter 6. If the cases are complete, you can determine a maximal likelihood model through simple counting. If a case contains missing values, the EM algorithm can be used; equivalently, disregard that case for the attributes that are missing.

All methods for learning classifiers from data have a problem with very rare cases, which may not be represented in the data set. Assume, for example, that the data set for learning a poker classifier does not contain a case in which I have lost with a hand with *3v*. If one is not careful, the classifier would deem this impossible regardless of the pattern of card changes. For a Bayesian network classifier, this problem corresponds to incorrectly setting a parameter to zero. To avoid zero values for parameters, you may simulate Bayesian learning by introducing virtual cases. An easy way of handling this is initially to give all parameters a small positive count.

Since NBCs are easy to learn, and easy to use as classifiers, and since they are very flexible with respect to missing values, they are very widespread. As mentioned in Section 3.1.5, NBCs assume the features to be independent given the class, and even though this is rarely the case, NBCs have proved surprisingly precise. A reason for this is that when doing classification we are interested only in the class of maximal probability and not in the exact probability distribution over the classes.

8.2 Evaluation of Classifiers

Assume that you have a classifier, *Clsf*, and a data set of cases covering the feature variables and the class variable. We wish to characterize the quality of *Clsf*. A way of characterizing *Clsf* is to calculate its *classification accuracy*: the fraction of correctly classified cases.

A more detailed description of a classifier would be to calculate the *confusion matrix*, $P^\#$(Classified value, Correct value). In addition to the confusion matrix you can also introduce a value for how bad a misclassification is, and thereby establish a *loss matrix*, describing a punishment for the various kinds of misclassification.

To illustrate this, consider again the poker game. Assume that you have established a classifier *Pcl*, and you have the set of cases in Table 8.1. Since 12 out of 20 cases are classified correctly, the classification accuracy is 0.6.

Case number:	BH	MH	FC	SC	Pcl
1	op	no	3	1	op
2	op	1a	2	1	op
3	draw	2 v	1	1	op
4	me	2 a	1	1	me
5	draw	fl	1	1	me
6	me	st	3	2	me
7	me	3 v	1	1	me
8	me	sfl	1	0	me
9	op	no	0	0	op
10	op	1 a	3	2	me
11	draw	2 v	2	1	op
12	me	2 v	3	2	draw
13	op	2 v	1	1	draw
14	op	2 v	3	0	op
15	me	2 v	3	2	me
16	draw	no	3	2	draw
17	draw	2 v	1	1	draw
18	op	fl	1	1	me
19	op	no	3	2	op
20	me	1 a	3	2	op

Table 8.1. Test cases for a poker classifier. The entry *Pcl* is the class value provided by the classifier.

The confusion matrix is given in Table 8.2, but it does not consider the stakes involved in the poker game. Let the situation be that both players initially have bet a euro, and you have to decide whether to *fold* (your opponent takes the pot) or to *call*. To simplify, assume that you place a euro when you call, and your opponent is forced to place a euro. The winner takes the pot,

	BH		
	me	draw	op
me	0.25	0.05	0.1
Plc draw	0.05	0.1	0.05
op	0.05	0.1	0.25

Table 8.2. Confusion matrix for the poker classifier. The sum of the diagonal elements is the classification accuracy.

and in the case of a draw you share the pot. The wins and losses in the various situations are given in Table 8.3.

		BH		
		me	draw	op
Action	fold	0	0	0
	call	3	1	−1

Table 8.3. Wins and losses in the poker game.

Based on Table 8.3, you decide on the strategy to call if and only if the classifier says *m* or *draw*. The loss matrix tells you what you lose by following the classifier compared to a situation with certainty on BH. It is given in Table 8.4.

	BH		
	me	draw	op
me	0	0	−1
Plc draw	0	0	−1
op	−3	−1	0

Table 8.4. Loss matrix for the poker classifier.

The confusion matrix and the cost matrix can now be used to calculate the expected loss of a strategy following the classifier (based on the data set $\mathcal{D}$):

Expected loss

$$= \sum_{\text{Classified,Correct}} P^{\#}(\text{Classified} \mid \text{Correct}) P^{\#}(\text{Correct})$$

$$\times \text{Loss}(\text{Classified}, \text{Correct})$$

$$= \sum_{\text{Classified,Correct}} P^{\#}(\text{Classified}, \text{Correct}) \text{Loss}(\text{Classified}, \text{Correct}).$$

That is, you first multiply the confusion matrix and the loss matrix term by term, and then you take the sum of all these elements.

The expected loss for the poker classifier is

$$\sum_{Plc,BH} P^{\#}(Plc, BH) \text{Loss}(Plc, BH) = -3.0.05 - 1.0.1 - 1.0.1 - 1.0.05 = -0.4.$$

A general problem in connection to machine learning is *overfitting*. What we are looking for is a classifier that can classify not-yet-seen cases. However, it may happen that the learned classifier is very accurate on the training data, but it is very poor when confronted with cases not represented there. To monitor overfitting, you usually divide the set data into training and test data, and you measure the classification accuracy on the test data set rather than on the training data set. A way of addressing overfitting in the choice of model is to reserve a part of the training set for validation and comparison of models only and not for establishing the models.

8.3 Extensions of Naive Bayes Classifiers

NBCs assume that the feature variables are independent given the class. Even though this assumption seldom holds, NBCs are surprisingly good with respect to classification accuracy. However, as described in the previous section, classification accuracy does not tell the full story. Often you are particularly interested in detecting a rare class. The class being rare also means that classification accuracy does not drop significantly if your classifier never identifies these cases.

A rare class is often identified through a set of feature values appearing together, where each value by itself does not point in that direction. NBCs cannot cope with that, since they assume the features to be independent given the class. Therefore, you may wish to extend NBCs to allow more elaborate dependency structure among feature variables. A simple extension of this kind is the *tree augmented naive Bayes classifier* (TAN): each feature variable has at most one feature variable as parent.

As opposed to the situation for NBCs, the structure is not given, and we have to look for a structure that with optimal parameter setting has maximal likelihood: out of the possible links between feature nodes we have to choose a set forming a tree. This is similar to the situation described in Section 7.3.3, and not surprisingly, the problem is solved through a slight modification of the Chow–Liu algorithm using conditional mutual information rather than mutual information (see equation (7.2), Page 237).

We give the construction without proof.

Theorem 8.1 (Learning TANs). *Let $\mathcal{D}$ be a data set over the variables $\{F_1, \ldots, F_m, C\}$. A TAN of maximal likelihood can be constructed as follows:*

1. *Calculate the conditional mutual information $MI(F_i, F_j \mid C)$ for each pair (F_i, F_j).*
2. *Consider the complete MI-weighted graph: the complete undirected graph over $\{F_1, \ldots, F_n\}$, where the links $F_i - F_j$ have the weight $MI(F_i, F_j \mid C)$.*
3. *Build a maximal-weight spanning tree for the complete MI-weighted graph.*
4. *Direct the resulting tree by choosing any variable as a root and setting the directions of the links to be outward from it.*
5. *Add the node C and a directed link from C to each feature node.*
6. *Learn the parameters.*

Running the TAN algorithm on the data for the poker domain resulted in the TAN in Figure 8.4.

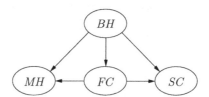

Fig. 8.4. A TAN for classifying poker.

Another extension is to introduce intermediate variables. For the poker example, the dependence between FC and SC can be mediated through a hidden variable C, as illustrated in Figure 8.5.

A problem with hidden variables is that even if you know how to connect the hidden variables introduced, you have to determine the number of states of the hidden variables. Let H be a hidden variable with n states and with children $\mathrm{ch}(H)$. If n is equal to the product of the number of states of the children, then H can represent any configuration of $\mathrm{ch}(H)$, and you cannot hope for a better fit. On the other hand, in that case, you should represent the product of $\mathrm{ch}(H)$ directly without a hidden variable. For the poker example it means that the number of states of C should be between 2 and 11. Now use the EM algorithm for these ten possible numbers of states. Since the likelihood increases with the number of states of C, the model of maximal likelihood has eleven hidden states, and that is not really what you are after. Therefore, you have to balance likelihood with size as described in Section 7.3.1.

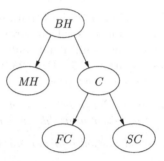

Fig. 8.5. The dependence between *FC* and *SC* is mediated by the hidden variable *C*.

8.4 Classification Trees

For the sake of completeness we shall in this section present a very popular method for doing classification. In the data mining literature the method is called a *decision tree*. However, since in this book we use this term differently (see Section 9.3), we shall call it a *classification tree*.

A classification tree is a directed tree whose internal nodes are feature variables. The links are labeled with values of the feature in question, and the leaves are labeled with class values (see Figure 8.6).

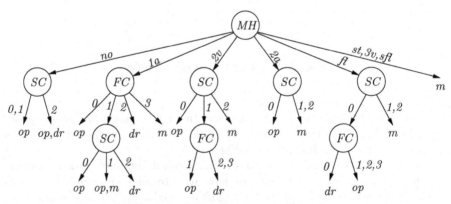

Fig. 8.6. A classification tree for poker

The tree in Figure 8.6 can be used to classify the situation with respect to *BH*. Classification is performed through processing the tree from the root toward the leaves. First you branch out based on the value of *MH*. Depending on the answer, you branch out according to the value of either *FC* or *SC*, and sometimes you also ask for the value of the other card change. When you reach a leaf, you read the classification.

To learn a classification tree, you first determine which feature variable to use as the root. Let C be the class variable with states $\{c_1, \ldots, c_n\}$, and let F be a feature variable with states $\{f_1, \ldots, f_k\}$. We wish to characterize how good a classifier F alone would be. That is, if we know the state of F, how close will we be at knowing the class value?

The values of F partitions $\mathcal{D}$ into the data sets $\mathcal{D}_1^1, \ldots, \mathcal{D}_k^1$, and for each data set $\mathcal{D}_i^1$ we have a distribution $P^\#(C|f_i)$. One way of measuring how close we are to knowing C in the data set $\mathcal{D}_i^1$ is to calculate the entropy for C. In general, for a variable X with distribution $P(X)$ (or $P^\#(X)$), the entropy is defined as

$$\text{Ent}(P(X)) = - \sum_{x \in \text{sp}(X)} P(x) \log_2(P(x)), \qquad (8.1)$$

where we let $0 \log_2(0) = 0$. If the probability of X being in a particular state approaches 1, then the entropy goes toward 0. On the other hand, the more dispersed the probability mass, the higher the entropy; in case we have a uniform distribution, the entropy attains its maximum value, $\log_2(|\text{sp}(X)|)$.

Now, if the entropy of each distribution $P^\#(C|f_i)$ is small, then knowing F brings us close to knowing C, but if the entropies are large, then knowing F does not give us much information about C. There are various ways of using the entropies as a score for ranking the variables. A method called ID3 uses the expected entropy as a measure of how good a feature is at predicting the class:

$$\mathbb{E}[\text{Ent}(F)] = \sum_F P^\#(F) \, \text{Ent}(P^\#(C \,|\, F)).$$

Actually, the algorithm uses information gain,

$$\text{Ent}(P^\#(C)) - \mathbb{E}[\text{Ent}(F)],$$

but since $\text{Ent}(P^\#(C))$ is independent of F, you look for a variable giving the lowest expected entropy.

Having chosen the feature F as the root, you continue recursively on the data sets $\mathcal{D}_1^1, \ldots, \mathcal{D}_k^1$.

As an illustration, the ID3 algorithm applied to the data set in Table 8.1 would first partition the data set for each variable. For the variable SC we have the sets $\{8, 9, 14\}$, $\{1, 2, 3, 4, 5, 7, 11, 13, 17, 18\}$, and $\{6, 10, 12, 15, 16, 19, 20\}$ corresponding to the states 0, 1, and 2, respectively. The set for state 0 has two cases with state op, and one with state m. This distribution has the entropy

$$-\frac{1}{3} \log_2\left(\frac{1}{3}\right) - \frac{2}{3} \log_2\left(\frac{2}{3}\right) = -\frac{1}{3}(2 - 2\log_2 3 - \log_2 3) = 0.918,$$

yielding a contribution of $3/20 \cdot 0.918 = 0.138$ to the expected entropy.

The following expected entropies are calculated (note that the maximal entropy for a distribution over three states is $\log_2 3 = 1.585$):

$$\mathbb{E}[\mathrm{Ent}(MH)] = 0.735, \quad \mathbb{E}[\mathrm{Ent}(FC)] = 1.351, \quad \mathbb{E}[\mathrm{Ent}(SC)] = 1.403.$$

Since MH has the lowest expected entropy, it is chosen as root. For each value of MH you now have a small data set, and you choose the best root for each. For $MH = no$ you have four cases, and since SC separates these cases better than FC, SC is chosen. The full tree is given in Figure 8.7; the ? indicates that no case covers the specified configuration, and for these situations you may take the majority class.

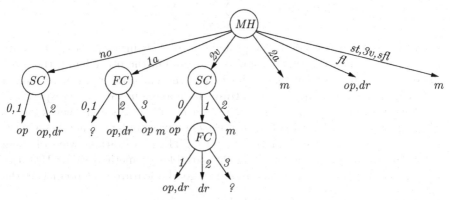

Fig. 8.7. The result of applying the ID3 algorithm on the data set in Table 8.1.

8.5 Summary

The Naive Bayes Classifier

In a naive Bayes classifier, each feature variable has the class variable as its only parent. This means that the structure is fixed, and learning a classifier therefore amounts to estimating the parameters.

Evaluating Classifiers

Two approaches for evaluating a classifier:

Classification accuracy: the fraction of correctly classified cases.

Expected loss:

Expected loss

$$= \sum_{\mathrm{Classified},\mathrm{Correct}} P^{\#}(\mathrm{Classified}, \mathrm{Correct})\mathrm{Loss}(\mathrm{Classified}, \mathrm{Correct}).$$

The Tree-Augmented Naive Bayes Classifier

In the tree-augmented naive Bayes classifier (TAN classifier), each feature variable has at most one other feature variable as parent in addition to the class variable.

Learning TANs: Let $\mathcal{D}$ be a dataset over the variables $\{F_1, \ldots, F_m, C\}$. A TAN of maximal likelihood can be constructed as follows:

1. Calculate the conditional mutual information $MI(F_i, F_j \mid C)$ for each pair (F_i, F_j).
2. Consider the complete MI-weighted graph: the complete undirected graph over $\{F_1, \ldots, F_n\}$, where the links $F_i - F_j$ have the weight $MI(F_i, F_j \mid C)$.
3. Build a maximal-weight spanning tree for the complete MI-weighted graph.
4. Direct the resulting tree by choosing any variable as a root and setting the directions of the links to be outward from it.
5. Add the node C and a directed link from C to each feature node.
6. Learn the parameters.

Classification Trees

A classification tree is a directed tree whose internal nodes are feature variables. The links are labeled with values of the feature in question, and the leaves are labeled with class values.

To learn a classification tree, you start with the empty tree and iteratively insert the node X that tells you the most about the class variable C. One possible measure is the expected entropy:

$$\mathbb{E}[\text{Ent}(X)] = \sum_X P^{\#}(X) \, \text{Ent}(P^{\#}(C \mid X)),$$

where

$$\text{Ent}(P(X)) = - \sum_{x \in \text{sp}(X)} P(x) \log_2(P(x)).$$

8.6 Bibliographical Notes

As mentioned, naive Bayes was used by de Dombal *et al.* (1972) and can be traced back at least to Minsky (1963). It was introduced to classification by Duda and Hart (1973). Its role in classification has been thoroughly studied in the last decade or so, with Domingos and Pazzani (1997) providing theoretical results on concepts that naive Bayes can classify better than any other classifier, and with empirical results that show how violations of the independence assumptions of the model are often of no consequence. Jaeger (2003) further clarifies the distinction between the concepts they can recognize, and the theoretical limits on the concepts that can be learned from data. Tree-augmented naive Bayes classifiers were introduced by Friedman *et al.* (1997). The ID3 algorithm for inferring classification trees was introduced by Quinlan (1979) and later improved in (Quinlan, 1986). For a general overview over classifiers, see (Mitchell, 1997).

8.7 Exercises

Exercise 8.1. Verify that the PC-algorithm results in the network in Figure 8.2 (or one of its equivalents) when run with an oracle based on the d-separation properties of the network in Figure 8.1, and with the variables *OH1* and *OH2* hidden.

Exercise 8.2. Learn the maximum likelihood parameters for the classifier in Figure 8.3 from the cases in Table 8.1. What class does your classifier assign to a case with *MH*=1a, *FC*=1, and *SC*=1?

Exercise 8.3. Verify that the TAN-algorithm constructs the classifier in Figure 8.4 and complete the classifier by learning the maximum likelihood parameters. What class does the classifier assign to the case with *MH*=1a, *FC*=1, and *SC*=1? What would the result be if you instead of maximum likelihood estimates used Bayesian parameter estimates?

Exercise 8.4. Consider the classification tree in Figure 8.6. How would this classifier classify the case with *MH*=1a, *FC*=1, and *SC*=1?

Exercise 8.5. Using the data in Table 7.4, construct a classification tree for classifying *A*. What class is assigned to $(B = 1, C = 2)$?

Part II

Decision Graphs

Graphical Languages for Specification of Decision Problems

A Bayesian network serves as a model for a part of the world, and the relations in the model reflect causal impact between events. The reason for building these computer models is to use them in taking decisions. In other words, the probabilities provided by the network are used to support some kind of decision making. In principle, there are two kinds of decisions, namely *test decisions* and *action decisions*.

A test decision is a decision to look for more evidence to be entered into the model, and an action decision is a decision to change the state of the world. In real life, this distinction is not very sharp; tests may have side effects, and by performing a treatment against a disease, evidence on the diagnosis may be acquired. In order to be precise, we should say that decisions have two *aspects*, namely a test aspect and an action aspect. The two aspects are handled differently in connection with Bayesian networks, and accordingly we treat them separately.

Although both observations and actions may change the probability distributions in the model, they are fundamentally different. To highlight this, consider the example in Figure 9.1.

A wheat type may be genetically resistant to mildew. If so, there will be no attack, and this has an impact on the quality of the crop. If you *observe* that there is no attack, the probabilities for *Resistance* and *Crop* are changed. If you, on the other hand, prevent an attack through spraying and thereby fix the state of *Attack* to *no*, then it has no impact on your belief about *Resistance*. That is, *the impact of actions can only follow the direction of the causal links*.

The example stresses the important point already made in Section 3.2.6 concerning the use of Bayesian networks. Using Bayes' theorem, it is easy to establish the model in Figure 9.2, which reflects a kind of diagnostic reasoning.

From the point of view of entering evidence and propagating probabilities, the two Bayesian networks in Figure 9.1 and Figure 9.2 represent the same joint probability distribution, so why bother emphasizing that the links in the network should be causal links? The difference becomes apparent when one

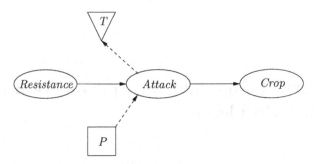

Fig. 9.1. A simple Bayesian network with an action and a test attached. The decision (Prevention) can by spraying fix the state of *Attack* to *no*. The test *T* can determine the state of *Attack*.

Fig. 9.2. A Bayesian network equivalent to the one in Figure 9.1.

sprays. In Figure 9.2, spraying will change the probability of resistance but it will have no impact on the crop.

In Section 9.1 we show how to extend a Bayesian network to cope with a single decision, and in Section 9.2 we describe fundamentals of rational decision making. Sections 9.3–9.5 present various graphical frameworks for modeling decision problems with several decisions involved, and in Section 9.6 we deal with problems that have an unbounded time horizon.

9.1 One-Shot Decision Problems

A Bayesian network provides a model of the world that can be used in making decisions. The typical situation is that we have observed some of the variables in the domain and based on these observations we make an inquiry to the Bayesian network about some other set of variables (probability updating). The result of the inquiry is in turn used in the subsequent decision-making process.

This type of application of Bayesian networks can be taken one step further, so that rather than keeping the model separated from the decision-making process, you could combine these two parts. That is, not only does the final model reveal the structure of the decision problem, but it can also be used to give advice about the decisions. In the simple situation in which only a single decision is to be made, the Bayesian network can readily be extended to reflect the structure of the decision problem.

9.1.1 Fold or Call?

Consider the poker example in Section 3.1.4 as extended in Exercise 3.13 with the variables *MH* ("my hand" having the same states as *OH2*) and *BH* ("best hand" with the states *me*, *opponent*, and *draw*), see Figure 9.3. The conditional probability distribution for *BH* is a deterministic function of *OH2* and *MH*.

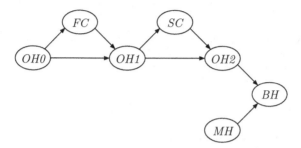

Fig. 9.3. The poker model extended with variables for my hand and best hand.

The reason I am interested in knowing which hand is best is that I shall take a decision on an action. For this game, the rules are that we both placed $1 on the table to get the initial hand, and after the rounds of card changing, my opponent places $1 extra (in this game she is forced to place $1 regardless of her hand). Now, I may either *fold* or *call*. If I fold, my opponent takes the pot, and if I call, I place $1 on the table, and we compare the hands. The player with the best hand takes the pot (in case of a draw we share).

My decision problem in deciding to fold or to call can be represented graphically by extending the Bayesian network with a couple of extra nodes. The decision options are represented by a rectangular node *D* with states *fold* and *call*. Another type of node, *U*, represents the possible outcomes in dollars. The node *U* is called a *utility node*, and the outcomes are called *utilities*. The variables determining the outcomes are *BH* and *D*, and this is shown graphically through directed links from *BH* and *D* to the diamond-shaped node *U*. See Figure 9.4. Note that in this example the utilities also include the initial $1 that I was forced to put on the table.

When I have extended the Bayesian network to the model in Figure 9.4, I can use the model to give advice on the decision *D*. I have observed my opponent's change of cards (for example, two cards and one card), and I know my own hand (for example, a flush). The probability for *BH* (best hand) is calculated, and it is used to calculate EU(*call*), the *expected utility* of calling: the sum of the various wins and losses weighted by their probability. The formula for EU(*call*) is

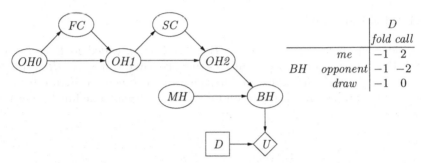

Fig. 9.4. Graphical representation of my decision problem of whether to fold or call. The variable D is a decision variable. The variable U represents the outcome in \$ (shown in the table), and the links into U indicate that the outcomes of the game (only) depend on D and BH.

$$\mathrm{EU}(call) = \sum_{BH} U(BH, call) P(BH \,|\, \mathrm{evidence})$$

$$= P(BH = me \,|\, FC = two, SC = one, MH = flush) U(BH = me, call)$$
$$+ P(BH = draw \,|\, FC = two, SC = one, MH = flush)$$
$$U(BH = draw, call)$$
$$+ P(BH = opponent \,|\, FC = two, SC = one, MH = flush)$$
$$U(BH = opponent, call).$$

If you use the probabilities found in Section 3.2.3, the expected utility of calling is

$$\mathrm{EU}(call) = 0.4 \cdot 2 + 0.054 \cdot (-2) + 0.546 \cdot 0 = 0.692,$$

and since the expected utility of folding is -1, I should call.

9.1.2 Mildew

Two months before the harvest of a wheat field, the farmer observes the state Q of the crop, and he observes whether it has been attacked by mildew, M. If there is an attack, he will decide on a treatment with fungicides.

There are five variables:

- Q with states fair (f), not too bad (n), average (a), and good (g);
- M with states no, little (l), moderate (m), and severe (s);
- H (state of the crop at time of harvest) with the states from Q plus rotten (r), bad (b), and poor (p) (farmers in all countries tend to describe their harvests in pessimistic terms);
- OQ (observation of Q) with the same states as Q;
- OM (observation of M) with the same states as M.

Furthermore, there is a decision node A with decision options no, light (l), moderate (m), and heavy (h) and a variable M' describing the mildew attack after the decision. We define a utility function $U(H)$ giving the utility of the outcome of the harvest for each state of the crop. The cost of the decisions is modeled as a utility function C attached to A (the values of C are either negative or zero). The total utility is $U + C$. Figure 9.5 gives a model.

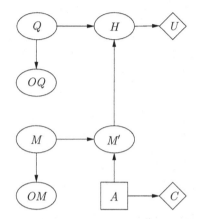

Fig. 9.5. A decision model for mildew.

With evidence e (statements on OQ and OM), the farmer wishes to determine an optimal decision (a decision of maximal expected utility). To do this, he needs to calculate the expected utility of the various options. That is, for each state a of A, we first calculate $P(H \mid A = a, e)$, and then

$$\text{EU}(A \mid e) = C(A) + \sum_H U(H)P(H \mid A, e).$$

9.1.3 One Decision in General

The general situation with one decision variable is as described in Figure 9.6. There is a Bayesian network structure with chance nodes and directed links. The network is extended with a single decision node D that may have an impact on the variables in the structure. In other words, there may be a link from D to some chance nodes. Furthermore, there is a set of utility functions, $U_1, \ldots, U_n$, over domains $\mathcal{X}_1, \ldots, \mathcal{X}_n$.

The task is to determine the decision that yields the highest expected utility. Thus, if none of the utility nodes contain D in the domain, then with evidence e we calculate

$$\text{EU}(D \mid e) = \sum_{\mathcal{X}_1} U_1(\mathcal{X}_1)P(\mathcal{X}_1 \mid D, e) + \cdots + \sum_{\mathcal{X}_n} U_n(\mathcal{X}_n)P(\mathcal{X}_n \mid D, e),$$

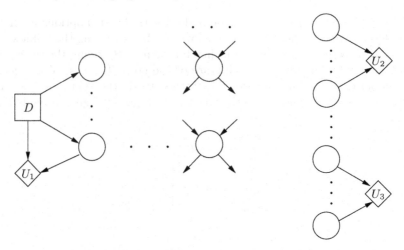

Fig. 9.6. A graphical representation of a one-action decision scenario.

and a state d maximizing $\mathrm{EU}(D = d \,|\, e)$ is chosen as an optimal decision. When D is contained in the domain of a utility node, such as U_1 in Figure 9.6, then we should perform the summation only over $\mathcal{X}_1 \setminus \{D\}$, and accordingly, we should use the probability distribution $P(\mathcal{X}_1 \setminus \{D\} \,|\, D, e)$.

A requirement of the method described above is that the decision problem contains only a single decision. When one is working with decision problems involving several decisions, things become a bit more complicated (we shall return to this issue in Sections 9.3 and 9.4).

9.2 Utilities

We treat decision problems in the framework of *theory*. Decisions are made because they may be of use in some way. Therefore, the various decisions should be evaluated on the basis of the usefulness of their consequences. We assume that "usefulness" is measured on a numerical scale called a *utility scale*, and if several kinds of utilities are involved in the same decision problem, then the scales have a common unit.

Management of Effort

> In your computer science studies you attend two courses, *Graph Algorithms* and *Machine Intelligence*. In the middle of the term, you realize that you cannot keep pace. You can either reduce your effort in both courses slightly or you can decide to attend one of the courses superficially. What is the best decision?

You have three possible actions:

Gm: Keep pace in Graph Algorithms and follow Machine Intelligence superficially.
SB: Slow down in both courses.
Mg: Keep pace in Machine Intelligence and follow Graph Algorithms superficially.

The results of the actions are your final marks for the courses. The marks
are integers between 0 and 5, where 0 and 1 are failing marks. You have certain
expectations for the marks given your effort in the rest of the term. They are
shown in Table 9.1.

	kp	sd	fs		kp	sd	fs
0	0	0	0.1	0	0	0	0.1
1	0.1	0.2	0.1	1	0	0.1	0.2
2	0.1	0.1	0.4	2	0.1	0.2	0.2
3	0.2	0.4	0.2	3	0.2	0.2	0.3
4	0.4	0.2	0.2	4	0.4	0.4	0.2
5	0.2	0.1	0	5	0.3	0.1	0

$P(GA \mid \textit{effort})$ $P(MI \mid \textit{effort})$

Table 9.1. The conditional probabilities of the final marks in Graph Algorithms
(GA) and Machine Intelligence (MI) given the efforts *keep pace (kp)*, *slow down
(sd)*, and *follow superficially (fs)*.

A way of solving your decision problem would be to say that the numeric
value of the mark is a utility, and you want to maximize the sum of the
expected marks. The calculations would then be

$$\mathrm{EU}(Gm) = \sum_{m \in GA} P(m \mid kp)m + \sum_{m \in MI} P(m \mid fs)m = 3.5 + 2.3 = 5.8,$$

$$\mathrm{EU}(SB) = \sum_{m \in GA} P(m \mid sd)m + \sum_{m \in MI} P(m \mid sd)m = 2.9 + 3.2 = 6.1,$$

$$\mathrm{EU}(Mg) = \sum_{m \in GA} P(m \mid fs)m + \sum_{m \in MI} P(m \mid kp)m = 2.3 + 3.9 = 6.2.$$

From this, you would conclude that you should follow Graph Algorithms superficially but keep pace in Machine Intelligence.

However, do the marks really reflect your utilities? If, for example, you
had the same number of marks but the numeric values were $0, 5, 6, 8, 9, 10$,
you would have come to another conclusion. The problem is that you cannot
expect that a difference of 1 in mark number always represents the same
difference in utility. Actually, in this case your subjective utility is probably
not increasing in the numeric value of the mark: the rule at your university

is that if you fail, you are given another chance, but if you pass, you are not allowed to try again to get a better mark. Therefore, you find that the worst mark to get is a 2 rather than a 0!

To overcome this problem, the mark scale is mapped into a utility scale going from 0 to 1. The best possible mark (5) is given the utility 1, and the worst possible mark (2) gets the utility 0.

The intermediate marks are given utilities between 0 and 1 by imagining that you have a choice between two games:

Game 1: You get for certain the mark x;
Game 2: You get mark 5 with probability p, and you get mark 2 with probability $1 - p$.

Which game would you prefer?

If $p = 0$, you would prefer Game 1, and for $p = 1$, Game 2 would be best. For some p between 0 and 1, you would be indifferent, and *this p is the utility for the mark x*. Specifically, if you should find a value for p that would make you indifferent between games 1 and 2, then it should hold that

$$\mathrm{EU}(\textit{Game 1}) = \mathrm{EU}(\textit{Game 2}).$$

This can be rewritten as $1 \cdot U(x) = (1 - p) \cdot U(2) + p \cdot U(5)$, and by exploiting that $U(2) = 0$ and $U(5) = 1$ we get $U(x) = p$.

In Table 9.2, we have performed the utility assessment for you. The utilities assessed are for only one course. We will now assume that the utility of marks for several courses is the sum of the individual utilities. Note that this is not evident (it might, for example, be that you prefer two 2's to failing both courses, which would delay your studies considerably), and an alternative could be to construct a single utility function for both courses.

Mark	0	1	2	3	4	5
Utility	0.05	0.1	0	0.6	0.8	1

Table 9.2. Utilities for the various marks (the same for both courses).

In Figure 9.7, the decision model is illustrated. To find an optimal decision, the calculations are

$$\mathrm{EU}(\textit{action}) = \sum_{m \in GA} P(m \mid \textit{action}) U_{GA}(m) + \sum_{m \in MI} P(m \mid \textit{action}) U_{MI}(m).$$

We get $\mathrm{EU}(Gm) = 1.015, \mathrm{EU}(SB) = 1.07, \mathrm{EU}(Mg) = 1.045$, and the optimal decision is therefore SB.

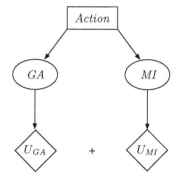

Fig. 9.7. A decision model for effort.

9.2.1 Instrumental Rationality

Beneath the principle of maximal expected utility there is a normative claim that rational decision making shall be represented as a task of calculating expected utilities and to choose an option of maximal expected utility. The question is whether this claim includes all kinds of human choice (private decisions, company decisions, political decisions, etc.). Does it cover choosing the dinner for tomorrow as well as whether to kill your husband or leave him? Does it include setting of tax rates and building dams for flood protection?

It is not claimed that humans/companies/politicians act in accordance with the principle of maximal expected utility (which can easily be disproved). The claim is that if the decision maker takes his time to analyze the situation to find out which choice seems the best, then it is irrational not to choose one of maximal expected utility.

In order not to enter into a circular argument, you need to be precise about the term *rational* without referring to utilities, and a way of doing so is to put up a set of rules that characterize rational choice. The rules need not be exhaustive or independent, but they should have the character that everybody agrees that it is irrational not to obey them.

Below we present the first such set of rules, presented by von Neuman and Morgenstern in 1947. The rules have been called *axioms of instrumentally rational choice*, and they are formulated in terms of preferences over lotteries. Formally, a *lottery* is a probability distribution over a set of outcomes/prices, denoted by X, where an outcome $X = x$ can be a bundle of commodities, services, resources, etc. The lottery with a certain outcome of the price x is denoted by $[x]$. The decision maker is supposed to rank the lotteries by preference. The notation $A \succeq B$ denotes that B is not preferred to lottery A, $A \succ B$ denotes that A is (strictly) preferred to B; and $A \sim B$ denotes that the decision maker is indifferent between A and B (shorthand for $A \succeq B$ and $B \succeq A$).

Construction of mixed lotteries. From two lotteries A and B we can construct compound lotteries. Let $\alpha \in [0, 1]$. Then $\alpha A + (1 - \alpha)B$ is a new lottery: with probability α, A is drawn, else B.

Axioms of instrumentally rational choice:

1. *Reflexivity.* For any lottery A, $A \succeq A$.
2. *Completeness.* For any pair (A, B) of lotteries, $A \succeq B$ or $B \succeq A$.
3. *Transitivity.* If $A \succeq B$ and $B \succeq C$, then $A \succeq C$.
4. *Preference increasing with probability.* If $A \succeq B$ then $\alpha A + (1 - \alpha)B \succeq \beta A + (1 - \beta)B$ if and only if $\alpha \geq \beta$.
5. *Continuity.* If $A \succeq B \succeq C$ then there exists $\alpha \in [0, 1]$ such that $B \sim \alpha A + (1 - \alpha)C$.
6. *Independence.* If $C = \alpha A + (1-\alpha)B$ and $A \sim D$, then $C \sim (\alpha D + (1-\alpha)B)$.

Theorem 9.1. *For an individual who acts according to a preference ordering satisfying rules 1–6 above, there exists a utility function over the outcomes so that the expected utility is maximized.*

Proof. Since the set of prices X is finite, there is a best price, x_B, and a worst price, x_W. Without loss of generality we set $U(x_B) = 1$ and $U(x_W) = 0$. The continuity axiom [5] then yields that for any price x there is an $\alpha \in [0, 1]$ such that $[x] \sim \alpha[x_B] + (1 - \alpha)[x_W]$. We set $U(x) = \alpha$.

Now let x_i denote prices and let t_i be probabilities. From standard probability calculus we have that if $A = \alpha B + (1 - \alpha)C$, $B = \sum_i t_i^B[x_i]$, and $C = \sum_i t_i^C[x_i]$, then $A = \sum_i (\alpha t_i^B + (1 - \alpha)t_i^C)[x_i]$. That is, any lottery A can be written in the form

$$A = \sum_i t_i[x_i],$$

and

$$\mathrm{EU}(A) = \sum_i t_i U(x_i).$$

Since $[x_i] \sim U(x_i)[x_B] + (1 - U(x_i))[x_W]$, we get (axiom [6])

$$A \sim \sum_i t_i(U(x_i)[x_B] + (1 - U(x_i))[x_W]).$$

Since $U(x_i)$ is independent of t_i, we have

$$A \sim \left(\sum_i t_i U(x_i)\right)[x_B] + \left(\sum_i t_i(1 - U(x_i))\right)[x_W].$$

Hence, for all lotteries A we have (axiom [3])

$$A \sim \alpha[x_B] + (1 - \alpha)[x_W],$$

where $\alpha = \text{EU}(A)$. Now let $A \sim \alpha[x_B] + (1 - \alpha)[x_W]$ and $B \sim \beta[x_B] + (1 - \beta)[x_W]$. By axiom [4] we have that $A \succeq B$ if and only if $\alpha \geq \beta$ if and only if $\text{EU}(A) \geq \text{EU}(B)$.

$\square$

The theorem says that if you agree that rules 1–6 apply for your decision problem, then you have to choose a decision that maximizes your expected utility. If you do not wish to follow the recommendation of a perfect max-EU analysis of your problem, your only way out is to attack the rules.

To illustrate this point, consider the following example (Allais' paradox). You have a choice between two lotteries:

- Lottery $A = [\$1\text{mill.}]$,
- Lottery $B = 0.1[\$5\text{mill.}] + 0.89[\$1\text{mill.}] + 0.01[\$0]$.

Most probably you would strictly prefer A to B because your life would be completely changed if you got \$1 million, and in B there is a risk of this not happening. This reasoning is perfectly rational. It reflects only that your subjective utility of \$1 million is very close to your utility of \$5 million. This must also be the case in other situations. Assume that you are faced with a new choice between two lotteries:

- Lottery $C = 0.11[\$1\text{mill.}] + 0.89[\$0]$,
- Lottery $D = 0.1[\$5\text{mill.}] + 0.9[\$0]$.

In turns out that if you chose D (as many people would do) you would not maximize expected utility. In other words, if you seriously mean that the difference in utility between \$1 million and \$5 million is very small, you must take the extra 1% chance of winning \$1 million.

The following calculations show that choosing D does not maximize your expected utility. Let $U(\$5\text{mill.}) = 1, \text{U}(0) = 0, \text{U}(\$1\text{mill.}) = u$. If you prefer A to B, you have

$$u > 0.1 + 0.89u.$$

Hence

$$u > \frac{10}{11}$$

and now

$$\text{EU}(C) = 0.11u > 0.11\frac{10}{11} = 0.1 = \text{EU}(D).$$

The rules presented here cover a simple type of decision problem. There is an extensive scientific debate about how wide the scope is for the principle of maximizing expected utilities in a world assigned with subjective probabilities. Axioms similar to the axioms presented here have been devised, and theorems similar to Theorem 9.1 have been proved.

9.3 Decision Trees

A classical way of representing decision problems with several decisions is with *decision trees*. A decision tree is a model that encodes the structure of the decision problem by representing all possible sequences of decisions and observations explicitly in the model.

The nonleaf nodes in a decision tree are decision nodes (rectangular boxes) or chance nodes (circles or ellipses), and the leaves are utility nodes (diamond shaped). The links in the tree have labels. A link from a decision node is labeled with the action chosen, and a link from a chance node is labeled by a state.

Example 9.1 (The Two-Test Milk Problem). Consider the infected milk scenario from Figure 3.1 and Section 3.2.1 (to keep things simple, we assume that the infections and tests are independent between the days). The farmer has 50 cows, and the milk from each cow is poured into a common container and transported to the dairy. The value of the milk is $2 per cow. The dairy checks the milk carefully, and if it is infected it is thrown away. After having milked a cow, the farmer may perform two different tests of the milk, T_A and T_B, before pouring it into the container. The price of the first test is 6 cents and it has a false positive/negative rate of 0.01, and the price of the second test is 20 cents and it has a false positive/negative rate of 0.001.

To establish the utilities, let us assume that the farmer has clean milk from the 49 other cows. If the farmer pours the milk into the container, he will gain $100 if it is not infected, and he will gain nothing if it is infected. If he throws the milk away, he will gain $98 regardless of the state of the milk.

The question is whether he should perform the tests and in which order. Figures 9.8 and 9.9 show the graphical part of a decision tree for the milk example with two tests.

A decision tree is read from the root downward. When you pass a decision node, the label tells you what the decision is, and when you pass a chance node, the label tells you the state of the node. If a decision node follows a chance node, then the chance node is observed before the decision is made. Hence the sequence in which we visit the nodes corresponds to the sequence of observations and decisions. We assume *no-forgetting*: when a decision is to be taken, the decision maker knows all the labels on the path from the root down to the current position in the decision tree. We adopt the shorthand *past* for the set of labels from the root to a position in the tree.

Each path from the root to a leaf specifies a complete sequence of observations and decisions, and we call such a sequence a *decision scenario*. Furthermore, we require decision trees to be complete: from a chance node there must be a link for each possible state, and from a decision node there must be a link for each possible decision option. This also means that a decision tree specifies all the possible scenarios in the decision problem.

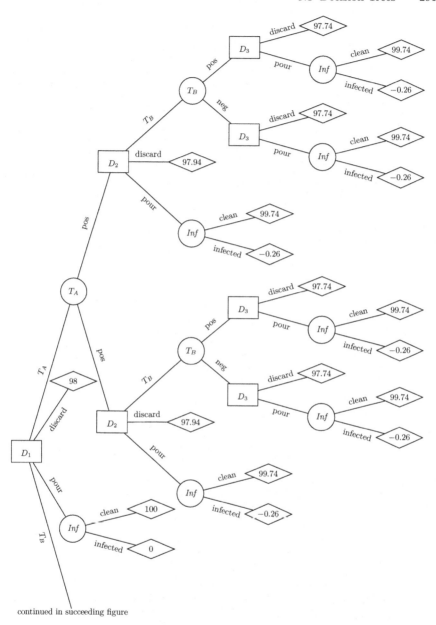

continued in succeeding figure

Fig. 9.8. The graphical part of a decision tree for the milk problem from Example 9.1. The tree reflects that no test is performed when the milk has been poured or discarded. Note that nodes in a decision tree may share names.

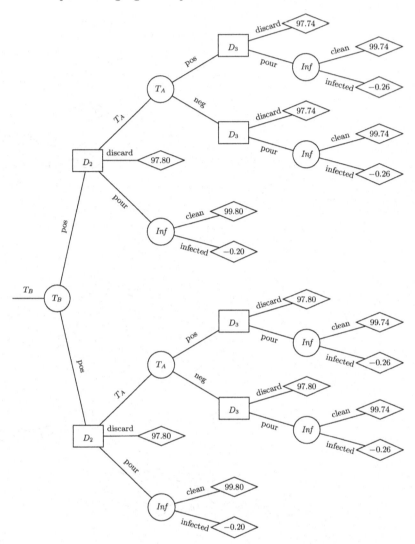

Fig. 9.9. Continuation of diagram in Figure 9.8.

The quantitative part of a decision tree consists of utilities and probabilities. Each leaf has a utility value attached to it. This utility reflects the utility of the decision scenario identified by the path from the root to the leaf in question. For the chance nodes, we associate a probability with each of the links emanating from them. See Figure 9.11 for an example. Let A be a chance node at a particular position in the tree with past o, and let l be an outgoing link labeled with a. We then associate $P(A = a \mid o)$ with this link. Either you can have the probabilities explicitly attached to the links (which can be rather impractical to work with), or you can use your Bayesian network model as a reference. You can, for example, complement the graphical part in Figures 9.8 and 9.9 with the Bayesian network in Figure 9.10 and then use the Bayesian network to calculate the required probabilities.

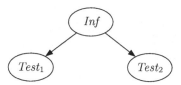

Fig. 9.10. A Bayesian network for calculating the probabilities for the decision tree in Figures 9.8 and 9.9.

9.3.1 A Couple of Examples

We now give two other examples of decision problems involving a sequence of decisions.

Example 9.2 (The Car Start Problem). In the morning, my car will not start. There are three possible faults: the *spark plugs* may be dirty, with probability 0.3; the *ignition system* may be malfunctioning, with probability 0.2; or there is some *other* cause, with probability 0.5. I can perform two repair actions myself: SP, which at the cost of 4 minutes always fixes spark plugs; and IS, which takes 2 minutes and fixes the ignition system with probability 0.5. I can also perform a test T, namely to check the charge on the spark plugs when starting. It takes half a minute, and it says *ok* if and only if the ignition system is okay. Finally, I can call road service RS, which at the cost of 15 minutes fixes everything. The car was okay yesterday evening, so I assume that there is at most one fault.

To work with utilities rather than costs, let us say that I have 30 minutes to fix the car and arrive at work, and I want to find a test–repair sequence that expectedly gives me as much time as possible for getting to work. Therefore, the utility of a test–repair sequence is the remaining time for getting to work.

A decision tree for this Car Start Problem is shown in Figure 9.11. The probabilities for the decision tree are calculated from the model in Figure 9.12, where the technique from Section 3.3.9 is used.

Example 9.3 (The Reactor Problem).

An electric utility firm must decide whether to build (B) a reactor of advanced design (a), a reactor of conventional design (c), or no reactor (n) at all. If the reactor is successful, an advanced reactor is more profitable, but it is also more risky.

If the firm builds a conventional reactor, the profits are \$8B if it is a success (cs), and −\$4B if there is a failure (cf). If the firm builds an advanced

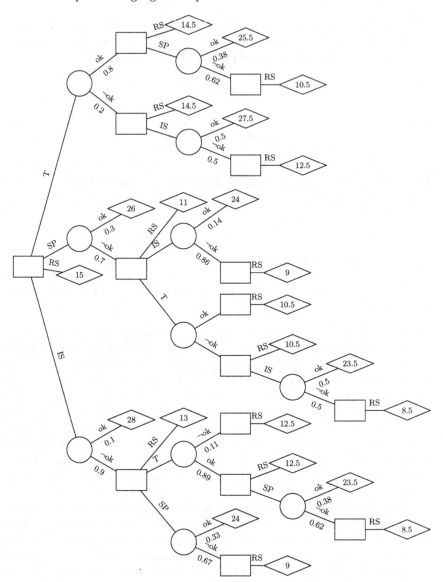

Fig. 9.11. A decision tree for the Car Start Problem in Example 9.2.

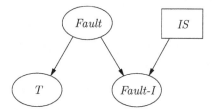

Fig. 9.12. A model for calculating the probabilities for a decision tree for the Car Start Problem in Example 9.2. Due to the assumption of exactly one fault, the faults are collected in the node *Fault* with states *is, sp,* and *other*.

reactor, the profits are \$12B if it is a success (*as*), −\$6B if there is a limited accident (*al*), and −\$10B if there is a major accident (*am*). The firm's utility is assumed to be linear in dollars. Before making the decision to build, the firm has the option to conduct a test ($T = t$) or not (*nt*) of the components (*Cp*) of the advanced reactor. The test results (R) can be classified as either bad (*b*), good (*g*), or excellent (*e*). The cost of the test is \$1B. If the test results are bad, then the Nuclear Regulatory Commission (NRC) will not permit the construction of an advanced reactor.

Figure 9.17 shows a decision tree representation of the problem, where the probabilities can be found from the Bayesian network in Figure 9.14.

The specification of the quantitative part (Figure 9.14) can be extended with decision nodes and utility nodes as shown in Figure 9.15, which can also be considered a model of the relevant world.

9.3.2 Coalesced Decision Trees

The main drawback of decision trees is that they grow exponentially with the number of decision and chance variables, and – as illustrated in the two examples – even very small decision problems require a relatively large decision tree. There are, however, methods for reducing the complexity by exploiting symmetries in the decision problem.

The idea is that when a decision tree contains identical subtrees, they can be collapsed. In the milk problem, if both tests are negative, the situations will be the same regardless of the order in which the tests are performed. The succeeding parts of the decision tree must therefore be the same, both in terms of structure and numerical information (probabilities and utilities); hence we can have the links to these parts meet in a common decision node. Figure 9.16 shows the structure of a coalesced decision tree for the milk problem, and Figure 9.17 shows the coalesced decision tree for the reactor problem.

The procedure for solving a coalesced decision tree is the same as the procedure for normal decision trees (see the next section).

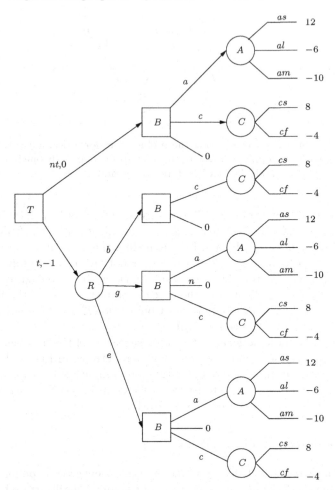

Fig. 9.13. A decision tree for the Reactor Problem. Note that the cost of the test is attached to the link $T = t$, indicating that the cost will be the same for all ensuing scenarios.

9.3.3 Solving Decision Trees

A solution to a decision tree is a *strategy* that specifies how we should act at the various decision nodes. An example of a strategy is illustrated in Figure 9.18 by the boldfaced links. Strategies are compared based on their expected utilities, and finding an *optimal strategy* amounts to finding a strategy with highest expected utility; such a strategy is not necessarily unique.

By assigning to each node in the decision tree a value corresponding to the maximum expected utility achievable at that node, an optimal strategy will pick an action leading to a child of maximum value. Looking at the end

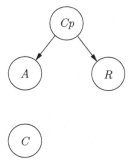

Fig. 9.14. A Bayesian network providing probabilities for the decision tree representation of the Reactor Problem shown in Figure 9.13.

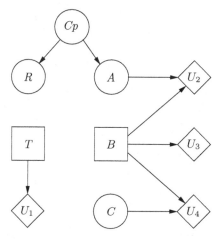

Fig. 9.15. A model of the world relevant for the reactor problem.

of the decision tree, one sees that the value of a leaf node is simply the utility assigned to that node. If we go one step further up the tree, then the value of a decision node D is the maximum value associated with its children/leaves, since D is under our full control. For a chance node, its value corresponds to the utility you can expect to achieve from that point in the decision tree: the value is the sum of the utilities of the leaves weighted with the probabilities of their outcomes. When all children of a node N have been assigned a value, we can calculate the value to assign to N. If N is a decision node, we assign it the maximum of the children's values, and if N is a chance node, we assign the weighted sum.

These observations form the basis for a procedure known as "average-out and fold-back" for calculating an optimal strategy and the maximum expected utility: start with nodes that have only leaves as children. If the node is a

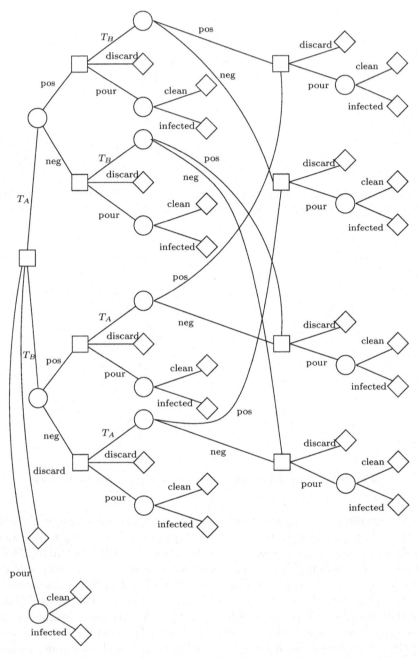

Fig. 9.16. The structure of a coalesced decision tree for the milk problem.

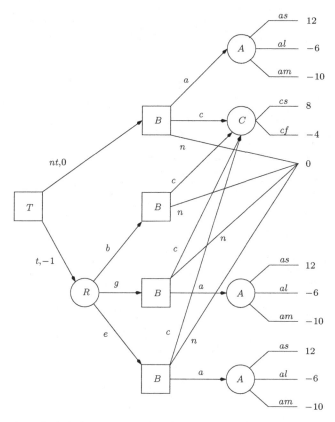

Fig. 9.17. A coalesced decision tree for the reactor problem. If we decide to build a conventional reactor the resulting subtrees will be the same regardless of our previous decisions and observations.

chance node A, the expected utility for A is calculated. Each child of A is an outcome o and has a utility $U(o)$ attached, and the link has a probability $P(A = a)$. We calculate the product $U(o) \cdot P(A = a)$ from each child, and their sum is attached to A. If the node is a decision node D, each child of D has an (expected) utility attached. Choose a child with maximal expected utility, highlight the link, and attach the value to D.

This is done repeatedly until the root is reached. The resulting value for the root is the expected utility if you adhere to the strategy of always maximizing the expected utility, and the paths from root to leaves following highlighted links when possible represent an optimal strategy for the decision problem.

Example 9.4 (The Car Start Problem, continued).

Figure 9.18 illustrates the calculations for solving the troubleshooting problem.

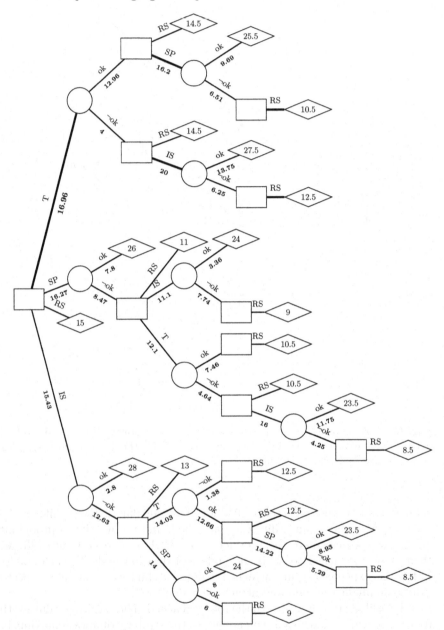

Fig. 9.18. Results when solving the decision tree from Figure 9.11. The boldfaced links indicate the optimal strategy.

As can be seen from Figure 9.18, the maximum expected utility is 16.96. A strategy close to the optimal one (in terms of expected utility) is to start performing SP and if unsuccessful to follow with T.

More formally, if we use $N(X = x)$ to denote the node following X by the link labeled x, then the "average-out and fold-back" algorithm can be specified recursively as follows.

Algorithm 9.1 [Expected-Utility (EU)] *Let X be a node in a decision tree T. To calculate an optimal strategy and the maximum expected utility for the subtree rooted at X, do:*

1. *If X is a utility node, then return $U(X)$.*
2. *If X is a chance node, then return*

$$EU(X) = \sum_{x \in sp(X)} P(X = x \mid past(X)) \, EU(N(X = x)).$$

3. *If X is a decision node, then return*

$$EU(X) = \max_{x \in sp(X)} EU(N(X = x)),$$

and mark the arc labeled:

$$x' = \arg \max_{x \in sp(X)} EU(N(X = x)).$$

□

By unfolding the calculations in the algorithm, we see that the expected utility of an optimal strategy Δ is the sum of the utilities of the possible outcomes o (the leaves in the decision tree) weighted by the probability of the path down to o under the strategy Δ:

$$EU(\Delta) = \sum_o U(o) P(o \mid \Delta).$$

The probability $P(o \mid \Delta)$ is the product of the probabilities attached to the arcs on the path from the root to o, where arcs emanating from decision nodes contribute 1 if they are part of Δ and 0 otherwise. For example, the strategy in Figure 9.18 is first to perform the test T, and if it says ok then follow with SP and possibly RS. If T says $\neg ok$, then follow with IS and possibly RS. The strategy has four possible outcomes, and the expected utility is

$$
\begin{aligned}
EU(\Delta) ={}& 25.5 \cdot P(T = ok, SP = ok \mid \Delta) + 10.5 \cdot P(T = ok, SP = \neg ok \mid \Delta) + \\
& 12.5 \cdot P(T = \neg ok, IS = \neg ok \mid \Delta) + 27.5 \cdot P(T = \neg ok, IS = ok \mid \Delta) \\
={}& 25.5 \cdot 0.8 \cdot 0.38 + 10.5 \cdot 0.8 \cdot 0.62 + 12.5 \cdot 0.2 \cdot 0.5 + 27.5 \cdot 0.2 \cdot 0.5 \\
={}& 16.96.
\end{aligned}
$$

In general, this procedure can be used for calculating the expected utility of any strategy; hence the identification of an optimal strategy could also be formulated as

$$\Delta = \arg\max_{\Delta'} \text{EU}(\Delta').$$

This approach, however, clearly has a complexity problem, since we should explore all possible strategies. The reason that this problem is not as apparent in the algorithm above is that it exploits a general principle known as *dynamic programming*. The idea is that the contribution from, say, the subtree rooted at $T = \neg ok$ is independent of the subtree rooted at $T = ok$; hence a strategy that is optimal for the subtree at $T = \neg ok$ will be part of an optimal strategy for the full decision tree.

9.4 Influence Diagrams

Decision trees are very easy to use, but they have a serious drawback: the number of decisions and observations need not be large before it becomes an inhuman task to specify the problem. We therefore look for other modeling frameworks that in a much more compact way can be used to represent decision problems with several decisions and observations.

In this section we present the *influence diagram* framework. It is particularly well suited for so-called *symmetric* decision problems.

In the decision tree framework, we used two models for describing a decision problem: a Bayesian network for calculating probabilities and a decision tree for representing the sequence of decisions and observations. In the influence diagram framework the approach is different: the Bayesian network is extended with syntactic features that will allow it to encode the probability model as well as the structure of the decision problem.

9.4.1 Extended Poker Model

In the poker problem described in Section 3.2.3, the final decision is whether to call or fold. When taking this decision I have information about my own hand (*MH*) as well as the number of cards my opponent has discarded in the first and in the second round of changing cards. However, before I come that far I would also have had to decide on my first change of cards (*MFC*) and my second change of cards (*MSC*). In order to make these two decisions explicit in the representation, you can extend the model in Figure 9.3 with *MFC* and *MSC* as well as two variables representing my initial hand (*MH0*) and my hand after the first change of cards (*MH1*). The resulting model is shown in Figure 9.19.

Looking at Figure 9.19 we see that even though all relevant variables are included in the model, it does not convey the order in which the decisions

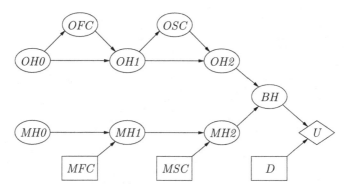

Fig. 9.19. The poker model in Figure 9.3 extended with variables for my initial hand (*MH0*), my first change of cards (*MFC*), my second hand (*MH1*), and my second change of cards (*MSC*).

are taken; nor does it specify the variables that are observed before a particular decision: before deciding on the first decision *MFC* I observe *MH0*; then I observe my opponent's first change of cards *OFC* as well as my second hand *MH1* before I decide on *MSC*; and finally, I observe both *MH2* and my opponent's second change of cards *OSC* prior to deciding on *D*.

An immediate way to encode this information directly in the model is to extend the model with so-called *information arcs*. An information arc is a directed arc $X \to D$ going into a decision node D from either a chance node or another decision X. Semantically it specifies that X is either observed (if it is a chance node) or decided on (if it is a decision node) before we decide on D. By extending the model in Figure 9.19 with information arcs we get the model in Figure 9.20, where we can see, for example, that when deciding on *MSC* we know the state of *OFC*, *MH0*, *MFC*, and *MH1*.

Now assume that we adopt the no-forgetting assumption from the decision tree framework, i.e., the decision maker remembers all previous observations and decisions. Given this assumption, we see that the model in Figure 9.20 contains redundant information arcs. For example, the arc $MFC \to MSC$ indicates that we decide on *MFC* before deciding on *MSC*, and the two arcs from *MH0* into *MFC* and *MSC* specify that the state of *MH0* is known when we decide on both *MFC* and *MSC*. However, under the no-forgetting assumption the link from *MH0* to *MSC* is redundant and it can therefore be removed. Similarly, *MFC* has an impact on *MH1*, which is observed before *MSC*. Therefore, *MFC* must precede *MSC*, and the link from *MFC* to *MSC* can be removed. By iteratively removing all redundant information arcs we obtain the model in Figure 9.21.

A model such as the one shown in Figure 9.21 is also called an *influence diagram*, and it encodes information about the probability model as well as the relevant information about the structure of the decision problem: the directed

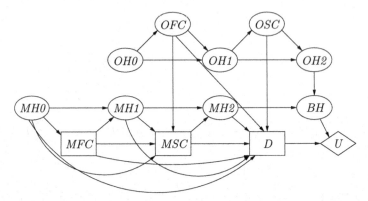

Fig. 9.20. The poker model in Figure 9.19 extended with information arcs into the decision variables.

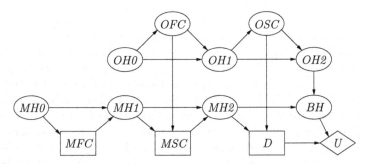

Fig. 9.21. The poker model in Figure 9.19, where the redundant information arcs have been removed.

path going through all the decision variables specifies the sequence in which the decisions are made, and the chance variables appearing as parents of a decision variable are the set of chance variables observed immediately before that decision. For example, since *MH2* and *OSC* are parents of *D*, they are observed immediately before *D* but after the decisions *MFC* and *MSC*. Note that we do not specify the sequence in which *MH2* and *OSC* are revealed, but their ordering will not affect the solution of the influence diagram (see also Section 9.3.3 and Section 10.1). In summary, the sequence of observations and decisions can be described as follows:

$$\{MH0\} \prec MFC \prec \{MH1, OFC\} \prec MSC \prec \{MH2, OSC\} \prec D$$
$$\prec \{OH0, OH1, OH2, BH\}.$$

For the last set of variables it should be noted that whether a variable will eventually be observed depends on the semantics of that variable and cannot be deduced from the syntax of the influence diagram. Finally, we also see that due to the no-forgetting assumption we can read that at the time of deciding on D, I will know the states of the parents $MH2$ and OSC, and by assuming that I do not forget my past, I will also know the states of $MH0$, MFC, $MH1$, OFC, and MSC.

9.4.2 Definition of Influence Diagrams

In the previous section we exemplified the influence diagram framework as an alternative to the decision tree framework. Historically, influence diagrams were invented as a compact representation of decision trees for symmetric decision problems (see Section 9.5). Now they are seen more as a decision tool extending Bayesian networks, and below we formally introduce the influence diagram framework in this way.

Syntax

An *influence diagram* consists of a directed acyclic graph over chance nodes, decision nodes, and utility nodes with the following structural properties:

- there is a directed path comprising all decision nodes;
- the utility nodes have no children;
- the decision nodes and the chance nodes have a finite set of states;
- the utility nodes have no states.

An influence diagram is *realized* when the following quantities have been specified:

- a conditional probability table $P(A \mid \mathrm{pa}(A))$ is attached to each chance node A;
- a real-valued function over $\mathrm{pa}(U)$ is attached to each utility node U.

Unless the context requires a distinction we let the term "influence diagram" include a specification of probabilities and utilities.

Figure 9.22 shows an example of an influence diagram (the states of the variables are not specified).

Semantics

Links into a decision node yield no quantitative requirements. They are called *information links*, and they indicate that the states of the parents are known prior to taking the decision. On the other hand, links into chance nodes or utility nodes represent functional relations.

The structural requirement that there be a path comprising all decision nodes ensures that the influence diagram defines a temporal sequence of decisions. This yields a partitioning of the chance variables into disjoint subsets according to the time of observation. The set $\mathcal{I}_0$ is the set of variables observed before any decision is taken. The set $\mathcal{I}_1$ is the set of variables observed after

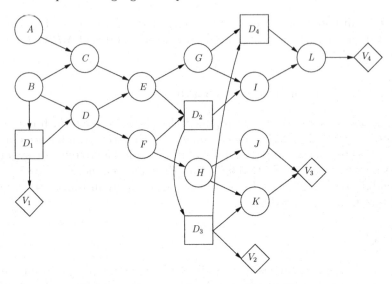

Fig. 9.22. An example of an influence diagram.

the first decision D_1 is taken but before the second decision D_2, and the set $\mathcal{I}_i$ is the set of chance variables observed after decision D_i but before decision D_{i+1}. If there are n decisions, $\mathcal{I}_n$ is the set of variables that are observed after D_n or not at all:

$$\mathcal{I}_0 \prec D_1 \prec \mathcal{I}_1 \prec \ldots \prec \mathcal{I}_{n-1} \prec D_n \prec \mathcal{I}_n.$$

For example, in Figure 9.22 we have $\mathcal{I}_0 = \{B\}$, $\mathcal{I}_1 = \{E, F\}$, $\mathcal{I}_2$ is empty, $\mathcal{I}_3 = \{G\}$, and $\mathcal{I}_4 = \{A, C, D, H, I, J, K, L\}$. The ordering $\prec$ therefore specifies a *partial temporal ordering* over the variables in the influence diagram; the ordering is partial since we do not have an ordering over the variables in each of the sets $\mathcal{I}_i$.

There is a hidden assumption behind the semantics of influence diagrams, namely *no-forgetting*: the decision maker remembers the past observations and decisions. Thus, at D_i we know the state of the variables appearing before D_i under $\prec$.

In some decision problems, two decisions may be independent in the sense that they can be taken in any order without changing the expected utilities. In Figure 9.22, the two decisions D_2 and D_3 are independent. Therefore, the link from D_2 to D_3 puts an unnecessary restriction on the decision maker. It could be removed and the representation would still be meaningful, although the first structural requirement would be violated. Unfortunately, it is not always easy to characterize situations in which decisions are independent, and we will keep the first structural requirement, which ensures a well-specified decision problem. We shall, however, return to this issue in Section 9.5.2 and Section 11.2.

If there is more than one utility node, then the entire utility can be either the sum or the product of the individual utilities. Due to the intuitive appeal, local utilities are usually treated as components in a sum. For instance, in the mildew example (Section 9.1.2) we have two local utility functions: C, which represents the cost of the various treatments, and U, which represents the utility of the harvest for each state of the crops. The total utility is the sum of C and U, and if we assume that both C and U are the actual costs and payoffs, then the sum simply encodes the overall monetary value of the different scenarios as described by the parents of C and U. Should it happen that the total utility is the product rather than the sum of the local utilities, then taking the logarithm of the utilities will transform the problem into an influence diagram in which the total utility is the sum of the transformed utilities.

Solving an Influence Diagram

An influence diagram provides a description of a decision problem and should subsequently be used to aid the decision maker in the decision process. This amounts to prescribing an action for each decision variable conditioned on the previous observations and decisions. A way of doing the prescription is to transform the influence diagram into a decision tree and then apply the "average-out and fold-back" algorithm. The influence diagram's decision tree representation has the property that each node representing a decision D has the same variables in the past. Let past(D) denote the variables in D's past. Thus, if in the decision tree we have an action for each such decision node, these actions will collectively specify an action for each possible configuration past(D). Such a specification is called a *policy* (denoted by δ) for D:

$$\delta_D : \mathrm{sp}(\mathrm{past}(D)) \to \mathrm{sp}(D).$$

If we have a policy for each decision variable in an influence diagram, we call it a *strategy*. For example, a strategy for the influence diagram in Figure 9.21 will consist of three policies:

$$\delta_{MFC} : \mathrm{sp}(MH0) \to \mathrm{sp}(MFC);$$
$$\delta_{MSC} : \mathrm{sp}(MH0, MFC, MH1, OFC) \to \mathrm{sp}(MSC);$$
$$\delta_D : \mathrm{sp}(MH0, MFC, MH1, OFC, MH2, OSC) \to \mathrm{sp}(D).$$

If the strategy encodes the solution of the "average-out and fold-back" algorithm (i.e., the strategy maximizes the expected utility), then the strategy is called an *optimal strategy* and each of its policies is called an *optimal policy*.

Definition 9.1. *A* policy *for decision D_i is a mapping δ_i that for any configuration of the past of D_i yields a decision for D_i. That is,*

$$\delta_i(\mathcal{I}_0, D_1, \ldots, D_{i-1}, \mathcal{I}_{i-1}) \in \mathrm{sp}(D_i).$$

A strategy for an influence diagram is a set of policies, one for each decision. A solution to an influence diagram is a strategy maximizing the expected utility.

By transforming the influence diagram into a decision tree in order to solve it, the complexity problem inherent in the decision tree framework is still present in the solution phase. However, solution methods working directly on the influence diagram have also been developed (see Chapter 10).

9.4.3 Repetitive Decision Problems

Fishing in the North Sea

Every year, the European Union undertakes very delicate political and biological negotiations to determine a volume of fishing for most kinds of fish in the North Sea. Simplified, you can say that each year the EU has a test for the volume of fish, and based on this test the volume of allowable catch is decided. This decision has an impact on the volume for next year (note that the decision on volume does not mean that only this volume is actually caught – quotas have a status similar to speed limits on highways). Figure 9.23 gives an influence diagram for a five-year strategy, where each variable is given ten states.[1]

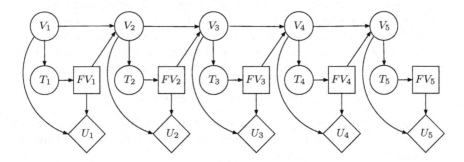

Fig. 9.23. An influence diagram for a five-year strategy for fishing volumes of herring in the North Sea.

The fishing model above has a complexity problem. For the fifth decision, all the past is relevant. Because there are nine ten-state variables in the past, the domain of the policy function for FV_5 has 10^9 elements.

---

[1] The model in Figure 9.23 is an example of a partially observable Markov decision process (POMDP), which we shall consider further in Section 9.6.2.

This does not mean that whenever the past is intractably large, the computer must give up. It fortunately often happens that not all information from the past is relevant (see Section 11.2).

Sometimes solving even fairly small influence diagrams represents an intractable task, and then you must use various approximation methods. One method is *blocking*. The principle in information blocking is to introduce variables that when observed, d-separate most of the past from the present decision.

Fishing Again

The problem with the model in Figure 9.23 is that all information from the past has an impact on how we will estimate the current volume of fish. We can make an approximation by allowing only this year's test and fishing volume to be used for estimating next year's volume of fish. In the model, we delete the arrows $V_i \rightarrow V_{i+1}$ and instead introduce the arrows $T_i \rightarrow V_{i+1}$ (see Figure 9.24).

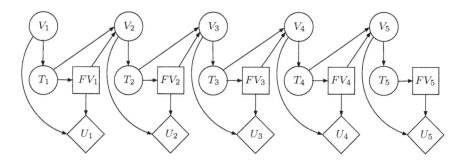

Fig. 9.24. The influence diagram from Figure 9.23 approximated through information blocking.

To establish the potential $P(V_{i+1} \mid T_i, FV_i)$, we can use the model in Figure 9.23.

$$P(V_2, T_1 \mid FV_1) = \sum_{V_1} P(V_1)P(T_1 \mid V_1)P(V_2 \mid V_1, FV_1),$$

$$P(T_1 \mid FV_1) = \sum_{V_2} P(V_2, T_1 \mid FV_1),$$

$$P(V_2 \mid T_1, FV_1) = \frac{P(V_2, T_1 \mid FV_1)}{P(T_1 \mid FV_1)}.$$

This last potential is used for all time slices.

The trick just shown is an example of a general information-blocking technique whereby you abstract the past into a *history variable* and allow only temporal links from observed variables and from the history variable (see Figure 9.25 for another example).

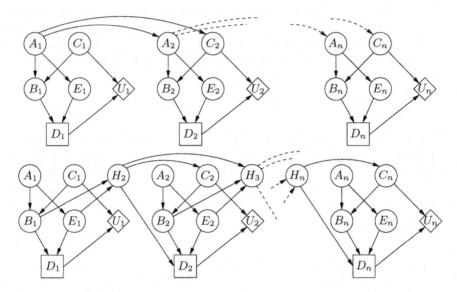

Fig. 9.25. In the top figure we have to take the entire past into account when deciding on D_n. In the lower figure, history variables have been introduced to perform information blocking.

9.5 Asymmetric Decision Problems

From the specification of the syntax for the influence diagram we see that the sequence in which the nodes are observed and decided on is the same in all possible scenarios (up to a permutation of the chance nodes in the sets $\mathcal{I}_i$). For instance, in the poker example we always start by observing *MH0*, and regardless of the outcome we then decide on *MFC*, etc. These types of decision problems are also called *symmetric* decision problems, because they can be represented by a decision tree that is completely symmetric (see Figure 9.26 for an example). If a decision problem is not symmetric we call it *asymmetric*.

Definition 9.2. *A decision problem is said to be* symmetric *if:*

- *in all of its decision tree representations, the number of scenarios is the same as the cardinality of the Cartesian product of the state spaces of all chance and decision variables, and*
- *in at least one decision tree representation, the sequence of chance and decision variables is the same in all scenarios.*

In particular, the first requirement ensures that the possible outcomes and decision options for a variable do not depend on previous observations and decisions. Moreover, the reason why the definition deals with several decision tree representations for a decision problem is that two consecutive observations (without intermediate decisions) or two consecutive decisions can be swapped without affecting the solution to the decision problem. For example, in Figure 9.26 the cardinality of the product of the state spaces of all variables is $2 \cdot 2 \cdot 2 \cdot 2 = 16$. This is also the number of scenarios in the decision tree, and since the decision tree also adheres to the second condition in the definition above, the underlying decision problem is symmetric.

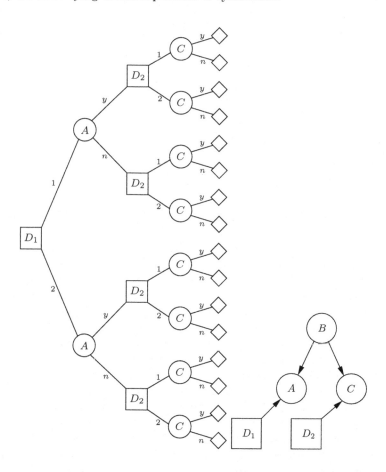

Fig. 9.26. A symmetric decision tree and the associated probability model.

The influence diagram corresponding to the decision problem shown in Figure 9.26 is given in Figure 9.27. From this example we see that the influence diagram provides a much more compact representation of the decision problem than does the decision tree. However, this holds only for symmetric problems: in the (asymmetric) decision tree shown in Figures 9.8 and 9.9 we observe only the result of the first test $Test_1$ if we decide to actually perform the test ($D_1 = T_1$ or $D_2 = T_1$). That is, the sequence in which we make observations and decisions may vary in the different scenarios, but the influence diagram does not provide an immediate mechanism for representing such types of conditional orderings.

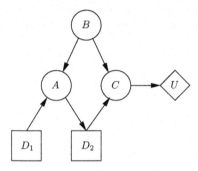

Fig. 9.27. An influence diagram representation corresponding to the decision tree from Figure 9.26.

The use of test decisions (like the ones in Figure 9.8 and 9.9 and in Example 9.1) is a frequent causes of asymmetry in decision problems: if you decide to perform a test, you will eventually observe the test result, but if you decide not to perform the test then a result will never be observed. Influence diagrams do not contain a special representation of test decisions. However, there is a general way of representing test decisions as ordinary decision variables. Assume, in the crop example in Figure 9.1, that I am in the situation that I can test the severity of the mildew attack before I decide on whether to spray. The node *Attack* represents the severity of the attack before spraying, so to model the impact of spraying we introduce a new chance node, *A-Attack*, representing the attack after the spray decision *P*. The decision is connected to the model by inserting a link from *P* to *A-Attack*. To model the test decision we insert a decision node *T*. This decision is basically a decision on whether the state of the chance node *Attack* should be revealed before deciding on *P* (we assume the test to be accurate). One way to model this situation is to introduce an additional node *Attack'* with the same states as *Attack* and with the additional state, *unobserved*, for handling the situation in which we decide not to perform the test. Next we add an arc from *T* and *Attack* to *Attack'* and an informational arc from *Attack'* to *P*. The final model is shown in Figure 9.28.

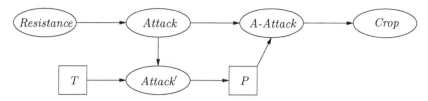

Fig. 9.28. An influence diagram representation (without utility nodes) of the crop problem: should you investigate the severity of the mildew attack before deciding on spraying against mildew?

The table for *Attack'* given T and *Attack* is specified so that the state is *unobserved* if T is *no*, and if T is *yes*, then *Attack'* is in the same state as *Attack* (see Table 9.3 and 9.4).

		\multicolumn{2}{c}{Attack}	
		y	n
T	y	$(1,0,0)$	$(0,1,0)$
	n	$(0,0,1)$	$(0,0,1)$

Table 9.3. The probability table $P(Attack' = (y, n, unobserved) \mid Attack, T)$ associated with *Attack'* in Figure 9.28.

This construction is general, and it is illustrated in Figure 9.29 and Table 9.4. In this way, methods developed for computing decision strategies can also be used for decision scenarios containing test decisions.

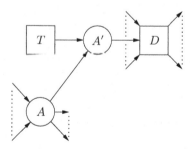

Fig. 9.29. A general way to model a decision on whether to observe A before deciding on D.

The construction can be made a bit simpler by extending the node A with the extra state *unobserved* and thereby avoiding the extra node A'. However, usually it is preferable not to change the nodes of the initial (causal) model.

$$
\begin{array}{c|ccc}
 & & \multicolumn{3}{c}{A} \\
 & & a_1 & \cdots & a_n \\
\hline
 & y & (1,\ldots,0,0) & \cdots & (0,\ldots,1,0) \\
T & n & (0,\ldots,0,1) & \cdots & (0,\ldots,0,1) \\
\end{array}
$$

Table 9.4. The probability table $P(A' = (a_1, \ldots, a_n, unobserved) \mid A, T)$ associated with A' in Figure 9.29.

As the modeling technique illustrates, influence diagrams can be used to model decision problems even when the decision problem is not completely symmetric. However, this comes at a cost since we need to introduce artificial states (e.g. the state *unobserved*) and in some situations it may also be necessary to introduce artificial nodes. In the extreme case in which the decision problem does not contain any symmetric substructures, the decision tree will provide a more compact representation than the influence diagram.

9.5.1 Different Sources of Asymmetry

As we have discussed above, influence diagrams are not really suitable for modeling asymmetric decision problems. However, decision trees are not really an alternative either when there are many observations and decisions. Therefore, much research has been directed at finding specification languages that much more compactly can represent the information needed for describing the decision problem. The following two examples shed additional light on some of the problems we face when constructing such languages.

Example 9.5 (The Diagnosis Problem). Consider a two-test problem like the one in Example 9.1, Page 290; after an initial observation I you have two tests, T_A and T_B, and a decision *Pour?*. The decision on pouring is the last decision, but the two tests can be performed in any order.

To represent this problem by an influence diagram we have to represent the unspecified ordering of the tests as a linear ordering of decisions. Introduce two decision nodes, $Test_1$ and $Test_2$, with options, t_A, t_B, and *no-test*; introduce two chance nodes, O_1 and O_2, as children of *Inf?* with states pos_A, neg_B, pos_A, neg_B, and *no-test*. To specify that two consecutive tests of the same type will give the same results, you introduce a link from O_1 to O_2 (See Figure 9.30).

Example 9.6 (The Dating Problem). Joe needs to decide whether he should ask (*Ask*) Emily for a date for Friday evening. He is not sure whether Emily likes him (*LikesMe*). If he decides not to ask Emily or if he decides to ask and she turns him down, he will then decide whether to go to a nightclub or watch a movie on TV at home (*NClub?*). Before making this decision, he will consult the TV guide to see whether there are any movies he would like to see (*TV*). If he decides to go to a nightclub, he will have to pay a cover charge and pay for drinks. His overall nightclub experience (*NCExp*) will depend on whether he

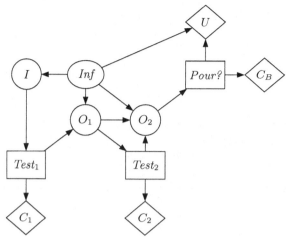

Fig. 9.30. An influence diagram representation of two tests and a decision on pouring. The *Test* nodes have three options, t_A, t_B, and *no-test*. The O nodes have five states, pos_A, pos_B, neg_A, neg_B, *no-test*. The arc $O_1 \rightarrow O_2$ indicates that repeating a test will give identical results.

meets his friends (*MeetFr*), the quality of the live music, etc (*Club*). If Emily accepts (*Accept*), then he will ask her whether she wishes to go to a restaurant or to a movie (*ToDo*); Joe cannot afford to do both. If Emily decides on a movie, Joe will have to decide (*Movie*) whether to see an action movie he likes or a romantic movie that he does not really care for, but which may put Emily in the right mood (*mMood*) to enhance his post movie experience with Emily (*mExp*). If Emily decides on a restaurant, he will have to decide (*Rest*) whether to select a cheap restaurant or an expensive restaurant. He knows that his choice will have an impact on his wallet and on Emily's mood (*rMood*), which in turn will affect his post restaurant experience with Emily (*rExp*).

From the examples above we can identify three types of asymmetry:

Functional asymmetry: The possible outcomes or decision options of a variable may vary depending on the past. We saw this in the reactor problem, where the options of the build decision are dependent on the result of a test.

Structural asymmetry: The very occurrence of an observation or a decision depends on the past. In the Dating Problem, for example, the restaurant options exist only if Emily accepts the invitation.

Order asymmetry: The ordering of the decisions and observations is not settled at the time the model is specified. For instance, in the Diagnosis Problem the ordering of the two tests is unspecified.

9.5.2 Unconstrained Influence Diagrams

In this section we shall look at a particular class of decision problems in which only order asymmetry is present.

Example

Consider again the two-test problem from Example 9.1 (Page 290) and its influence diagram representation shown in Figure 9.30. A much more direct specification would be to use decision nodes representing each test explicitly. If we knew, for example, that $Test_A$ comes before $Test_B$, it can done with an influence diagram (see Figure 9.31(a)). However, in practice this is rarely the case.

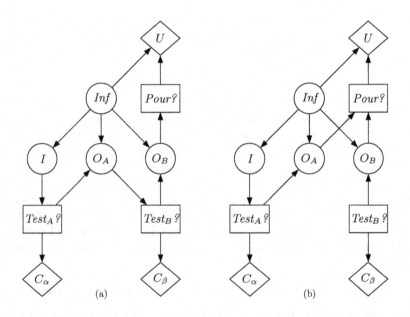

Fig. 9.31. (a) An ID representing the scenario in which you first decide on $Test_A$? and next on $Test_B$?. (b) An attempt to remove the temporal constraint on the test decisions.

To relax the temporal constraint on the test decisions, you may remove the link from O_A to $Test_B$? (Figure 9.31(b)). However, now there is no specification that the result of the first test is known when deciding on the next test. To specify this we introduce a new type of chance variables, *observables*. They are drawn as double circles, and they are observed when all preceding decision nodes have been decided (Figure 9.32). In that case we say that the observable is *free* and that the last preceding decision *released* the observable.

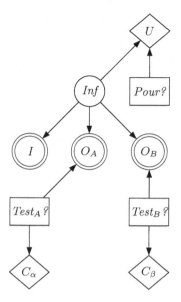

Fig. 9.32. A graphical representation of two tests and a decision of pouring. Here I is observed prior to any decision, O_A is observed when $Test_A$? has been decided, and O_B is observed when $Test_B$? has been decided.

Looking at Figure 9.32 it may seem that we have not specified that O_A is actually observed immediately after deciding on $Test_A$?. However, since the expected utility cannot increase by delaying an observation free of cost, we can safely introduce the rule that an observable chance node is observed immediately after it has been released. This means that the decision problem has been uniquely specified, and the rest can be left to a computer. The specification in Figure 9.32 yields that solving the decision problem boils down to solving two influence diagrams (one for each order of the test decisions) and choosing the order and strategy from the one giving the highest expected utility. This also means that while the influence diagram encoded the possible sequences of observations and decisions at the graphical level, this new framework has postponed it to the solution phase.

Next, consider a more complex situation. A patient may suffer from two different diseases. After an initial observation O_I, there are two possible tests, T_A and T_B, and each disease has a specific treatment, Tr_1 and Tr_2. After each treatment, the new state of the disease is observed (cost free). In Figure 9.33 the problem is specified graphically.

Even for a simple problem like the one above it is extremely cumbersome to draw a decision tree, and it is rather tricky to squeeze the scenario into the ID straightjacket; the problem is that all possible sequences must be represented explicitly.

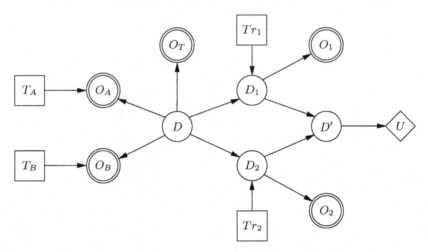

Fig. 9.33. A graphical representation of a situation with two tests and two treatments.

Definition of UIDs

As the examples above illustrate, we can meaningfully relax the linear temporal order constraint for influence diagrams without getting an ambiguous representation.

Definition 9.3. *An* unconstrained influence diagram *(UID) is an acyclic directed graph over decision variables (rectangular shaped), chance variables (circular shaped), and utility variables (diamond shaped). Utility variables have no children. There are two types of chance variables, observables (doubly circled) and nonobservables (singly circled). A nonobservable cannot have a decision as a child.*

Let U be a UID. The set of decision variables is denoted by $\mathcal{D}_U$, and the set of observables is denoted by $\mathcal{O}_U$. The partial temporal order induced by U is denoted by $\prec_U$. When obvious from the context we avoid the subscript.

The quantitative specification required is similar to the specification for influence diagrams: conditional probabilities and utility functions. We add the convention that each decision variable D has a cost. If this cost depends only on D, it is not represented graphically. We say that a UID is *realized* when the structure has been extended with the required quantitative specifications.

The semantics of a UID are similar to the semantics of an ID. A link into a decision variable represents temporal precedence; a link into a chance variable represents causal influence; a link into a utility variable represents functional dependence. We assume *no-forgetting*: at each point of the decision process the decision maker knows all previous decisions and observations.

An observable can be observed when all its antecedent decision variables have been decided on. In that case we say that the observable is *free*, and we *release* an observable when the last decision in its ancestral set is taken.

The structural specification of a UID yields a partial temporal ordering of the decisions and observations. An extension to a linear ordering is called an *admissible order*. Any admissible order yields an influence diagram.

S-DAGs and Strategies

As for decision trees and influence diagrams, the graphical language and its suitability as a language supporting human modeling are the most important properties. Having constructed an adequate model, you can hand it over to a computer, which may then unfold the model to a decision tree and compute an optimal strategy.

In dealing with UIDs, the concept of *strategy* is more complex than in the case of IDs (see Section 9.4.2). In principle we look for a set of rules telling us what to do given the current information, where "what to do" is to choose the next action as well as to choose a decision option if the next action is a decision. That is, a strategy consists of a function prescribing the next step and a set of functions for choosing decisions. The structure of the step function can be represented in a graphical structure, called an *S-DAG* (strategy DAG).

Definition 9.4. *Let U be a UID. An* S-DAG *is a directed acyclic graph G. The nodes are labeled with variables from $\mathcal{D}_U \cup \mathcal{O}_U$ such that each maximal directed path in G represents an admissible ordering of $\mathcal{D}_U \cup \mathcal{O}_U$. For notational convenience we add two unary nodes* Source, *and* Sink. Source *is the only node with no parents and* Sink *is the only node with no children.*

Note that an S-DAG need not contain all admissible orderings. Figure 9.34 gives an example of an S-DAG for the two-tests-two-treatments problem.

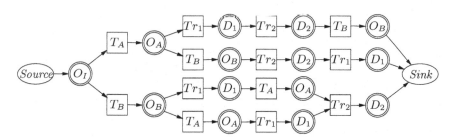

Fig. 9.34. An example of an S-DAG for the UID in Figure 9.33.

For a node N in an S-DAG G, the *history* of N is defined as the union of the labels of N and its ancestors, denoted by $\mathrm{hst}_G(N)$. When the S-DAG is obvious from the context we drop the subscript. For example, the O_B-node at the bottom path in Figure 9.34 has the history $\{O_I, T_B, O_B\}$ and the children $\{T_A, Tr_1\}$; the set of labels of N's children is denoted by $\mathrm{ch}(N)$. A *step policy* for node N is now defined as a function

$$\sigma : \mathrm{sp}(\mathrm{hst}(N)) \to \mathrm{ch}(N).$$

Recall that $\mathrm{sp}(\mathrm{hst}(N))$ denotes all possible configurations of the variables in $\mathrm{hst}(N)$.

A *step strategy* for a UID U is a pair $(G, \mathcal{S})$, where G is an S-DAG for U and $\mathcal{S}$ is a set of step policies, one for each node in G (except for *Sink*); when a node has only one child, the step policy is trivial. For a decision node N a *decision policy* is a function

$$\delta : \mathrm{sp}(\mathrm{past}(N)) \to \mathrm{sp}(N).$$

A *strategy* for U is a step strategy together with a decision policy for each decision node.

Example

Consider the UID in Figure 9.35. A strategy may have the structure illustrated by the S-DAG and the simple policy rules in Figure 9.36. Note that the policies combine step policies and decision policies.

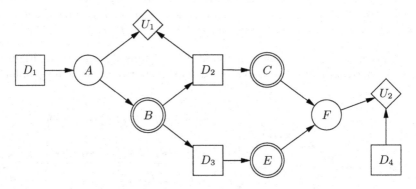

Fig. 9.35. An example UID.

The strategy represented in Figure 9.36 can be unfolded to the *strategy tree* in Figure 9.37. The expected utility from following the strategy can be calculated in the same way as for decision trees, where the UID is used for calculating the probabilities.

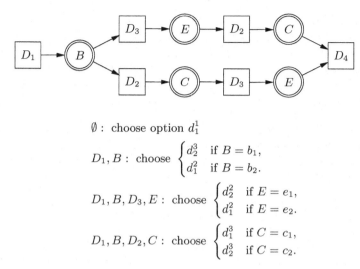

$\emptyset$: choose option d_1^1

D_1, B : choose $\begin{cases} d_2^3 & \text{if } B = b_1, \\ d_1^2 & \text{if } B = b_2. \end{cases}$

D_1, B, D_3, E : choose $\begin{cases} d_2^2 & \text{if } E = e_1, \\ d_1^2 & \text{if } E = e_2. \end{cases}$

D_1, B, D_2, C : choose $\begin{cases} d_1^3 & \text{if } C = c_1, \\ d_2^3 & \text{if } C = c_2. \end{cases}$

Fig. 9.36. The structure of a strategy for the UID in Figure 9.35.

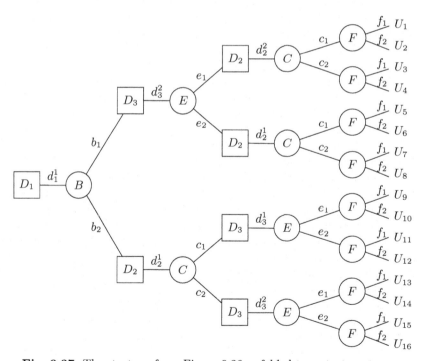

Fig. 9.37. The strategy from Figure 9.36 unfolded to a strategy tree.

Definition 9.5. *Let Δ be a strategy for the UID U. The* expected utility *of Δ is the expected utility of the corresponding unfolded strategy tree for Δ with respect to U. A* solution *to U is a strategy of maximal expected utility. Such a strategy is called an* optimal *strategy. The S-DAG for an optimal strategy is called* optimal, *and the step policies as well as the decision policies are also called* optimal.

Rather than trying out all possible strategy trees in looking for an optimal strategy, there are efficient solution algorithms that exploit dynamic programming and work on a (single) S-DAG representation of the UID (see Section 10.4).

9.5.3 Sequential Influence Diagrams

There is no widely recognized graphical language that compactly can cope with all types of asymmetry. Here we shall indicate only one attempt, called *sequential influence diagrams* (SIDs). The SID framework has its source in a (causal) world model like the one in Figure 9.15. To extend this world model to also represent the structure of the decision problem we need to specify the order of the decisions and observations as well as any asymmetry constraints. There are various ways of doing so. In the case of influence diagrams, the order is specified in the same graph through information links, but you may also have a separate specification (as in decision trees).

The SID framework takes the former approach by extending the world model with features specifying order and asymmetry constraints. This is done in Figure 9.38. The world model is extended with dashed arrows (*structural links*) indicating informational precedence. A label on a link is a *guard* reflecting asymmetry constraints. A guard consists of two parts. The first part takes care of structural asymmetry, and the second part describes functional asymmetry. That is, the first part describes the condition for following the link. If the condition is satisfied we say that the link is *open*. For example, if we decide to perform the test $T = t$ in Figure 9.38, then the next node will be R. If there are constraints on the choices at a decision node, then this is specified in the second part of the guard (this part is empty when there are no constraints). In Figure 9.38 the choice a in B can be taken only if a test is not performed or a test is performed and the result is either good or excellent (i.e., the scenario satisfies $(T = nt) \vee (T = t \wedge (R = e \vee R = g))$).

The specification can be unfolded to a decision tree by iteratively following the open arcs from a source node (a node with no incoming structural arcs) until a node is reached with no open outgoing arcs.

An SID specification of the Dating Problem is shown in Figure 9.39. The framework partly adopts the UID method of representing order asymmetry by introducing *clusters* of nodes (encapsulated in a dashed ellipse). In terms of information precedence, we can think of a cluster $\mathbf{C}$ of nodes as a single node in the sense that a structural arc going into $\mathbf{C}$ from a node X indicates

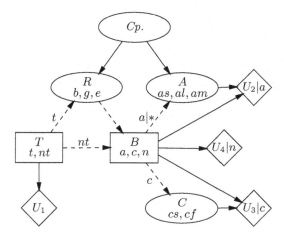

Fig. 9.38. A graphical representation of the Reactor Problem; the $*$ denotes that the choice $B = a$ is allowed only in scenarios that satisfy $(T = nt) \lor (T = t \land (R = e \lor R = g))$.

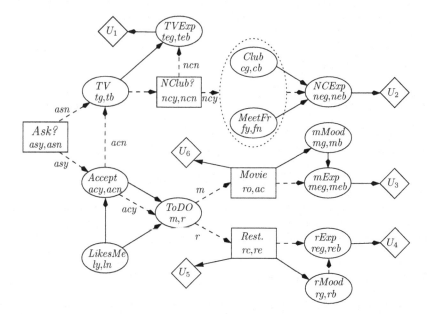

Fig. 9.39. An SID representation of the Dating Problem.

that when X has been observed or decided on, the next node is a node in **C**. A structural arc from **C** to a node Y indicates that Y will be the next node in the ordering on leaving **C**. Figure 9.39 illustrates the use of clusters for representing the partial temporal ordering over the chance nodes *Club* and *MeetFr*. From the model we see that these two nodes will be observed only after a decision on *NClub?* but before *NCExp* is observed.

A sequential influence diagram can be solved by unfolding it into a decision tree. There are, however, more efficient ways, which identify symmetric subtrees and solve them as influence diagrams, but that is outside the scope of this book.

9.6 Decision Problems with Unbounded Time Horizons

Consider a problem of robot navigation in which a robot is placed in some environment and its task is to find a path from its current position to a certain goal position. Each time the robot moves from one position to another it incurs a loss (fuel expenditure), but when it reaches the goal state it receives a reward and the navigation task ends. The aim is now to find a sequence of moves that will maximize the robot's expected reward (and minimize its expected loss):

The problem above is an example of a general type of problem called *planning under uncertainty*:

- at each step we are faced with the same type of decision,
- at each step we are given a certain reward (possibly negative) determined by the chosen decision and the state of the world,
- the outcome of a decision may be uncertain,
- the time horizon of the decision problem is unbounded.

Examples of other problems of this type include factory process control and transportation logistics.

In Section 9.4.3 we discussed a related type of decision problem, namely repetitive decision problems with a bounded time horizon. In what follows we extend this discussion to unbounded time horizons.

9.6.1 Markov Decision Processes

In the robot navigation problem above, the robot's process can roughly be described as an unbounded loop over the following events:

1. observe the state of the world (for example the robot's position in the world),
2. decide on the next action and collect the reward (possibly negative),
3. perform the action.

Using the influence diagram modeling language, we can represent the qualitative part of this problem by the structure in Figure 9.40. The node S_i represents the state of the world at step i; D_i is the ith decision of the robot; and R_i is the reward received when action D_i is performed in state S_i. The dashed arcs indicate that the future time horizon may be unbounded.

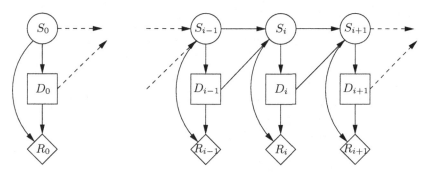

Fig. 9.40. A snapshot of a model of a Markov decision process. S_i represents the state at step i; D_i represents the ith decision; and R_i represents the reward of taking decision D_i in state S_i.

In order to specify the quantitative part of the model we need some additional information about the problem domain. Specifically, we shall assume that the robot is placed in the 3×3 grid environment shown in Figure 9.41. The robot can move *north, east, south,* and *west,* and for each move it incurs a loss of 0.1. If the robot decides to move, say, *north,* then this move will succeed with probability 0.7, and with probability 0.3 the robot will "slip" and move in one of the other three directions with equal probability; if the robot moves into a wall it will remain at its current position. At any point in time the robot can observe its exact position, and the aim is now to find a sequence of moves that will take it to the goal state at position $(3, 1)$ in the upper right corner. At the goal state it will receive a reward of 10, and from this state it cannot exit. Such a state is called a *terminal state*. At positions $(2, 2)$ and $(3, 2)$ two obstacles are placed that will incur a loss of 5 and 1, respectively. Although the environment is bounded, the decision problem is in principle unbounded. The robot may, for example, cycle between two positions an indefinite number of times before entering the goal state.

Returning to the model in Figure 9.40, we see that the variable S_i has a state for each possible position of the robot (a total of nine), and based on the description above, the associated transition function can (for $D_i = $ *north*) be specified as in Table 9.5; the structure of the transition function is similar for the other actions. For this particular example, the reward function is independent of the chosen decision, and $R(S_i, D) (= R(S_i))$ specifies a value for each position.

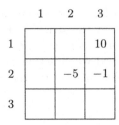

Fig. 9.41. A 3×3 grid world.

S_{i+1}		$(1,1)$	$(1,2)$	$(1,3)$	$(2,1)$	$(2,2)$	$(2,3)$	$(3,1)$	$(3,2)$	$(3,3)$
	$(1,1)$	0,8	0,7	0	0,1	0	0	0	0	0
	$(1,2)$	0,1	0,1	0,7	0	0,1	0	0	0	0
	$(1,3)$	0	0.1	0.2	0	0	0.1	0	0	0
	$(2,1)$	0.1	0	0	0.7	0.7	0	0	0	0
S_{i+1}	$(2,2)$	0	0.1	0	0.1	0	0.7	0	0.1	0
	$(2,3)$	0	0	0.1	0	0.1	0.1	0	0	0.1
	$(3,1)$	0	0	0	0.1	0	0	1	0.7	0
	$(3,2)$	0	0	0	0	0.1	0	0	0.1	0.7
	$(3,3)$	0	0	0	0	0	0.1	0	0.1	0.2

Table 9.5. The transition function $P(S_{i+1} \mid north, S_i)$ for the robot navigation problem.

The robot navigation problem is an example of a *Markov decision process* (MDP). In general, in a Markov decision process:

- the world is *fully observable*, i.e., the agent can observe the true state of the world at any point in time,
- the uncertainty in the system is a result of the consequences of the actions being nondeterministic (when performing an action we make a state transition with a certain probability), and
- for each decision we get a reward (which may be negative) that may depend on the current world state.

More formally:

Definition 9.6 (Markov decision Processes). *An MDP consists of an unbounded set of identical time steps. Each time step i consists of:*

1. *A finite set of states of the world represented by the chance variable S_i.*
2. *A finite set of actions represented by the decision variable D_i.*
3. *A transition function $P(S_{i+1} = s' \mid S_i = s, D_i = a)$ specifying the probability that the next state is s' when action a is taken in state s.*
4. *A reward function $R(S_i = s, D_i = a)$ specifying the reward of taking action a in state s, for each $a \in \mathrm{sp}(D_i)$ and $s \in \mathrm{sp}(S_i)$.*

5. *An* initial state $s_0 \in S_0$.

The transition function and the reward function are the same for all time steps.

In the definition above, we require that at any given point in time the world state be represented by a single variable. This means that when specifying the transition function we need to elicit $|\mathrm{sp}(S) \times \mathrm{sp}(S)|$ probabilities for each decision. In order to make this elicitation task easier, you may exploit the internal structure of the world and represent S as a Bayesian network.

Types of Strategies

A policy for a decision variable is in general a function that returns a decision option for each possible configuration of the variables previously observed and decided on. In dealing with MDPs, however, the past is irrelevant in determining the optimal decision. More precisely, from the d-separation properties of the MDP model in Figure 9.40 we see that the future is independent of the past given the current state of the world S_i (this is also called the *Markov property*). Hence, instead of considering the past for decision D_i, it is sufficient to include only S_i:

$$\delta_{D_i} : \mathrm{sp}(S_i) \rightarrow \mathrm{sp}(D_i).$$

In decision problems with a bounded time horizon we have previously defined an optimal strategy as a collection of optimal polices, one for each decision. However, in dealing with unbounded time horizons the situation is a bit different. To illustrate the difference, consider again the model in Figure 9.24 approximating the fishing in the North Sea decision problem (described in Section 9.4.3). Strictly speaking, according to Definition 9.6, this model is not an MDP, but by marginalizing out the unobserved variables we obtain the equivalent MDP structure in Figure 9.42.

In looking for an optimal strategy for this model it is obvious that the optimal policy for FV_1, say, is not necessarily the same as the optimal policy for FV_5 ($\delta_{FV_1}(T_1) \neq \delta_{FV_5}(T_5)$); even though the tests conducted at year 1 and year 5 produce the same results, the decisions at these two points in time will in general be different. For example, at year 1 the optimal policy may set the allowable catch to a conservative number to ensure that there will be enough fish in the forthcoming years. On the other hand, at year 5 these concerns are irrelevant, since the time horizon stops at that year, and the optimal policy may set the allowable catch to a higher volume. To take another example, in the robot navigation problem we look for a strategy for arriving at the goal state from some starting position, say $(2, 3)$. Suppose now that we have a finite time horizon and require that the robot should reach the goal state within 4 steps. With this constraint we do not have time to follow the route left around the center state $(2, 2)$ corresponding to the relative sequence of positions (*west, north, north, east, east*). Instead we would have to pass either $(2, 2)$ or $(3, 2)$, both of which incur a loss.

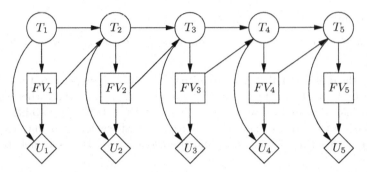

Fig. 9.42. The approximated model for fishing in the North Sea obtained from the original model in Figure 9.24 by marginalizing out the unobserved state variables V_i.

In general, we can say that optimal decisions at the end will be different from the ones at the beginning. For these situations we say that the optimal strategy is a *nonstationary strategy*.

Consider now the case in which we have an unbounded time horizon. At any time step, the optimal decision can depend only on the current state and what may happen in the future. If two time steps are in the same state, then they also have the same possibilities in the future, and therefore the optimal decision must be the same.

In the fishing in the North Sea example with unbounded time horizon, the optimal policy for deciding on the allowable catch at year 1 will not be any different from the policy at year 5. Similarly, in the robot example the optimal policy at state $(2, 3)$ does not depend on the point in time at which the robot entered that state. That is, when there is no fixed time horizon there is no reason to change the optimal policy for a given state at different points in time. For the robot example, this allows us to represent the optimal policy as in Figure 9.43.

More formally, an optimal strategy Δ consists of a set of identical policies, which are functions of only the current state. Such a strategy is called *stationary*, since it can be completely described by a single policy. We will not distinguish between a stationary strategy and a policy, and these terms will also be used interchangeably.

Optimality in Markov Decision Process

When evaluating a strategy for a decision problem with an unbounded time horizon, you might be tempted to simply consider the expected utilities/rewards for each time step and sum them up over time. However, if the process never stops, the sum may not be bounded, and you cannot compare two strategies with an expected reward of $+\infty$. This is not a problem for the

Fig. 9.43. A strategy for the robot in the 3 × 3 grid world.

robot example, since it has a terminal state in which the robot will eventually end up. However, in the fishing example, any catching policy that at each time step gives a positive reward will have an infinite sum. An immediate approach for handling this problem could be to specify some fixed horizon k so that the utility of a state sequence $s_0, s_1, s_2, \ldots$ is simply the sum of the rewards obtained at the first k states. For notational convenience we shall in this section assume that the reward is independent of the chosen action:

$$U(s_0, s_1, s_2, \ldots) = R(s_0) + R(s_1) + \cdots + R(s_k).$$

However, this raises the question of how to choose k, and, more importantly, it has the effect of postponing unpleasant decisions to after the horizon; in the extreme case in which $k = 0$ we care only about the immediate reward. The bounded fishing model in Figure 9.42 illustrates this point. With a fixed time horizon, you will be very greedy, in the end not caring about the volume of fish in later years.

Another approach is to weigh rewards in the immediate future higher than rewards in the distant future. This can be done by introducing a *discounting factor* γ, $0 \leq \gamma \leq 1$, so that the utility of a state sequence $s_0, s_1, s_2, \ldots$ is the accumulated discounted reward of each of the states:

$$U(s_0, s_1, s_2, \ldots) = R(s_0) + \gamma R(s_1) + \gamma^2 R(s_2) + \cdots.$$

In the extreme case that $\gamma = 0$, the agent considers all future rewards as being insignificant (corresponding to $k = 0$ above), and if $\gamma = 1$ then the discounted utility corresponds to having additive rewards as in the robot navigation problem. When $\gamma < 1$ the utility of an infinite sequence is always finite:

$$U(s_0, s_1, s_2, \ldots) = \sum_{i=0}^{\infty} \gamma^i R(s_i) \leq \sum_{i=0}^{\infty} \gamma^i \mathrm{max}R = \frac{\mathrm{max}R}{1 - \gamma}, \tag{9.1}$$

where $\mathrm{max}R$ is the maximum reward we can achieve in any state. A problem domain in which the discounted reward model has been applied is economics; here the discounting factor has been used, for example, to represent inflation or an interest rate. Discounted rewards have also been used to model unbounded decision problems, in which the decision process may terminate at any point in time with probability $(1 - \gamma)$. This could, for example, be used to model that there is a risk of $(1 - \gamma)$ that the robot will break down after it has performed a move.

Some decision problems cannot naturally be modeled using discounted rewards. The robot navigation problem with no terminal state is an example of such a decision problem: the navigation task is not only to reach the goal state but also to avoid the obstacles, and if we disregard the potential problem of the robot breaking down, then there is no real justification for using discounted rewards (why should it be worse to hit an obstacle now than in the future?). In this situation, the *average reward* may be a more appropriate model:

$$U(s_0, s_1, s_2, \ldots) = \lim_{N \to \infty} \frac{1}{N} \sum_{i=0}^{N-1} R(s_i) \leq \max R.$$

No matter whether we use discounted reward or average reward, we should take into account that each strategy Δ corresponds to a set of different state sequences due to the actions being nondeterministic. For example, if the robot starts at $(1, 3)$, then a performed action sequence (*north, north, east, east*) will result in the state sequence $[(1, 2), (1, 1), (2, 1), (3, 1)]$ with probability $0.7^4 = 0.2401$. Thus, we evaluate strategies based on their *expected reward*. Let $P(S_i \mid \Delta, s_0)$ be the probability distribution for S_i given that we start in s_0 and follow the strategy Δ. Then

$$\sum_{S_i} R(S_i) P(S_i \mid \Delta, s_0)$$

is the expected reward at step i, and $\gamma^i \sum_{S_i} R(S_i) P(S_i \mid \Delta, s_0)$ is the discounted expected reward. The expected reward of Δ is defined as

$$U^*(s, \Delta) = \lim_{N \to \infty} \sum_{i=0}^{N} \gamma^i \left(\sum_{S_i} R(S_i) P(S_i \mid \Delta, s_0) \right).$$

A standard notation for $U^*(s_0, \Delta)$ is also

$$\mathbb{E} \left[\sum_{i=0}^{\infty} \gamma^i R(s_i) \,\middle|\, \Delta, s \right].$$

In Section 10.6 we shall return to the actual calculation of these expectations.

9.6.2 Partially Observable Markov Decision Processes

In many decision problems, the assumption that the environment is fully observable is not realistic. For example, the sensors used by a robot for positioning may be inaccurate, and they will therefore provide only a blurred picture of the state of the world. We call such an environment *partially observable*, and in the Bayesian framework we can encode this uncertainty with a probability distribution over the possible world states. For bounded horizons, we have actually encountered such a decision problem before, namely in the form of the more exact model for the fishing in the North Sea decision problem specified in Figure 9.23.

In general, we can model that type of decision problem as a so-called *partially observable Markov decision process* (POMDP) illustrated in Figure 9.44. In the POMDP model the node O_i represents the observation at step i, and the conditional probability distribution attached to this node encodes the uncertainty associated with the observation; the information arc from O_i to D_i specifies that only O_i is observed immediately before decision D_i. More formally, a POMDP consists of:

1. A set of states and actions as in the MDP framework.
2. A transition function and a reward function as specified for the MDP.
3. A set of possible observations represented by the chance variable O_i at time step i.
4. An *observation function* $P(O_i \mid S_i, D_{i-1})$ that specifies the probability of the possible observations conditioned on the current state of the world and the last decision.

Observe that as for the MDP we use a single variable to represent the observation and the state at the ith time step. However, as for the MDP, we can consider these variables as being the products of several variables, so that both the transition function and the observation function can be specified more compactly using a Bayesian network. To simplify the model, we will stick to the single-variable representations.

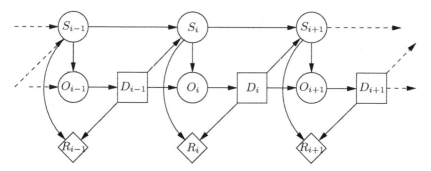

Fig. 9.44. A snapshot of a model of a partially observable Markov decision process. The state of the world S_i is observed only indirectly through the observation node O_i.

When the world is only partially observable we can no longer execute an action based on the current state of the world. In fact, based on the d-separation properties of the model in Figure 9.44, we see that when decision D_i is taken, all previous observations and decisions are d-connected to the current and future state variables, hence the entire past is relevant when the decision is taken. Another way of interpreting this situation is that all our previous observations and decisions have an impact on our current beliefs about the state of the world, and our ensuing action is based on these beliefs. This also means that while for MDPs we specified a policy conditionally on the observed state of the world, we should now specify a policy conditionally on our belief of the state of the world. Since the actual state of the world is not observed, our belief will in general not point to any specific state but will rather be a probability distribution over the possible states. That is, our belief can be expressed as a probability distribution $P(S_i \mid D_1, O_1, \ldots, D_{i-1}, O_i)$, and an optimal policy for step i will therefore specify an action for each possible probability distribution over S_i. This implies that if $P(S_i|past_i) = P(S_j|past_j)$, then the optimal decisions for D_i and D_j are the same.

9.7 Summary

One Action

Decision D, utility functions $U_1, \ldots, U_n$ over domains $X_1, \ldots, X_n$, evidence e. The expected utility is

$$\mathrm{EU}(D \mid e) = \sum_{X_1} U_1(X_1)P(X_1 \mid D, e) + \cdots + \sum_{X_n} U_n(X_n)P(X_n \mid D, e),$$

and a state d maximizing $\mathrm{EU}(D \mid e)$ is chosen as an optimal action.

Instrumental Rationality

For an individual who acts according to a preference ordering satisfying the rules below, there exists a utility function so that the individual maximizes the expected utility.

1. *Reflexivity.* For any lottery A, $A \succeq A$.
2. *Completeness.* For any pair (A, B) of lotteries, $A \succeq B$ or $B \succeq A$.
3. *Transitivity.* If $A \succeq B$ and $B \succeq C$, then $A \succeq C$.
4. *Preference increasing with probability.* If $A \succeq B$ then $\alpha A + (1 - \alpha)B \succeq \beta A + (1 - \beta)B$ if and only if $\alpha \geq \beta$.
5. *Continuity.* If $A \succeq B \succeq C$ then there exists $\alpha \in [0, 1]$ such that $B \sim \alpha A + (1 - \alpha)C$.
6. *Independence.* If $C = \alpha A + (1-\alpha)B$ and $A \sim D$, then $C \sim (\alpha D + (1-\alpha)B)$.

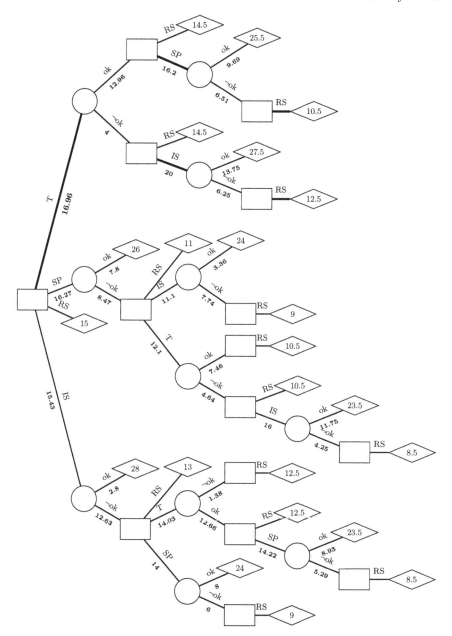

Fig. 9.45. An example of a decision tree. The probabilities may be taken from a Bayesian network. The bold links indicate an optimal strategy.

Decision Trees

An example is shown in Figure 9.45.
To calculate an optimal strategy and the maximum expected utility for the subtree rooted at node X, do:

1. If X is a utility node, then return $U(X)$.
2. If X is a chance node, then return

$$\text{EU}(X) = \sum_{x \in \text{sp}(X)} P(X = x \,|\, \text{past}(X)) \, \text{EU}(N(X = x)).$$

3. If X is a decision node, then return

$$\text{EU}(X) = \max_{x \in \text{sp}(X)} \text{EU}(N(X = x)),$$

and mark the arc labeled

$$x' = \arg \max_{x \in \text{sp}(X)} \text{EU}(N(X = x)).$$

Influence Diagrams

An *influence diagram* consists of a directed acyclic graph over chance nodes, decision nodes, and utility nodes with the following structural properties:

– there is a directed path comprising all decision nodes;
– the utility nodes have no children.

For the quantitative specification, we require that:

– the decision nodes and the chance nodes have a finite set of mutually exclusive states;
– the utility nodes have no states;
– to each chance node A there be attached a conditional probability table $P(A \,|\, pa(A))$;
– to each utility node V there be attached a real-valued function over $pa(V)$.

Figure 9.46 gives an example of the structural part of an influence diagram.

A *policy* for decision D_i is a mapping δ_i that for any configuration of the past of D_i yields a decision for D_i. That is

$$\delta_i(\mathcal{I}_0, D_1, \ldots, D_{i-1}, \mathcal{I}_{i-1}) \in \text{sp}(D_i).$$

A *strategy* for an influence diagram is a set of policies, one for each decision. A *solution* to an influence diagram is a strategy maximizing the expected utility.

Methods for determining optimal strategies from influence diagrams are given in Chapter 10.

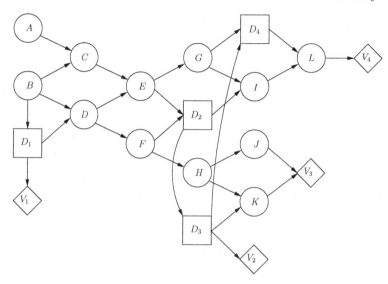

Fig. 9.46. An example of the structure of an influence diagram. We have $\mathcal{I}_0 = \{B\}, \mathcal{I}_1 = \{E, F\}, \mathcal{I}_2$ is empty, $\mathcal{I}_3 = \{G\}, \mathcal{I}_4 = \{A, C, D, H, I, J, K, L\}$.

Asymmetric Decision Problems

A decision problem is said to be *symmetric* if:

- in all of its decision tree representations, the number of scenarios is the same as the cardinality of the Cartesian product of the state spaces of all chance and decision variables, and
- in at least one decision tree representation, the sequence of chance and decision variables is the same in all scenarios.

There are three types of asymmetry:

Functional asymmetry: The possible outcomes or decision options of a variable may vary depending on the past.

Structural asymmetry: The very occurrence of an observation or a decision depends on the past.

Order asymmetry: The ordering of the decisions and observations is not settled at the time the model is specified.

Unconstrained Influence Diagrams

Unconstrained influence diagrams are used to model order asymmetry. Compared to influence diagrams there need not be a total ordering of the decisions, and the chance variables are partitioned into two sets: observable chance variables and nonobservable chance variables. An observable chance variable is released (for observation) when all its antecedent decision variables have been decided on.

Solving an unconstrained influence diagram involves finding the next action as well as finding an optimal policy if the next action is a decision (that is, finding the conditional sequence of action and observations maximizing the expected utility). The solution is specified in terms of an S-DAG:

An S-DAG is a directed acyclic graph G. The nodes are labeled with variables from $\mathcal{D}_U \cup \mathcal{O}_U$ such that each maximal directed path in G represents an admissible ordering of $\mathcal{D}_U \cup \mathcal{O}_U$.

A *step policy* for a node N in an S-DAG G is a function

$$\sigma : \mathrm{sp}(\mathrm{hst}(N)) \to \mathrm{ch}(N).$$

A *step strategy* for U is a pair $(G, \mathcal{S})$, where G is an S-DAG for U and $\mathcal{S}$ is a set of step policies, one for each node in G (except for *Sink*). A *policy* for N is a function

$$\delta : \mathrm{sp}(past(N)) \to \mathrm{ch}(N).$$

A *strategy* for U is a step strategy together with a policy for each node.

Decision Problems with an Unbounded Time Horizon

An MDP consists of an unbounded set of identical time steps. Each time step i consists of:

1. A finite set of states of the world (represented by the chance variable S_i).
2. A finite set of actions (represented by the decision variable D_i).
3. A *transition function* $P(S_{i+1} = s' \mid S_i = s, D_i = a)$ specifying the probability that the next state is s' when taking action a in state s.
4. A *reward function* $R(S_i = s, D_i = a)$ specifying the reward of taking action a in state s, for each $a \in \mathrm{sp}(D_i)$ and $s \in \mathrm{sp}(S_i)$.
5. An *initial state* $s_0 \in S_0$.

The transition function and the reward function are the same for all time steps.

There are three standard ways to ensure that the utility of an unbounded state sequence $s_0, s_1, s_2, \ldots$ is bounded:

Fixed time horizon: The sum of the rewards obtained at the first k states:

$$U(s_0, s_1, s_2, \ldots) = R(s_0) + R(s_1) + \cdots + R(s_k).$$

Discounted reward: The accumulated discounted reward of each of the states:

$$U(s_0, s_1, s_2, \ldots) = R(s_0) + \gamma R(s_1) + \gamma^2 R(s_2) + \cdots,$$

where $0 \geq \gamma < 1$.

Average expected reward: The accumulated average reward at each of the states:

$$U(s_0, s_1, s_2, \ldots) = \lim_{N \to \infty} \frac{1}{N} \sum_{i=0}^{N-1} R(s_i).$$

A POMDP consists of:

1. A set of states and actions as in the MDP framework.
2. A transition function and a reward function as specified for the MDP.
3. A set of possible observations (represented by the chance variable O_i at time step i).
4. An *observation function* $P(O_i \mid S_i, D_{i-1})$ that specifies the probability of the possible observations conditioned on the current state of the world and the last decision.

9.8 Bibliographical Notes

Decision theory has a long history but achieved a breakthrough in the work of von Neumann and Morgenstern (1944), who laid down the axioms for instrumental rationality. Decision trees were introduced by Raiffa and Schlaifer (1961). Influence diagrams were proposed by Howard and Matheson (1981), and were adapted to allow for additive decompositions of utility functions in (Tatman and Shachter, 1990). Unconstrained influence diagrams were introduced in (Jensen and Vomlelova, 2002), and sequential influence diagrams in (Jensen *et al.*, 2006). The latter is a fusion of the valuation networks of Shenoy (1996) and the asymmetric influence diagrams of Nielsen and Jensen (2003a). The study of Markov decision processes can be traced back at least to Howard (1960). A good starting point for further reading is (Puterman, 1994). Partially observed Markov decision processes originate with Drake (1962) and Åström (1965). The reactor problem, as presented here, is due to Covaliu and Oliver (1995).

9.9 Exercises

Exercise 9.1. Consider the management of effort example in Section 9.2.

(i) Let the marks be $0, 5, 6, 8, 9, 10$. What is the optimal decision if the numerical values are used as utilities?
(ii) Consider the approach in which the marks are given subjective utilities. Show that action Gd can be optimal only if the mark 0 is given higher utility than mark 3.

Exercise 9.2. Prove that if U is a utility function for a decision maker and if a $(a > 0)$ and b are real numbers, then $aU + b$ is an equivalent utility function.

Exercise 9.3. [E] Extend the model from Exercise 3.14 to a model for folding or calling.

Exercise 9.4. [E] Extend Exercise 3.18 with the following:

In golf, the task is to use as few strokes as possible at each hole. I am driving at a hole 260 m long. If the drive is 265 m, I will on average use 1.8 strokes to finish the hole. If the drive is 240 m, on average 2 extra strokes are needed; 220 m requires 2.5 extra strokes; 200 m requires 2.7; 180 m 2.9 extra strokes; 160 m 3.1; 145 m 3.3; a drive of 290 m will carry the ball into a sand trap, requiring 3.5 extra strokes; if the drive is misshit, the ball will drop into a lake, and it will require 4.5 extra strokes to finish the hole.

Construct a system that helps me decide whether to use the 3-wood or the driver in the drive.

Exercise 9.5. [E] Consider the stud farm example from Section 3.2.2. Extend the model to be an aid for deciding for each horse whether it should be taken out of breeding. Table 9.6 gives the utilities.

	Carrier	Pure		Carrier	Pure
Out	−10	−10	*Out*	−3	−3
In	−40	100	*In*	−10	40
	Stallions			*Mares*	

Table 9.6. Tables for Exercise 9.5.

Exercise 9.6. Let the hypothesis variable H have n states. Introduce an action variable A with the same states as H; let the utility table be as follows:

$$U(h, a) = \begin{cases} 1 \text{ if } h \text{ and } a \text{ are the same,} \\ 0 \text{ otherwise.} \end{cases}$$

Show that a value function based on U corresponds to selecting a hypothesis state of highest probability.

Exercise 9.7. Construct a decision tree for the mildew decision problem in Section 9.1.2. How many numbers would you need to specify to render it complete?

Exercise 9.8. Solve the decision tree in Figure 9.47.

Exercise 9.9. Consider an altered version of the poker decision problem in which each player is now allowed three rounds of changing hands. What would an influence diagram look like for this altered problem? What is the past for each decision variable in the diagram?

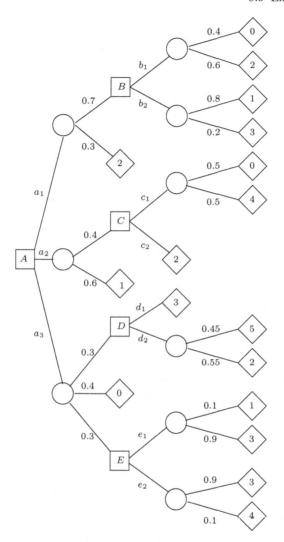

Fig. 9.47. Figure for Exercise 9.8.

Exercise 9.10. What is the partial temporal ordering of observations and decisions in the influence diagrams in Figures 9.23 and 9.24?

Exercise 9.11. [E] (The oil wildcatter's problem)

An oil wildcatter must decide whether to drill or not to drill. The cost of drilling is $70,000. If he decides to drill, the hole may be soaking (with a return of $270,000), wet (with a return of $120,000), or dry (with a return of $0). The prior probabilities for soaking, wet, and dry are (0.2, 0.3, 0.5). At the cost of $10,000, the oil wildcatter could decide to take seismic soundings of the geological structure at the site. The specifics of the test are given in Table 9.7.

$T \setminus S$	dr	wt	so
n	0.6	0.3	0.1
o	0.3	0.4	0.4
c	0.1	0.3	0.5

$P(Test \mid Structure)$

Table 9.7. Table for Exercise 9.11. The states n, o, and c are the outcomes of the test.

(i) Solve the problem with a decision tree.

(ii) Solve the problem with an influence diagram.

Exercise 9.12. (The used car buyer's problem)

Joe is considering buying a used car from a dealer for $1,000. The market price of similar cars with no defects is $1,100. Joe is uncertain whether the particular car he is considering is a "peach" or a "lemon." Of the ten major subsystems in the car, a peach has a serious defect in only one subsystem, whereas a lemon has a serious defect in six subsystems. The probability that the used car under consideration is a lemon is 0.2. The cost of repairing one defect is $40, and the cost of repairing six defects is $200.

For an additional $60, Joe can buy the car from the dealer with an "antilemon guarantee." The antilemon guarantee will normally pay for 50% of the repair cost, but if the car is a lemon, then the guarantee will pay 100% of the repair cost.

Before buying the car, Joe has the option of having the car examined by a mechanic for an hour. In this period, the mechanic offers three alternatives t_1, t_2, t_3 as follows:

t_1: test the steering subsystem alone at a cost of $9,

t_2: test the fuel and electrical subsystems for a total cost of $13,

t_3: do a two-test sequence in which Joe can authorize a second test after the result of the first test is known. In this alternative, the mechanic will first test the transmission subsystem at a cost of $10 and report the results to Joe. If Joe approves, the mechanic will then proceed to test the differential subsystem at an additional cost of $4.

All tests are guaranteed to find a defect in the subsystem if a defect exists. We assume that Joe's utility for profit is linear in dollars.

(i) Solve the problem with a decision tree.

(ii) Consider how to represent the problem as an influence diagram (you may add dummy states and variables as you wish).

Exercise 9.13. Draw an influence diagram for the decision problem in Section 9.1.2.

Exercise 9.14. Solve the decision tree in Figures 9.8 and 9.9 (the probabilities can be taken from the model in Figure 9.10).

Exercise 9.15. Complete the reduced decision tree from Figure 9.16 and solve it.

Exercise 9.16. ^{E} Solve Exercise 3.16 as a decision problem.

Exercise 9.17. ^{E} Solve the example in Section 11.1.1 as an influence diagram.

Exercise 9.18. ^{E} Extend the poker model from Exercise 9.3 to the influence diagram in Figure 9.21.

Exercise 9.19. ^{E} Represent the Car Start Problem in Section 9.3.1 as an influence diagram. (What are the decision options at each step?)

Exercise 9.20. Unfold the sequential influence diagram in Figure 9.38 with the following probabilities: A conventional reactor (C) has probability 0.980 of being successful (cs), and a probability 0.020 of a failure (cf). An advanced reactor (A) has probability 0.660 of being successful (as), probability 0.244 of a limited accident (al), and probability 0.096 of a major accident (am).

Exercise 9.21. Consider the Dating Problem in Example 9.6. What are the asymmetries in the decision problem? Which of them are functional asymmetries/structural asymmetries/order asymmetries?

Exercise 9.22. Construct an S-DAG for the UID in Figure 9.48.

Exercise 9.23. Consider the two-player turn-taking game of tic-tac-toe in which each player has three game pieces, and the objective is to place all your pieces in a straight line on a 3×3 board. The players take turns placing a piece in one of the free slots on the board, and when a player has no more pieces off the board, he must take one of his pieces already on the board and place it somewhere else. Formalize the game as a Markov decision process, seen from the point of view of one of the players.

Exercise 9.24. Consider the example of the possibly infected milk from a single cow introduced in Sections 3.1.1 and 3.2.1. Add to that the daily decision of throwing the milk out or pouring it into the tank, and associate the utility of

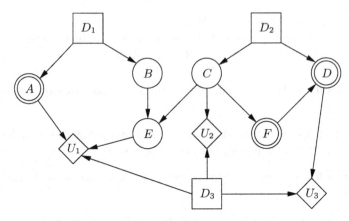

Fig. 9.48. A UID.

- 0 with pouring infected milk into the tank,
- 98 with throwing the milk out, and
- 100 with pouring noninfected milk into the tank.

Formalize the setting as a POMDP.

10

Solution Methods for Decision Graphs

In Chapter 9 we presented graphical languages for modeling decision problems. The languages ease the burden of specifying the problem and transfer the complexity of the problem to the computer. For problems with a finite time horizon, the computer may fold out the specification to a decision tree and determine an optimal strategy by averaging out and folding back as described in Section 9.3.3. However, the calculations may be intractable, and in this chapter we present alternative methods exploiting symmetries in the decision problem. Sections 10.1–10.3 are devoted to solution methods for influence diagrams. Section 10.4 presents a method for solving unconstrained influence diagrams. In Section 10.5 we consider decision theoretic troubleshooting, which has next to no temporal ordering, and for which the decision trees tend to be intractably large. In Section 10.6 we present two methods for solving MDPs, and a method for solving POMDPs is indicated. The last section presents LIMIDs, which is a way of approximating influence diagrams by limiting the memory of the decision maker.

10.1 Solutions to Influence Diagrams

An influence diagram has three types of nodes: *chance nodes*, *decision nodes*, and *utility nodes*. The set of chance nodes is denoted by $\mathcal{U}_C$, the set of decision nodes is denoted by $\mathcal{U}_D$ and the set of utility nodes is denoted by $\mathcal{U}_V$. The *universe* is $\mathcal{U} = \mathcal{U}_C \cup \mathcal{U}_D$. We shall also refer to the members of $\mathcal{U}$ as the variables of the influence diagram.

The decision nodes have a temporal order, $D_1, \ldots, D_n$, and the chance nodes are partitioned according to when they are observed: $\mathcal{I}_0$ is the set of chance nodes observed prior to any decision, $\ldots, \mathcal{I}_i$ is the set of chance nodes observed after D_i is taken and before the decision D_{i+1} is taken. Finally, $\mathcal{I}_n$ is the set of chance nodes never observed or observed after the last decision. That is, we have a partial temporal ordering $\mathcal{I}_0 \prec D_1 \prec \mathcal{I}_1 \prec \ldots \prec D_n \prec \mathcal{I}_n$.

Recall that an influence diagram is constructed so that if $A \prec D_i$, then there is a directed path from A to D_i.

We shall in this chapter use the influence diagram DI in Figure 10.1 as a standard example, where $\mathcal{I}_0 = \emptyset$, $\mathcal{I}_1 = \{T\}$, and $\mathcal{I}_2 = \{A, B, C\}$. In order not to make things unnecessarily complicated, all variables in DI are binary.

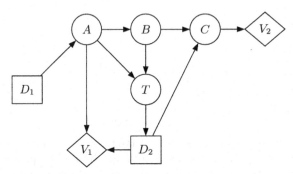

Fig. 10.1. The example influence diagram, DI.

As in Bayesian networks, the graphical representation of influence diagrams supports an analysis of conditional independence. However, d-separation for influence diagrams is performed slightly differently from the way it is done for Bayesian networks: ignore the utility nodes, and since the links into decision nodes encode only information precedence, they shall also be ignored.

For the DI example, we can perform d-separation analysis on Figure 10.1. We get, for example, that C is d-separated from T given B (note that you need not condition on D_2, since the link from T to D_2 is ignored). Also, A and T are d-separated from D_2. This means that if I *perform an action* from D_2, then this action has no impact on T. Note that this is different from: if I am told what action from D_2 was performed, what can I infer about T? If, for example, I know that the decision maker maximizes expected utilities, I may be able to infer a great deal about T.

Decision variables play a different role from that played by chance variables. For chance variables you ask the question, may information about node A change my belief about node B? For decision variables the question is, may an action from D have consequences for node B? Although the two concepts are different, they are in the case of influence diagrams not in conflict. In general, effects of decisions cannot "go back in time":

Proposition 10.1. *Let $A \in \mathcal{I}_i$ and let D_j be a decision variable with $i < j$. Then*

(i) A and D_j are d-separated and hence

$$P(A \mid D_j) = P(A).$$

(ii) Let W be any set of variables prior to D_j in the temporal ordering. Then A and D_j are d-separated given W and hence

$$P(A \mid D_j, W) = P(A \mid W).$$

Proof.

(i) Since D_j has no parents, any impact from D_j must follow the direction of a link from D_j. The only way the impact can start going in the opposite direction from that of a link is if it meets a converging connection at a chance variable B, and then it can do so only if either B or one of its children C has received evidence. Since D_j is the only variable we condition on, this cannot happen. Hence if D_j and A are not d-separated, there must be a directed path from D_j to A. Since $A \prec D_j$ in the temporal ordering, there is a directed path from A to D_j, and since the graph is acyclic, there cannot be a directed path from D_j to A.

(ii) We argue in the same way as for *(i)*. By following directions of links from D_j, we can only start going opposite to the direction by meeting evidence. Since all evidence is prior in the temporal ordering, we know from *(i)* that we cannot meet it.

$\square$

10.1.1 The Chain Rule for Influence Diagrams

For Bayesian networks we have that $P(\mathcal{U})$ is the product of all probability potentials attached to the variables in the network. For influence diagrams we have a similar theorem. Again, decision variables act differently from chance variables. since a decision variable eventually will come under my control, it requires no prior probabilities. Also, it has no meaning to attach a probability distribution to a chance variable A effected by a decision variable D, unless a decision has been taken and the action performed. So in Figure 10.1 it has no meaning to consider $P(A)$ or $P(A, D)$. What is meaningful is $P(A \mid d)$ for all $d \in D$, and we may lump the probabilities for all decisions of D together in the expression $P(A \mid D)$.

Theorem 10.1 (The chain rule for influence diagrams). *Let ID be an influence diagram with universe $\mathcal{U} = \mathcal{U}_C \cup \mathcal{U}_D$. Then*

$$P(\mathcal{U}_C \mid \mathcal{U}_D) = \prod_{X \in \mathcal{U}_C} P(X \mid \mathrm{pa}(X)).$$

Proof. Let us first look at the influence diagram DI. From the fundamental rule we have

$$P(C, T, B, A \mid D_1, D_2) = P(C \mid T, B, A, D_1, D_2) P(T, B, A \mid D_1, D_2)$$
$$= P(C \mid T, B, A, D_1, D_2) P(T \mid B, A, D_1, D_2)$$
$$\times P(B \mid A, D_1, D_2) P(A \mid D_1, D_2). \qquad (10.1)$$

Since C is d-separated from A, T, and D_1 given B and D_2, we have

$$P(C \mid T, B, A, D_1, D_2) = P(C \mid B, D_2).$$

We also have

$$P(T \mid B, A, D_1, D_2) = P(T \mid B, A),$$
$$P(B \mid A, D_1, D_2) = P(B \mid A),$$
$$P(A \mid D_1, D_2) = P(A \mid D_1).$$

Substituting in equation (10.1) yields

$$P(C, B, T, A \mid D_1, D_2) = P(C \mid B, D_2) P(T \mid B, A) P(B \mid A) P(A \mid D_1,),$$

which is the product of the probability potentials for DI.

A general proof can follow another line of reasoning. Let $\mathbf{d}$ be a particular configuration of decisions. By inserting them in the influence diagram ID, you get a Bayesian network representing $P(\mathcal{U}_C \mid \mathbf{d})$, the joint probability of $\mathcal{U}_C$, under the condition that the decisions $\mathbf{d}$ are taken. Using the chain rule for Bayesian networks, you infer that $P(\mathcal{U}_C \mid \mathbf{d})$ is the product of all probability potentials attached to the decision variables instantiated to $\mathbf{d}$. Since this holds for all instantiations of $\mathcal{U}_D$, you get the result. $\qquad \square$

10.1.2 Strategies and Expected Utilities

To solve an influence diagram, you may unfold it into a decision tree and solve it. In Figure 10.2 we have unfolded DI from Figure 10.1.

When solving the decision tree in Figure 10.2, we start at the leaves and work toward the root (see Section 9.3.3). Consider the path (d_1^1, t_1). We wish to compute the expected utility of performing action d_1^2 given (d_1^1, t_1). We have

$$\mathrm{EU}(d_1^2 \mid d_1^1, t_1) = \sum_{A,C} P(A, C \mid d_1^1, t_1, d_1^2)(V_1(A, d_1^2) + V_2(C)).$$

For the action d_2^2, we have

$$\mathrm{EU}(d_2^2 \mid d_1^1, t_1) = \sum_{A,C} P(A, C \mid d_1^1, t_1, d_2^2)(V_1(A, d_2^2) + V_2(C)).$$

Taken together, we write

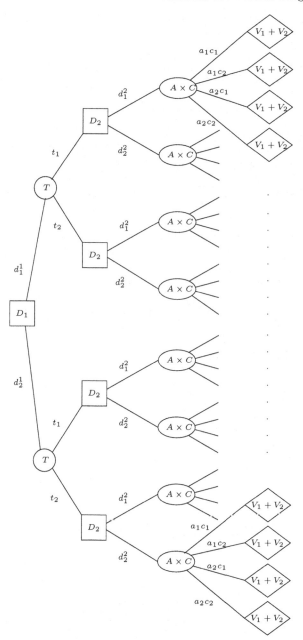

Fig. 10.2. *DI* from Figure 10.1 unfolded into a decision tree. Note that to reduce the size of the decision tree the last chance node in each path is defined as the Cartesian product of A and C, and that the utilities in the leaves are the sums of V_1 and V_2.

$$\mathrm{EU}(D_2 \,|\, d_1^1, t_1) = \sum_{A,C} P(A, C \,|\, d_1^1, t_1, D_2)(V_1(A, D_2) + V_2(C)).$$

We choose the action of maximal expected utility, and we get a decision rule for D_2 with $D_1 = d_1^1$ and $T = t_1$

$$\delta_2(d_1^1, t_1) = \arg\max_{D_2} \mathrm{EU}(D_2 \,|\, d_1^1, t_1).$$

If there are several decisions yielding the maximum, either of them will do. The maximal expected utility from D_2 given (d_1^1, t_1) is

$$\rho_2(d_1^1, t_1) = \max_{D_2} \sum_{A,C} P(A, C \,|\, d_1^1, t_1, D_2)(V_1(A, D_2) + V_2(C)).$$

Generalizing these two formulas to any path over D_1, T, we get a policy for D_2

$$\delta_2(D_1, T) = \arg\max_{D_2} \mathrm{EU}(D_2 \,|\, D_1, T)$$

$$= \arg\max_{D_2} \sum_{A,C} P(A, C \,|\, D_1, T, D_2)(V_1(A, D_2) + V_2(C)),$$

and a new utility function

$$\rho_2(D_1, T) = \max_{D_2} \sum_{A,C} P(A, C \,|\, D_1, T, D_2)(V_1(A, D_2) + V_2(C)), \quad (10.2)$$

which gives the expected utilities when we know the values of (D_1, T). The decision tree in Figure 10.2 can now be reduced to the one in Figure 10.3.

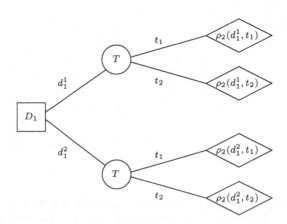

Fig. 10.3. The decision tree from Figure 10.2 with D_2 replaced by a utility function reflecting that the policy δ_2 for D_2 is followed.

Next, look at the decision D_1 as in Figure 10.3. If we take the action d_1^1, we get the expected utility

$$\text{EU}(d_1^1) = P(t_1 \mid d_1^1)\rho_2(d_1^1, t_1) + P(t_2 \mid d_1^1)\rho_2(d_1^1, t_2),$$

which can also be written

$$\text{EU}(D_1) = \sum_T P(T \mid D_1)\rho_2(D_1, T).$$

The policy for D_1 is

$$\delta_1 = \arg\max_{D_1} \sum_T P(T \mid D_1)\rho_2(D_1, T),$$

and the expected utility of performing optimal decisions is

$$\rho_1 = \max_{D_1} \sum_T P(T \mid D_1)\rho_2(D_1, T). \tag{10.3}$$

So far we have written various expressions without really connecting them to the potentials from the influence diagram. In principle, all probabilities in the expressions can be calculated from the influence diagram by inserting and propagating evidence. However, by taking a closer look at equation (10.3) we can make a much tighter connection between the specification of the influence diagram and its solution: By combining equation (10.2) and equation (10.3), we get

$$
\begin{aligned}
\rho_1 &= \max_{D1} \sum_T P(T \mid D_1) \max_{D_2} \sum_{A,C} P(A, C \mid D_1, T, D_2)(V_1(A, D_2) + V_2(C)) \\
&= \max_{D_1} \sum_T \max_{D_2} \sum_{A,C} P(T \mid D_1)P(A, C \mid D_1, T, D_2)(V_1(A, D_2) + V_2(C)) \\
&= \max_{D_1} \sum_T \max_{D_2} \sum_{A,C} P(T \mid D_1, D_2)P(A, C \mid D_1, T, D_2)(V_1(A, D_2) + V_2(C)) \\
&= \max_{D_1} \sum_T \max_{D_2} \sum_{A,C} P(A, C, T \mid D_1, D_2)(V_1(A, D_2) + V_2(C)) \\
&= \max_{D_1} \sum_T \max_{D_2} \sum_{A,B,C} P(A, B, C, T \mid D_1, D_2)(V_1(A, D_2) + V_2(C)) \\
&= \max_{D_1} \sum_T \max_{D_2} \sum_{A,B,C} P(\mathcal{U}_C \mid \mathcal{U}_D)(V_1(A, D_2) + V_2(C)).
\end{aligned}
$$

The formula for δ_1 is

$$\delta_1 = \arg\max_{D_1} \sum_T \max_{D_2} \sum_{A,B,C} P(\mathcal{U}_C \mid \mathcal{U}_D)(V_1(A, D_2) + V_2(C)).$$

For the policy δ_2 we have

$$\delta_2(D_1, T) = \arg\max_{D_2} \sum_{A,C} P(A, C \mid D_1, T, D_2)(V_1(A, D_2) + V_2(C)).$$

We can multiply inside "$\arg\max_{D_2}$" with anything not varying with D_2:

$$
\begin{aligned}
\delta_2(D_1, T) &= \arg\max_{D_2} P(T \mid D_1) \sum_{A,C} P(A, C \mid D_1, T, D_2)(V_1(A, D_2) + V_2(C)) \\
&= \arg\max_{D_2} \sum_{A,C} P(T \mid D_1, D_2) P(A, C \mid D_1, T, D_2)(V_1(A, D_2) + V_2(C)) \\
&= \arg\max_{D_2} \sum_{A,C} P(A, T, C \mid D_1, D_2)(V_1(A, D_2) + V_2(C)) \\
&= \arg\max_{D_2} \sum_{A,B,C} P(\mathcal{U}_C \mid \mathcal{U}_D)(V_1(A, D_2) + V_2(C)),
\end{aligned}
$$

and the similarity with the formula for δ_1 is transparent. Similar calculations yield for ρ_2,

$$\rho_2(D_1, T) = \frac{1}{P(T \mid D_1)} \max_{D_2} \sum_{A,B,C} P(\mathcal{U}_C \mid \mathcal{U}_D)(V_1(A, D_2) + V_2(C)).$$

Theorem 10.2. *Let ID be an influence diagram over $\mathcal{U} = \mathcal{U}_C \cup \mathcal{U}_D$ and $\mathcal{U}_V = \{V_i\}$. Let the temporal order of the variables be described as $\mathcal{I}_0 \prec D_1 \prec \mathcal{I}_1 \prec \cdots \prec D_n \prec \mathcal{I}_n$ and let $V = \sum_i V_i$. Then:*

(i) An optimal policy for D_i is

$$\delta_i(\mathcal{I}_0, D_1, \ldots, \mathcal{I}_{i-1}) = \arg\max_{D_i} \sum_{\mathcal{I}_i} \max_{D_{i+1}} \cdots \max_{D_n} \sum_{\mathcal{I}_n} P(\mathcal{U}_C \mid \mathcal{U}_D)V.$$

(ii) The expected utility from following the policy δ_i (and acting optimally in the future) is

$$
\begin{aligned}
\rho_i(\mathcal{I}_0, D_1, \ldots, \mathcal{I}_{i-1}) = \ &\frac{1}{P(\mathcal{I}_0, \ldots, \mathcal{I}_{i-1} \mid D_1, \ldots, D_{i-1})} \\
&\max_{D_i} \sum_{\mathcal{I}_i} \max_{D_{i+1}} \cdots \max_{D_n} \sum_{\mathcal{I}_n} P(\mathcal{U}_C \mid \mathcal{U}_D)V,
\end{aligned}
$$

and the strategy for ID consisting of an optimal policy for each decision yields the maximum expected utility:

$$\mathrm{MEU}(ID) = \sum_{\mathcal{I}_0} \max_{D_1} \sum_{\mathcal{I}_1} \max_{D_2} \cdots \max_{D_n} \sum_{\mathcal{I}_n} P(\mathcal{U}_C \mid \mathcal{U}_D)V.$$

Proof. We start with the last decision D_n. We have for the expected utility given the past

$$\text{EU}(D_n \mid \mathcal{I}_0, D_1, \ldots, D_{n-1}, \mathcal{I}_{n-1})$$

$$= \sum_{\mathcal{I}_n} P(\mathcal{I}_n \mid \mathcal{I}_0, D_1, \ldots, D_{n-1}, \mathcal{I}_{n-1}, D_n)V$$

$$= \sum_{\mathcal{I}_n} \frac{1}{P(\mathcal{I}_0, \ldots, \mathcal{I}_{n-1} \mid D_1, \ldots, D_n)} P(\mathcal{I}_n, \mathcal{I}_0, \ldots, \mathcal{I}_{n-1} \mid D_1, \ldots, D_n)V$$

$$= \frac{1}{P(\mathcal{I}_0, \ldots, \mathcal{I}_{n-1} \mid D_1, \ldots, D_{n-1})} \sum_{\mathcal{I}_n} P(\mathcal{U}_C \mid \mathcal{U}_D)V.$$

In the last expression we used that $P(\mathcal{I}_0, \ldots, \mathcal{I}_{n-1} \mid D_1, \ldots, D_n) = P(\mathcal{I}_0, \ldots, \mathcal{I}_{n-1} \mid D_1, \ldots, D_{n-1})$. We now get

$$\rho_n(\mathcal{I}_0, D_1, \ldots, \mathcal{I}_{n-1})$$

$$= \frac{1}{P(\mathcal{I}_0, \ldots, \mathcal{I}_{i-1} \mid D_1, \ldots, D_{n-1})} \max_{D_n} \sum_{\mathcal{I}_n} P(\mathcal{U}_C \mid \mathcal{U}_D)V,$$

and

$$\delta_n(\mathcal{I}_0, D_1, \ldots, \mathcal{I}_{n-1})$$

$$= \arg\max_{D_n} \frac{1}{P(\mathcal{I}_0, \ldots, \mathcal{I}_{n-1} \mid D_1, \ldots, D_{n-1})} \sum_{\mathcal{I}_n} P(\mathcal{U}_C \mid \mathcal{U}_D)V$$

$$= \arg\max_{D_n} \sum_{\mathcal{I}_n} P(\mathcal{U}_C \mid \mathcal{U}_D)V.$$

Next, assume the theorem to hold for $i+1, \ldots, n$ and consider decision D_i. We have

$$\text{EU}(D_i \mid \mathcal{I}_0, D_1, \ldots, D_{n-1}, \mathcal{I}_{i-1})$$

$$= \sum_{\mathcal{I}_i} P(\mathcal{I}_i \mid \mathcal{I}_0, D_1, \ldots, D_{i-1}, \mathcal{I}_{i-1}, D_i)\rho_{i+1}(\mathcal{I}_0, D_1, \ldots, \mathcal{I}_i)$$

$$= \sum_{\mathcal{I}_i} \frac{1}{P(\mathcal{I}_0, \ldots, \mathcal{I}_{i-1} \mid D_1, \ldots, D_i)} P(\mathcal{I}_i, \mathcal{I}_0, \ldots, \mathcal{I}_{i-1} \mid D_1, \ldots, D_i)$$

$$\frac{1}{P(\mathcal{I}_0, \ldots, \mathcal{I}_i \mid D_1, \ldots, D_i)} \max_{D_{i+1}} \sum_{\mathcal{I}_{i+1}} \cdots \max_{D_n} \sum_{\mathcal{I}_n} P(\mathcal{U}_C \mid \mathcal{U}_D)V$$

$$= \sum_{\mathcal{I}_i} \frac{1}{P(\mathcal{I}_0, \ldots, \mathcal{I}_{i-1} \mid D_1, \ldots, D_i)} \max_{D_{i+1}} \sum_{\mathcal{I}_{i+1}} \cdots \max_{D_n} \sum_{\mathcal{I}_n} P(\mathcal{U}_C \mid \mathcal{U}_D)V$$

$$= \frac{1}{P(\mathcal{I}_0, \ldots, \mathcal{I}_{i-1} \mid D_1, \ldots, D_{i-1})} \sum_{\mathcal{I}_i} \max_{D_{i+1}} \sum_{\mathcal{I}_{i+1}} \cdots \max_{D_n} \sum_{\mathcal{I}_n} P(\mathcal{U}_C \mid \mathcal{U}_D)V,$$

and we get the formulas in *(i)* and *(ii)*.

Since we have repeatedly determined a policy maximizing the expected utility regardless of the past, no other set of policies can give a higher expected utility. The formula for $\text{MEU}(ID)$ is the formula from *(ii)* for ρ_0. It is calculated by taking $\rho_1(D_1)$, multiplying by $P(\mathcal{I}_0)$, and marginalizing $\mathcal{I}_0$ out:

$$\text{MEU}(ID) = \sum_{\mathcal{I}_0} P(\mathcal{I}_0)\rho_1(\mathcal{I}_0)$$

$$= \sum_{\mathcal{I}_0} P(\mathcal{I}_0)\frac{1}{P(\mathcal{I}_0)} \max_{D_1} \sum_{\mathcal{I}_1} \max_{D_2} \cdots \max_{D_n} \sum_{\mathcal{I}_n} P(\mathcal{U}_C \,|\, \mathcal{U}_D)V$$

$$= \sum_{\mathcal{I}_0} \max_{D_1} \sum_{\mathcal{I}_1} \max_{D_2} \cdots \max_{D_n} \sum_{\mathcal{I}_n} P(\mathcal{U}_C \,|\, \mathcal{U}_D)V.$$

□

Since $P(\mathcal{U}_C \,|\, \mathcal{U}_D)$ is the product of all probability distributions attached to ID, we have a method for calculating ρ_i as well as δ_i. The method specifies that you may start with the product of all probability potentials and then marginalize out in reverse temporal order where chance variables are sum-marginalized and decision variables are max-marginalized. Each time an $\mathcal{I}_i$ is marginalized out, the result is used to determine a policy for D_i.

The method has the same problem as the method for Bayesian networks, namely that $P(\mathcal{U}_C \,|\, \mathcal{U}_D)$ may be an intractably large table, and we therefore have to look for methods that reduce the size of the domains to deal with. We shall consider this task in detail in Section 10.2.

10.1.3 An Example

The influence diagram DI in Figure 10.1 has the potentials in Table 10.1. Using Theorem 10.2 we get $\delta_2(D_1, T)$ and $\rho_2(D_1, T)$ as listed in Table 10.2.

$A \setminus D_1$	d_1^1	d_2^1
y	0.2	0.8
n	0.8	0.2

$P(A \,|\, D_1)$

$B \setminus A$	y	n
y	0.8	0.2
n	0.2	0.8

$P(B \,|\, A)$

$A \setminus B$	y	n
y	(0.9, 0.1)	(0.5, 0.5)
n	(0.5, 0.5)	(0.1, 0.9)

$P(T \,|\, A, B)$

$B \setminus D_2$	d_1^2	d_2^2
y	(0.9, 0.1)	(0.5, 0.5)
n	(0.5, 0.5)	(0.9, 0.1)

$P(C \,|\, D_2, B)$

$A \setminus D_2$	d_1^2	d_2^2
y	3	0
n	0	2

$V(A, D_2)$

$$V_2(C) = (10, 0)$$

Table 10.1. Potentials for DI.

$T \setminus D_1$	d_1^1	d_2^1
y	d_1^2	d_1^2
n	d_2^2	d_2^2

$T \setminus D_1$	d_1^1	d_2^1
y	9.51	11.29
n	10.34	8.97

$\delta_2(D_1, T)$ $\rho_1(D_1, T)$

Table 10.2. $\delta_2(D_1, T)$ and $\rho_2(D_1, T)$ for DI.

Finally, we get $\delta_1 = d_2^1$ and MEU(DI) = 10.58. Note that $\delta_2(D_1, T)$ has the property that the state of T alone determines the decision to choose, hence we can remove D_1 from the domain of δ_2. This phenomenon can sometimes be determined from the d-separation properties of the influence diagram (see Figure 10.4 and Section 11.2), and we say that this part of the past is not *required* for the decision in question. For DI it cannot be deduced from the structure; the potentials happened to cause it.

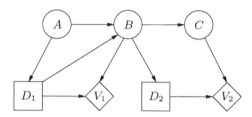

Fig. 10.4. An influence diagram in which D_1 is not required for D_2.

10.2 Variable Elimination

The method for solving influence diagrams has many similarities with the junction tree propagation algorithm for Bayesian networks: you start off with a set of potentials, and you eliminate one variable at a time. There are, however, differences. First of all, the elimination order is constrained by the temporal order. Since max-marginalization and sum-marginalization do not commute, you have to do it in an order whereby you first sum-marginalize $\mathcal{I}_n$, then max-marginalize D_n, sum-marginalize $\mathcal{I}_{i-1}$, etc. This type of elimination order is called a *strong elimination order*. Furthermore, you have two types of potentials to deal with. Also, you need to eliminate in only one direction; this corresponds to COLLECTEVIDENCE.

We shall first analyze the calculations in eliminating a variable. Let Φ be a set of probability potentials and Ψ a set of utility potentials. The two sets represent the expression $\prod \Phi(\sum \Psi)$, the product of all probability potentials multiplied by the sum of all utility potentials.

Now assume that we shall calculate $\sum_X \prod \Phi(\sum \Psi)$ for some chance variable X. To do that we partition Φ into two sets: Φ_X, which is the set of potentials with X in the domain, and $\Phi^* = \Phi \setminus \Phi_X$. The set Ψ is in the same way divided up in the two sets Ψ_X and Ψ^*. Set $\phi_X = \sum_X \prod \Phi_X$ and $\psi_X = \sum_X \prod \Phi_X(\sum \Psi_X)$. Using the distributive law we get

$$
\begin{aligned}
\sum_X \prod \Phi\left(\sum \Psi\right) &= \prod \Phi^* \sum_X \left(\prod \Phi_X \left(\sum \Psi^* + \sum \Psi_X\right)\right) \\
&= \prod \Phi^* \left(\left(\sum \Psi^*\right) \sum_X \left(\prod \Phi_X\right) + \sum_X \prod \Phi_X \left(\sum \Psi_X\right)\right) \\
&= \prod \Phi^* \left(\left(\sum \Psi^*\right) \phi_X + \psi_X\right) \\
&= \prod \Phi^* \phi_X \left(\sum \Psi^* + \frac{\psi_X}{\phi_X}\right).
\end{aligned}
$$

We see that the result of eliminating the chance variable X is that Φ_X is removed from the set of probability potentials and substituted with ϕ_X. For the set of utility potentials, Ψ_X is removed and $\frac{\psi_X}{\phi_X}$ is added.

Let D be a decision variable. We again divide the potentials into Φ_D, Φ^* and Ψ_D, Ψ^*. Since all variables coming after D in the temporal ordering have been eliminated when we are about to eliminate D, it follows that $\prod \Phi_D$ does not vary with D (See Exercise 10.3). So taking $\max_D$ of $\prod \Phi_D$ is an almost empty operation; it only removes D from the domain. Using the distributive law for max, setting $\phi_D = \max_D \prod \Phi_D$ and $\psi_D = \max_D \prod \Phi_D(\sum \Psi_D)$, and exploiting that $\prod \Phi_D(\sum \Psi^*)$ does not vary with D, we get

$$
\begin{aligned}
\max_D \prod \Phi\left(\sum \Psi\right) &= \prod \Phi^* \max_D \left(\prod \Phi_D \left(\sum \Psi^* + \sum \Psi_D\right)\right) \\
&= \prod \Phi^* \left(\max_D \prod \Phi_D \left(\sum \Psi^*\right) + \max_D \prod \Phi_D \left(\sum \Psi_D\right)\right) \\
&= \prod \Phi^* \left(\phi_D \left(\sum \Psi^*\right) + \psi_D\right) \\
&= \prod \Phi^* \phi_D \left(\sum \Psi^* + \frac{\psi_D}{\phi_D}\right).
\end{aligned}
$$

The result is similar to the result for sum-elimination. To sum up:

Algorithm 10.1 [Variable elimination for influence diagrams] *You work with two sets of potentials: Φ, the set of probability potentials; Ψ, the set of utility potentials. When a variable X is eliminated, the potential sets are modified in the following way:*

1.

$$
\begin{aligned}
\Phi_X &:= \{\phi \in \Phi \mid X \in \mathrm{dom}\,(\phi)\}; \\
\psi_X &:= \{\psi \in \Psi \mid X \in \mathrm{dom}\,(\psi)\}.
\end{aligned}
$$

2. *If X is a chance variable, then*

$$\phi_X := \sum_X \prod \Phi_X;$$

$$\psi_X := \sum_X \prod \Phi_X \left(\sum \Psi_X \right).$$

If X is a decision variable, then

$$\phi_X := \max_X \prod \Phi_X;$$

$$\psi_X := \max_X \prod \Phi_X \left(\sum \Psi_X \right).$$

3.

$$\Phi := (\Phi \setminus \Phi_X) \cup \{\phi_X\}$$

$$\Psi := (\Psi \setminus \Psi_X) \cup \left\{ \frac{\psi_X}{\phi_X} \right\}.$$

□

The influence diagram is solved by repeatedly eliminating variables according to a strong elimination order.

10.2.1 Strong Junction Trees

The considerations on triangulated graphs and junction trees (see Section 4.4) can be applied when the method above is used for solving influence diagrams. The considerations shall not be repeated here. Consider now the influence diagram in Figure 10.5

When solving the influence diagram, you first establish the moral graph: for each potential you link all variables in the domain. For the graph it means that you remove information links, add a link for each pair of nodes with a common child (including a common utility node), and finally remove the directions and the utility nodes. It is done in Figure 10.6 for the influence diagram in Figure 10.5.

As opposed to Bayesian networks, we cannot choose any elimination order for the triangulation. We have to follow a strong elimination order: first eliminate $\mathcal{I}_n$ (in any order), then eliminate D_n, then $\mathcal{I}_{n-1}$ and so on (if some $\mathcal{I}_i$ is empty, we may permute the elimination of D_{i+1} and D_i). The resulting triangulation is called a *strong triangulation*. Figure 10.7 shows the strong triangulation resulting from eliminating the nodes in the moral graph in Figure 10.5 through the strong elimination order $A, L, I, J, K, H, C, D, D_4, G, D_3, D_2, E, F, D_1, B$.

If you use the method for constructing the join trees from Section 4.3.1, the result of a strong triangulation is called a *strong junction tree* with the

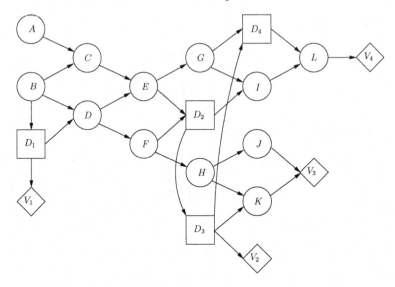

Fig. 10.5. The influence diagram from Figure 9.22.

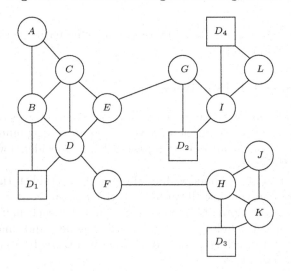

Fig. 10.6. The moral graph for the influence diagram in Figure 10.5.

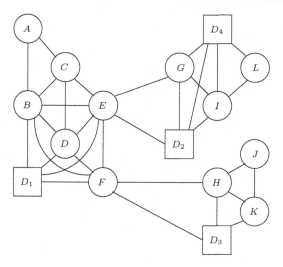

Fig. 10.7. A strong triangulation of the graph in Figure 10.6.

last clique constructed in the strong elimination order serving as a *strong root*. A junction tree with a strong root R has the following property: for any two neighboring cliques C, C' with separator S and C' closest to R, it holds that the variables in S do not appear after the variables in $C \setminus S$ in the temporal order. This property ensures that when COLLECTEVIDENCE(R) is called, then whenever a variable is eliminated the appropriate potentials are present. Figure 10.8 shows a strong junction tree for the graph in Figure 10.7.

Note: Although the influence diagram prescribes a specific order of the decisions, it happens that some decisions are independent such that the order may be altered without changing the strategy or the MEU. This is sometimes detected in constructing a strong junction tree. That is, if you follow the method from Section 4.3.1, you may get a tree in which the decision nodes are eliminated in two different branches (as is the case in Figure 10.8, where the elimination of D_3 can be done independently of D_2 and D_4).

From the strong junction tree, you can construct elimination sequences that do not meet the temporal constraints (the elimination sequence $J, K, H,$ $D_3, A, C, L, I, D_4, G, D_2, D, E, F, D_1, B$ is a perfect elimination sequence ending with B, but it does not follow the temporal order). Since the result of COLLECTEVIDENCE(R) is independent of the actual order of messages sent, all elimination sequences allowed by the strong junction tree give the same result (as long as the elimination order inside each clique obeys the temporal constraints). This means that the strong junction tree in Figure 10.8 discloses that D_3 and D_4 are independent, and the temporal order can be relaxed to a partial ordering of the decision nodes.

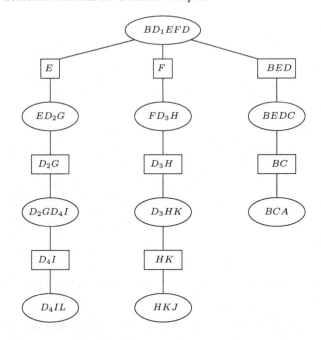

Fig. 10.8. A strong junction tree for the graph in Figure 10.7.

It may also happen that the strong junction tree does not allow for a strong elimination sequence when COLLECTEVIDENCE(R) is called. An example is given in Figure 10.9, where C and F are the first variables to be eliminated according to the temporal ordering, but in the strong junction tree, C is eliminated after D_4. However, this is not a problem, since C cannot affect the policy for D_4 (see Section 11.2).

10.2.2 Required Past

As noted previously, the domain of a policy for a decision variable D_1 is in general $(\mathcal{I}_0, D_1, \ldots, \mathcal{I}_{i-1})$, but a strong elimination order can reveal reductions of the domain: whenever D_i is eliminated, you consider only the potentials with D_i in the domain. The required past must therefore be a subset of the union of these domains, and thus part of the clique closest to the strong root containing D_i.

With the strong elimination ordering $A, L, I, J, K, H, C, D, D_4, G, D_3, D_2,$ E, F, D_1, B for the influence diagram in Figure 10.5, we get the following policies for the decision variables: $\delta_4(G, D_2)$, $\delta_3(F)$, $\delta_2(E)$, and $\delta_1(B)$. Here we see that the policy for D_4, say, contains only two variables as opposed to the seven variables that constitute the past for D_4.

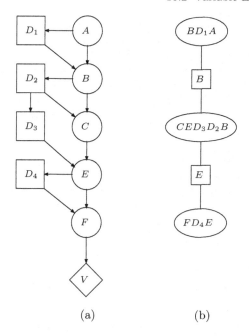

(a)   (b)

Fig. 10.9. An influence diagram (a) with a strong junction tree (b) for which COLLECTEVIDENCE(R) does not initiate a strong elimination sequence meeting the temporal constraints: C should be eliminated before D_4.

This analysis does not guarantee minimal policy domains, as can be seen from the influence diagram in Figure 10.10. We shall return to this issue in Section 11.2.

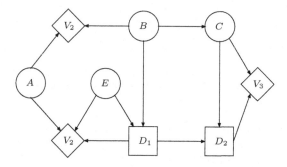

Fig. 10.10. The minimal domain of the policy for D_1 contains only the variable E, but a strong elimination ordering would produce a policy over E and B.

10.2.3 Policy Networks

When a strategy for an influence diagram has been determined, we have a policy δ_i for each decision node D_i. The domain of δ_i is $(\mathcal{I}_0, D_1, \ldots, \mathcal{I}_{i-1})$, but as shown above (and explained in Section 11.2) we may be able to reduce it so that it contains only the required variables, denoted by $\mathrm{req}(D_i)$.

A decision variable can together with its policy be represented in a Bayesian network.

Definition 10.1. *Let D be a decision variable with policy δ_D. The chance-variable representation of D is the result of the following construction: Substitute D with a chance variable D^* having parents $\mathrm{req}(D)$, and assign D^* the conditional probability distribution $P(D^* \mid \mathrm{req}(D))$:*

$$ P(d|\mathbf{r}) = \begin{cases} 1 & \text{if } \delta_D(\mathbf{r}) = d, \\ 0 & \text{otherwise.} \end{cases} $$

If all decision variables are substituted with their chance-variable representations, we obtain a so-called policy network for the influence diagram.

Definition 10.2. *Let I be an influence diagram over $\mathcal{U} = \mathcal{U}_C \cup \mathcal{U}_D$. A policy network for ID (denoted by I^*) is a Bayesian network over $\mathcal{U} = \mathcal{U}_C \cup \mathcal{U}_D^*$ in which all decision variables D_i have been substituted with their chance-variable representations. The probability potentials from I are kept (with D_js replaced by D_j^*).*

Figure 10.11 shows the policy network for the influence diagram in Figure 10.5 with the policy domains determined in Section 10.2.2.

Example 10.1. A farmer has a wheat field. Twice during the season, he observes the state of the field and decides on a possible treatment with fungicides. Later, he observes the state of the field to decide on the booking of machinery for the harvest. Figure 10.12 shows an influence diagram for his decision problem.

To make an advance booking of machinery and for booking plane tickets for his summer vacation, he wishes to know the time of harvest on which he may eventually decide.

Based on the influence diagram an optimal strategy is determined, and the policy network is constructed (see Figure 10.13).

From the policy network he can read the probabilities of his future decision as to the time of harvest. After the first observation and decision, he may enter this as evidence and now get a new probability distribution for the optimal time of harvesting.

Policy networks can be used in other ways. Assume that you know the farmer's influence diagram and observe some of his actions. Then the policy network can give you estimates on what he may have observed or done in

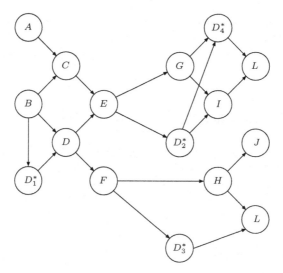

Fig. 10.11. A policy network for the influence diagram in Figure 10.5.

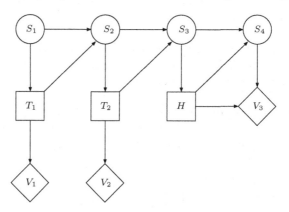

Fig. 10.12. An influence diagram for treatment and time of harvest.

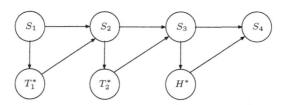

Fig. 10.13. A policy network for the influence diagram in Figure 10.12.

the past. Furthermore, policy networks can be used for analyzing the strategy proposed by the system: risk profile (what is the probability of losing \$X or going bankrupt?), probability of success (winning at least \$X), variance of the expected utility, etc.

10.3 Node Removal and Arc Reversal

In this section we present a method for solving influence diagrams by successively removing the nodes from the diagram. That is, the influence diagram is solved through the construction of a sequence of simpler and simpler influence diagrams. Actually, this method was historically the first, which worked directly on the influence diagram rather than unfolding it to a decision tree.

10.3.1 Node Removal

The method has four operations: removal of barren nodes, removal of chance nodes, removal of decision nodes, and *arc reversal*. The first three operations are rather straightforward.

Removal of barren nodes: A chance or decision node is *barren* if it has no children or if all its children are barren. Since a barren node plays no role for any decision, it can safely be removed.

Removal of chance nodes: Let the only children of the chance node C be the utility nodes $U_1, \ldots, U_k$. Then C and the utility nodes can be removed by integrating them into one utility node (see Figure 10.14) with the utility potential

$$U^* = \sum_C P(C \mid \text{pa}(C)) \left[\sum_{i=1}^{k} U_i \right].$$

Removal of decision nodes: Let the only children of the decision node D be the utility nodes $U_1, \ldots, U_k$. Assume that all parents of $U_1, \ldots, U_k$ are known at the time of deciding on D. Then the optimal policy for D is

$$\delta_D = \arg\max_D \left(\sum_{i=1}^{k} U_i \right),$$

and D and $U_1, \ldots, U_k$ can be removed by substituting them with a new utility node having the potential

$$U^* = \max_D \left(\sum_{i=1}^{k} U_i \right).$$

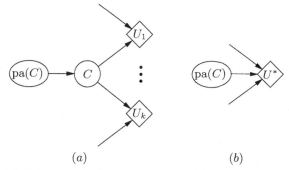

Fig. 10.14. (a) C has only utilities as children. (b) The result of removing C.

10.3.2 Arc Reversal

Consider now the influence diagram in Figure 10.15 (a). None of the removal operations can be applied. However, by applying Bayes' rule we can reverse the arrow from A to B, and now A can be removed.

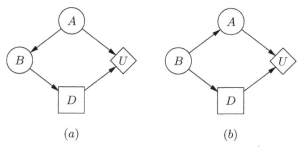

Fig. 10.15. (a) An influence diagram, where no nodes can be removed. (b) The arc has been reversed, and A can now be removed.

To generalize this operation, consider the node A with parents $C, \ldots, D$, and B with parents A and $E, \ldots, F$ (see Figure 10.16 (a)). Assume further that there is no other directed path between A and B.

Now, if the arc from A to B is reversed and the two nodes are given the same parents (see Figure 10.16 (b)), then all d-separation properties in the resulting Bayesian network also hold in the initial network (see Exercise 10.12).

Therefore, the resulting network can represent the probability distribution from the initial network. It is just a question of determining the new conditional probabilities. We substitute the potentials $P(A \mid C, \ldots, D)$ and $P(B \mid A, E, \ldots, F)$ with the potentials $P(A \mid B, C, \ldots, D, E, \ldots, F)$ and $P(B \mid C, \ldots, D, E, \ldots, F)$, and if the product of the new potentials is equal to the product of the old potentials, then the chain rule for Bayesian networks

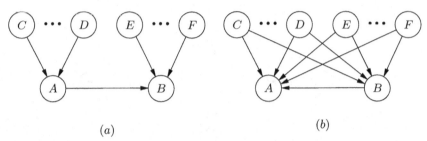

Fig. 10.16. (a) A part of a Bayesian network. (b) The arc from A to B has been reversed, and the two variables are given the same parents.

grants that the two networks represent the same probability distribution. Furthermore, we wish to use only the old potentials for the computation of the new. For this purpose we first establish the following proposition.

Proposition 10.2. *Let A be a node with parents $\mathrm{pa}(A)$ in a Bayesian network, and let X be a nonparent ancestor of A. Then X and A are d-separated given $\mathrm{pa}(A)$.*

Proof. An active path from A to X not containing parents of A must go from A to a child of A. Since there cannot be converging connections on this path, the path must be a directed path from A to X. Since X is an ancestor of A, this would create a directed cycle; hence the path cannot be active. $\square$

To establish the new potentials we look at $P(A, B \mid C, \ldots, D, E, \ldots, F)$. From the fundamental rule we have

$$P(A, B \mid C, \ldots, D, E, \ldots, F) = P(B \mid A, C, \ldots, D, E, \ldots, F)$$
$$P(A \mid C, \ldots, D, E, \ldots, F).$$

The proposition yields that B is independent of $C, \ldots, D$ given $A, E, \ldots, F$. Since there is no directed path between A and B (other than the directed link), A is independent of $E, \ldots, F$ given $C, \ldots, D$. Hence

$$P(A, B \mid C, \ldots, D, E, \ldots, F) = P(B \mid A, E, \ldots, F)P(A \mid C, \ldots, D),$$

and this can be calculated from the potentials in the Bayesian network. Then

$$P(B \mid C, \ldots, D, E, \ldots, F) = \sum_A P(A, B \mid C, \ldots, D, E, \ldots, F),$$

and

$$P(A \mid B, C, \ldots, D, E, \ldots, F) = \frac{P(A, B \mid C, \ldots, D, E, \ldots, F)}{P(B \mid C, \ldots, D, E, \ldots, F)}.$$

Note that the product of the new potentials is equal to the product of the old potentials.

Arc reversal: Let A and B be chance nodes so that A is a parent of B and there are no other directed path from A to B. Let $C, \ldots, D$ be the parents of A and let $A, E, \ldots, F$ be the parents of B. Then the arc from A to B can be reversed by assigning A and B the conditional probability distributions

$$P(B \mid C, \ldots, D, E, \ldots, F) = \sum_A P(B \mid A, E, \ldots, F) P(A \mid C, \ldots, D),$$

$$P(A \mid B, C, \ldots, D, E, \ldots, F) = \frac{P(B \mid A, E, \ldots, F) P(A \mid C, \ldots, D)}{P(B \mid C, \ldots, D, E, \ldots, F)},$$

respectively.

10.3.3 An Example

Consider the influence diagram in Figure 10.17(a). First we remove the barren node E, and we get the influence diagram in Figure 10.17(b).

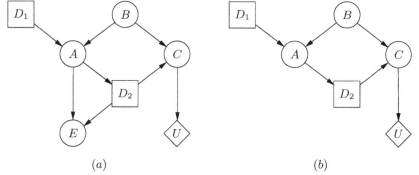

(a)   (b)

Fig. 10.17. (a) An example influence diagram. (b) The same influence diagram without barren nodes.

Next we can remove C, which has only the utility node as a child, and we get the new potential

$$U^1(D_2, B) = \sum_C U(C) P(C \mid D_2, B).$$

The resulting influence diagram is shown in Figure 10.18(a). Next, no node can be removed, and we look for an application of arc reversal. The node B cannot be removed since it has a chance variable as a child, and we fix this by arc reversal. The result is shown in Figure 10.18(b).
The new potentials are

$$P(A \mid D_1) = \sum_B P(A \mid B, D_1) P(B),$$

$$P(B \mid D_1, A) = P(A \mid B, D_1) P(B) / P(A \mid D_1).$$

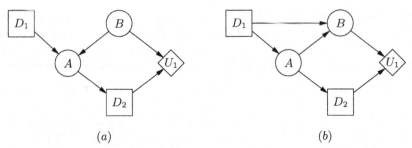

(a)   (b)

Fig. 10.18. (a) C has been removed from Figure 10.17(a). (b) The arc from B to A has been reversed.

Now we can remove B, and we get the new utility (see Figure 10.19(a))

$$U^2(D_1, A, D_2) = \sum_B U^1(B, D_2)P(B \mid A, D_1).$$

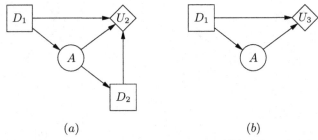

(a)   (b)

Fig. 10.19. (a) B has been removed from Figure 10.18(b). (b) D_2 has been removed.

In the influence diagram in Figure 10.19(a) we can determine the policy for D_2. We have a potential, which directly gives us the utility for each configuration of the relevant past and for each decision option. Hence the policy is achieved by determining the max-value

$$\delta_2(D_1, A) = \arg \max_{D_2} U^2(D_1, A, D_2),$$

and the maximum expected utility is

$$U^3(D_1, A) = \max_{D_2} U^2(D_1, A, D_2).$$

The result is the influence diagram in Figure 10.19(b). A chance-node removal followed by a decision-node removal does the rest.

Finally, we need to show that the four rules above are complete: all influence diagrams can be solved by successive application of the four rules. What we need to show is that if no variables can be removed, then arc reversal will bring us further. See Exercise 10.9.

10.4 Solutions to Unconstrained Influence Diagrams

A solution to an unconstrained influence diagram is an S-DAG together with optimal policies. An S-DAG containing all admissible orderings and all possible branchings after each observation can support all policies, and it could therefore be a candidate for a computational structure for the solution algorithm. However, this full S-DAG grows exponentially in the number of "holes" in the ordering, and there is a risk that it will become intractably large. Also, some nodes in the full S-DAG may never be visited by an optimal strategy, and the corresponding policy is superfluous. Therefore it is worthwhile to reduce the S-DAG under investigation.

Before presenting an algorithm for calculating optimal policies, we shall illustrate various ways of reducing the full S-DAG, while keeping it an S-DAG for an optimal strategy.

10.4.1 Minimizing the S-DAG

Consider the UID in Figure 10.20 with the full S-DAG shown in Figure 10.21; since nothing is gained by delaying a cost-free observation the observables are placed immediately after they have been released.

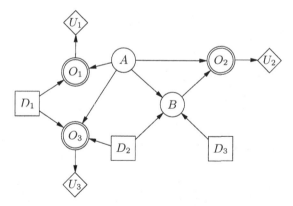

Fig. 10.20. An example UID.

In order to reduce the size of the S-DAG, you can merge paths at points where they have the same history. For example, the upper path in Figure 10.21 $D_1 - O_1 - D_2 \cdots$ shares history with the path $D_2 - D_1 - O_1 \cdots$, and from that

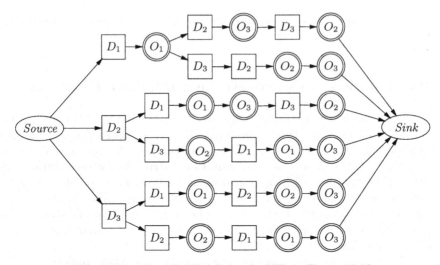

Fig. 10.21. The full S-DAG for the UID in Figure 10.20.

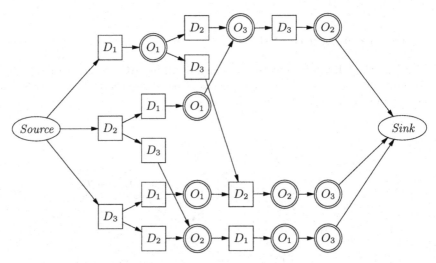

Fig. 10.22. The result of merging paths in the S-DAG from Figure 10.21.

point on, they follow the same routes. The result of merging paths according to this principle is shown in Figure 10.22.

Next, consider the path $D_2 - D_1 - O_1 \cdots$. Since the two decisions D_2 and D_1 can be swapped without changing the expected utilities, the path $D_1 - D_2 - O_1 \cdots$ will have the same expected utility as $D_2 - D_1 - O_1 \cdots$. However, on this path, the observation O_1 is not taken as soon as it has been released, and we say that O_1 is misplaced. Moving O_1 to the other side of D_2

cannot decrease the expected utilities, and we get the path $D_1 - O_1 - D_2 \cdots$. The conclusion is that the path $D_2 - D_1 - O_1 \cdots$ can never be better than $D_1 - O_1 - D_2 \cdots$, and it can therefore be removed from the S-DAG. We say that the path $D_1 - O_1 - D_2 \cdots$ *dominates* the path $D_2 - D_1 - O_1 \cdots$.

For the same reasons $D_1 - O_1 - D_3 \cdots$ dominates $D_3 - D_1 - O_1 \cdots$, $D_2 - D_3 - O_2 \cdots$ is the same as $D_3 - D_2 - O_2 \cdots$, $D_1 - O_1 - D_2 - O_3 - D_3 \cdots$ dominates $D_1 - O_1 - D_3 - D_2 - O_3 \cdots$. We end up with the S-DAG in Figure 10.23, and for this particular example the job is reduced to solving two different influence diagrams. The solution for the UID is then the optimal strategy of the one with highest expected utility.

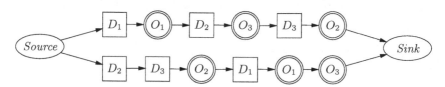

Fig. 10.23. The result of removing dominated paths from the S-DAG in Figure 10.22.

The reduction of the full S-DAG as performed above has the drawback that you start out with the full S-DAG, which may be intractably large. To avoid that, you can start from behind and build up a reduced S-DAG. The procedure is like a breadth-first search in which you go stepwise backward over the "cross section" of the S-DAG constructed so far. For the UID in Figure 10.20 you start with the sink, add all decisions that may come last, and finally add the observables released by each last decision (see Figure 10.24 (a)).

Consider the path with D_2 as the last decision. Then D_3 must be placed at some stage before (see Figure 10.24(b)). If the child of D_3 is a decision node, you can swap until you reach an observable, O. If O is not released by D_3, O is misplaced and it can be swapped with D_3. Since O_2 is the only observable released by D_3, you can move D_3 until it meets O_2, and then D_3 has passed D_2. To conclude, you can avoid D_2 as the last decision.

In general you have the following:

Proposition 10.3. *Let D be a decision node (or Sink) in an S-DAG, and let D_1 and D_2 be parents of D. If the set of observables released by D_1 is a subset of the set of observables released by D_2, then the path with D_2 as a parent of D can be removed without reducing the maximal expected utility.*

The proof goes along the same lines as the reasoning above in removing the path with D_2 as a parent of *Sink*.

To continue the construction of the reduced S-DAG, expand backward from D_1 and D_3. The result is shown Figure 10.25.

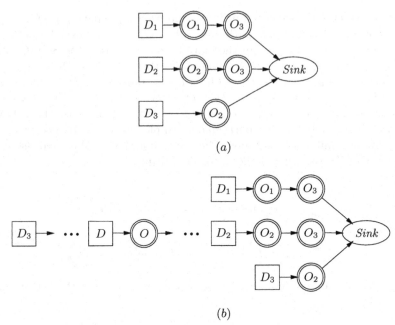

(a)

(b)

Fig. 10.24. (a) The first step in a roll-back construction of a reduced S-DAG. (b) An illustration showing why D_2 can be avoided as the last decision.

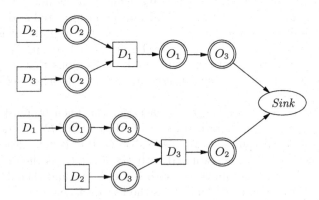

Fig. 10.25. The nodes D_1 and D_3 in Figure 10.24 are expanded backward.

Due to the proposition above, we can remove D_2 as a parent of D_1 as well as D_1 as a parent of D_3. The last expansions yield the S-DAG in Figure 10.23.

10.4.2 Determining Policies and Step Functions

A solution for a reduced S-DAG is determined in almost the same manner as for influence diagrams. We eliminate variables in reverse order; when a branching point is met, the elimination is branched out; when several paths meet, the probability potentials are the same, and the utility potentials are unified through maximization. To illustrate the method we use the UID in Figure 10.26 with the reduced S-DAG in Figure 10.27.

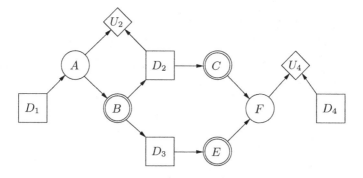

Fig. 10.26. A UID. Recall that each decision node has a hidden utility function.

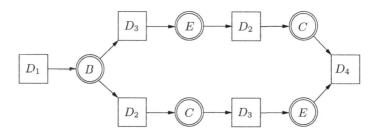

Fig. 10.27. A reduced S-DAG for the UID in Figure 10.26 (*Sink* and *Source* are ignored).

We start off with the two sets:

$$\Phi = \{P(A \mid D_1), P(B \mid A), P(C \mid D_2), P(E \mid D_3), P(F \mid C, E)\},$$
$$\Psi = \{U_1(D_1), U_2(A, D_2), U_3(D_3), U_4(F, D_4)\}.$$

First the nonobservables are eliminated. The actual variable elimination follows the same procedure as for influence diagrams (see Section 10.2). When A and F are eliminated, we get the sets

$$\Phi' = \{P(B \mid D_1), P(C \mid D_2), P(E \mid D_3)\},$$
$$\Psi' = \{U_1(D_1), U_2'(B, D_1, D_2), U_3(D_3), U_4'(C, E, D_4)\},$$

where

$$P(B \mid D_1) = \sum_A P(A \mid D_1)P(B \mid A);$$

$$U_2'(B, D_1, D_2) = \frac{1}{P(B \mid D_1)} \sum_A P(A \mid D_1)P(B \mid A)U_2(A, D_2);$$

$$U_4'(C, E, D_4) = \sum_F P(F \mid C, E)U_4(F, D_4).$$

Note that $\sum_F P(F \mid C, E) = 1$. When D_4 has been eliminated we have

$$\Psi^4 = \{U_1(D_1), U_2'(B, D_1, D_2), U_3(D_3), U_4''(C, E)\},$$

where

$$U_4''(C, E) = \max_{D_4} U_4'(C, E, D_4),$$
$$\delta_{D_4}(C, E) = \arg\max_{D_4} U_4'(C, E, D_4).$$

Next we branch and produce one set of potentials after elimination of C and another set after eliminating E:

$$\Phi^C = \{P(B \mid D_1), P(E \mid D_3)\},$$
$$\Psi^C = \{U_1(D_1), U_2'(B, D_1, D_2), U_3(D_3), U_4^C(E, D_2)\},$$

where $U_4^C(E, D_2) = \sum_C P(C \mid D_2)U_4''(C, E)$, and

$$\Phi^E = \{P(B \mid D_1), P(C \mid D_2)\},$$
$$\Psi^E = \{U_1(D_1), U_2'(B, D_1, D_2), U_3(D_3), U_4^E(C, D_3)\},$$

where $U_4^E(C, D_3) = \sum_E P(E \mid D_3)U_4''(C, E)$.

When eventually D_3 has been eliminated in the C-branch, and D_2 is eliminated in the E-branch, we have the two potential sets

$$\Phi^{Ce} = \{P(B \mid D_1)\},$$
$$\Psi^{Ce} = \{U_1(D_1), U^C(B, D_1)\};$$
$$\Phi^{Ec} = \{P(B \mid D_1)\},$$
$$\Psi^{Ec} = \{U_1(D_1), U^E(B, D_1)\}.$$

It is no coincidence that the two probability potential sets are identical. They are both the result of sum-marginalizing the same set of variables from the same set of potentials. Since sum-marginalizations can be commuted, the two branches must give the same result. Before marginalizing B we unify the utility function sets by taking the max for each entry in the utility functions:

$$\Psi = \{U_1(D_1), \max(U^C(B, D_1), U^E(B, D_1))\}.$$

The step function is

$$\sigma(b, d_1) = \begin{cases} D_3 \text{ if } U^C(b, d_1) \geq U^E(b, d_1), \\ D_2 \text{ otherwise.} \end{cases}$$

Finally, the eliminations of B and D_1 are standard.

10.5 Decision Problems Without a Temporal Ordering: Troubleshooting

A special subclass of decision problems is that of decision problems with no temporal ordering imposed on the decisions (an extreme type of order asymmetry). An important example is troubleshooting, whereby a fault causing a device to malfunction should be identified and eliminated through a sequence of troubleshooting steps. Some steps are *repair steps*, which may or may not fix the problem; some steps are *observation steps*, which cannot fix the problem but may give indications of the causes of the problem; and some steps have repair aspects as well as observation aspects. The task is to find the cheapest strategy for sequencing the troubleshooting steps. As a first attempt you might try to model the decision problem using the unconstrained influence diagram framework, but the lack of temporal constraints will quickly cause the S-DAG to become intractably large, thereby making inference prohibitive. The car start problems of Sections 2.1.1 and 9.3.1 are examples of troubleshooting tasks.

In this section, we shall consider a solution method for decision problems with no temporal ordering by focusing solely on troubleshooting problems. In addition we will deal with pure repair steps and pure observation steps only, and we will call them *actions* and *questions*, respectively.

10.5.1 Action Sequences

First we consider a set of steps consisting of actions only. An action A_i has two possible outcomes, namely "$A_i = yes$" (the problem was fixed) and "$A_i = no$" (the action failed to fix the problem). Each action A_i has a cost $C_{A_i}(e)$, which may depend on evidence e. We sometimes use $C_i(e)$ (or C_i) as shorthand for $C_{A_i}(e)$. Because there are no questions, a *troubleshooting strategy* is a sequence of actions $\mathbf{s} = \langle A_1, \ldots, A_n \rangle$ prescribing the process of repeatedly performing the next action until an action fixes the problem or the last action has been performed.

When solving a troubleshooting problem, we have some initial evidence e and in the course of executing actions in the troubleshooting sequence $\mathbf{s} = \langle A_1, \ldots, A_n \rangle$ we collect further evidence, namely that the previous actions have failed. We let e^i denote the evidence that the first i actions have failed, and we refer to a set of failed actions as *simple evidence*. In the following, we will not mention the initial evidence explicitly.

Definition 10.3. *The* expected cost of repair *(ECR) of a troubleshooting sequence* $\mathbf{s} = \langle A_1, \ldots, A_n \rangle$ *with costs* C_i *is the mean of the costs until an action succeeds or all actions have been performed:*

$$\mathrm{ECR}(\mathbf{s}) \equiv \sum_i \mathrm{ECR}_i(\mathbf{s}),$$

where

$$\mathrm{ECR}_i(\mathbf{s}) = C_i(e^{i-1})P(e^{i-1}).$$

Note that the term "expected cost of repair" may be misleading because we allow a situation in which all actions have been performed without having fixed the problem. If this happens, it will happen with the same probability regardless of the sequence, and therefore we need not estimate a cost for it. We may also extend the set of actions with a *call service* action, CS, that will fix the problem for sure. We return to this in Section 10.5.3.

Now consider two neighboring actions A_i and A_{i+1} in $\mathbf{s}$, and let $\mathbf{s}'$ be obtained from $\mathbf{s}$ by swapping the two actions. The contribution to $\mathrm{ECR}(\mathbf{s})$ from the two actions is

$$C_i(e^{i-1})P(e^{i-1}) + C_{i+1}(e^i)P(A_i = no, e^{i-1}), \qquad (10.4)$$

and the contribution to $\mathrm{ECR}(\mathbf{s}')$ from the two actions is

$$C_{i+1}(e^{i-1})P(e^{i-1}) + C_i(e^{i-1}, A_{i+1} = no)P(A_{i+1} = no, e^{i-1}). \qquad (10.5)$$

The difference between (10.5) and (10.4) equals $\mathrm{ECR}(\mathbf{s}') - \mathrm{ECR}(\mathbf{s})$, so we get

$$\mathrm{ECR}(\mathbf{s}') - \mathrm{ECR}(\mathbf{s}) = P(e^{i-1}) \cdot \left[C_{i+1}(e^{i-1}) - C_i(e^{i-1}) \right.$$
$$\left. + C_i(e^{i-1}, A_{i+1} = no)\, P(A_{i+1} = no \,|\, e^{i-1}) - C_{i+1}(e^i)\, P(A_i = no \,|\, e^{i-1}) \right].$$

If $\mathbf{s}$ is an optimal troubleshooting sequence, we must have $\mathrm{ECR}(\mathbf{s}) \leq \mathrm{ECR}(\mathbf{s}')$, and therefore

$$C_i(e^{i-1}) + C_{i+1}(e^i)P(A_i = no \,|\, e^{i-1}) \qquad (10.6)$$
$$\leq C_{i+1}(e^{i-1}) + C_i(e^{i-1}, A_{i+1} = no)P(A_{i+1} = no \,|\, e^{i-1}).$$

If it holds that the costs are independent of the actions taken previously, (10.6) can be rewritten as

$$\frac{P(A_i = yes \,|\, e^{i-1})}{C_i} \geq \frac{P(A_{i+1} = yes \,|\, e^{i-1})}{C_{i+1}}. \tag{10.7}$$

Definition 10.4. *Let A be a repair action and e be the evidence collected so far. The* efficiency *of A is defined as*

$$\mathrm{ef}(A \,|\, e) = \frac{P(A = yes \,|\, e)}{C_A(e)}.$$

The considerations above yield the following result:

Proposition 10.4. *Let* s *be an optimal sequence of actions for which the costs are independent of the actions taken previously. Then it must hold that*

$$\mathrm{ef}(A_i \,|\, e^{i-1}) \geq \mathrm{ef}(A_{i+1} \,|\, e^{i-1}), \quad \textit{for all } i.$$

10.5.2 A Greedy Approach

As remarked initially, it is not feasible to solve the troubleshooting problem using, for example, the decision tree framework or the unconstrained influence diagram framework. Alternatively, you might try to solve the troubleshooting problem by doing the sequencing in a greedy fashion: always choose an action with the highest efficiency. However, Proposition 10.4 does not guarantee that this approach will yield an optimal troubleshooting sequence.

As an example, consider Figure 10.28, where there are four possible causes, C_1, C_2, C_3, and C_4, for a malfunctioning device; we assume that exactly one of the causes is present, and that the prior probabilities are 0.2, 0.25, 0.40, and 0.15, respectively. Assume that all actions are perfect and have cost 1. Then, action A_2 has the highest efficiency, and if A_2 fails, then A_1 has higher efficiency than A_3. The sequence $\langle A_2, A_1, A_3 \rangle$ has ECR = 1.50. However, the sequence $\langle A_3, A_1 \rangle$ has ECR = 1.45.

To analyze why the decreasing-efficiency approach does not guarantee an optimal sequence, let $\langle A_1, \ldots, A_n \rangle$ be a sequence ordered by decreasing efficiency. If the sequence is not optimal, there must be two actions A_i and A_j, $i < j$, that in the optimal sequence are taken in reverse order. At the time at which A_i is chosen, we have

$$\frac{P(A_i = yes \,|\, e)}{C_i} \geq \frac{P(A_j = yes \,|\, e)}{C_j}.$$

In the optimal sequence, in which A_j is chosen before A_i, we have

$$\frac{P(A_i = yes \,|\, e')}{C_i} < \frac{P(A_j = yes \,|\, e')}{C_j},$$

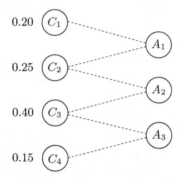

Fig. 10.28. An example of dependent actions. The C's are causes for the device failing. The A variables represent actions. An action will repair a parent if faulty. A single fault is assumed.

where e and e' are simple evidence (not involving A_i and A_j). From this we can infer that an action sequence $\langle A_1, \ldots, A_n \rangle$ is optimal if for all $i < j$ it holds that

$$\mathrm{ef}(A_j \mid e) \le \mathrm{ef}(A_i \mid e),$$

where e is simple evidence (not involving A_i and A_j).

Proposition 10.5. *Consider the following assumptions:*

- *The device has n different faults $F_1, \ldots, F_n$ and n different repair actions $A_1, \ldots, A_n$.*
- *Exactly one of the faults is present.*
- *Each action has a specific probability of repair, $p_i = P(A_i = yes \mid F_i)$, and $P(A_i = yes \mid F_j) = 0$ for $i \ne j$.*
- *The cost C_i of a repair action does not depend on the performance of any previous actions.*

If these assumptions hold, then $\mathrm{ef}(A_j) \le \mathrm{ef}(A_i)$ implies that $\mathrm{ef}(A_j \mid e) \le \mathrm{ef}(A_i \mid e)$, where e is simple evidence (not involving A_i and A_j).

Note that we do not assume the repair actions to be perfect. They may fail to fix a fault that they are supposed to fix.

Proof.   Let A_m be an action that has failed. We calculate $P(A_i = yes \mid A_m = no)$ (for notational convenience, we omit mention of the current evidence e). Due to the single-fault assumption, we have $P(A_m = no \mid A_i = yes) = 1$. Using Bayes' rule, we get

$$P(A_i = yes \mid A_m = no) = \frac{P(A_m = no \mid A_i = yes)P(A_i = yes)}{P(A_m = no)}$$

$$= \frac{P(A_i = yes)}{P(A_m = no)}.$$

In other words, $P(A_m = no)$ is a normalizing constant for the remaining actions, and the relative order of efficiencies is preserved.   □

Example 10.2 (Expansion of Example 9.2). On a cold and wet morning, my car will not start. Moisture may have affected the ignition system or the carburetor, the spark plugs may be dirty, there may be a lack of fuel, or there may be some other fault that I cannot fix myself.

Table 10.3 gives the initial probabilities and costs for the various causes. Because my car started yesterday evening, I assume that exactly one of the causes is present. I have one repair action for each possible cause, but the actions may not be perfect. The measure of precision is the probability of success given that the cause is present. Table 10.3 gives the precision and time requirement of the various actions.

	SP	*IS*	*Carb*	*Fu*	*Other*
Cost	4 min.	2 min.	3 min.	1 min.	n.a.
Prob.	0.3	0.1	0.1	0.1	0.4
Prec.	0.8	0.7	0.6	0.95	n.a.

Table 10.3. Initial probabilities of the causes, precision, and cost in terms of minutes for the various repair actions.

The efficiencies are calculated as

$$\text{ef}(SP) = \frac{0.3 \cdot 0.8}{4} = 0.060,$$

$$\text{ef}(IS) = \frac{0.1 \cdot 0.7}{2} = 0.035,$$

$$\text{ef}(Carb) = \frac{0.1 \cdot 0.6}{3} = 0.02,$$

$$\text{ef}(Fu) = \frac{0.1 \cdot 0.95}{1} = 0.095;$$

hence I should start with *Fu*. Assume now that *Fu* did not solve the problem. By updating the efficiencies of the remaining repair actions (as in the proof above), we get

$$\text{ef}(SP) = \frac{0.3 \cdot 0.8}{(1 - 0.1 \cdot 0.95) \cdot 4} = 0.066;$$

$$\text{ef}(IS) = \frac{0.1 \cdot 0.7}{(1 - 0.1 \cdot 0.95) \cdot 2} = 0.039;$$

$$\text{ef}(Carb) = \frac{0.1 \cdot 0.6}{(1 - 0.1 \cdot 0.95) \cdot 3} = 0.022,$$

which specify the same sequence of the remaining actions as before the update.

The following theorem concludes the considerations.

Theorem 10.3. *Let* $\mathbf{s} = \langle A_1, \ldots, A_n \rangle$ *be an action sequence for a troubleshooting problem fulfilling the conditions in Proposition 10.5. Assume that* $\mathbf{s}$ *is ordered according to decreasing initial efficiencies. Then* $\mathbf{s}$ *is an optimal action sequence and*

$$\text{ECR}(\mathbf{s}) = \sum_{i=1}^{n} C_i \left(1 - \sum_{j=1}^{i-1} p_j \right). \tag{10.8}$$

Proof. From the proof of Proposition 10.5, we have that the relative order of the efficiencies of the actions is preserved. For any action sequence $\mathbf{s}'$ that is not ordered according to $\text{ef}(A_i)$, there will be a j such that $\text{ef}(A_j) < \text{ef}(A_{j+1})$ and therefore $\text{ef}(A_j \,|\, e^{j-1}) < \text{ef}(A_{j+1} \,|\, e^{j-1})$. Hence, $\mathbf{s}'$ can be improved by swapping A_j and A_{j+1}. From the definition, we have

$$\text{ECR}(\mathbf{s}) = \sum_{i=1}^{n} C_i P(e^{i-1}).$$

Due to the single fault assumption, we have $P(e^{i-1}) = 1 - \sum_{j=1}^{i-1} p_j$.   □

10.5.3 Call Service

The action *call service* (CS) will always solve the problem. The cost of CS is not the unknown price of fixing the device but the possible overhead of having outsiders fixing a problem you could have fixed yourself. The efficiency of CS is $1/C_{CS}$ no matter what set of actions has been performed so far.

Let $\mathbf{s} = \langle A_1, \ldots, A_n \rangle$ be an optimal action sequence resulting from a situation meeting the assumptions in Proposition 10.5. It may be that the sequence should be broken before A_n and service called. According to Proposition 10.4, CS should be performed only after an action of higher efficiency. It is a good idea to perform the CS action as soon as it has maximal efficiency. However, this is not guaranteed to be optimal. The question of finding an optimal action sequence including CS is of higher combinatorial complexity: instead of looking for a sequencing of actions each of which must eventually be performed if the other actions fail, we now look for a subset of actions and a sequencing of them. We will not go further into this problem.

10.5.4 Questions

The outcome of a question may shed light on any of the possible faults, or it may be focused on a particular fault.

The troubleshooting task is to interleave actions and questions such that the expected cost is minimal. To do so, we must analyze the value of answers to questions.

Imagine that we are in the middle of a troubleshooting sequence; we have so far gained the evidence e, and now we have the option to ask the question Q with cost C_Q. For simplicity, we assume that Q has only two outcomes, "yes" and "no." Assume that regardless of the outcome of Q, we are able to calculate the minimal expected cost of repair for the remaining sequence. Therefore, let ECR be the minimal expected cost if Q is not performed, and let $\mathrm{ECR}_{Q=yes}$ and $\mathrm{ECR}_{Q=no}$ denote the same for the outcomes "yes" and "no," respectively.

Then, the value of observing Q is

$$V(Q) = \mathrm{ECR} - \left(P(Q = yes \,|\, e) \underset{Q=yes}{\mathrm{ECR}} + P(Q = no \,|\, e) \underset{Q=no}{\mathrm{ECR}} \right), \qquad (10.9)$$

and Q is performed if and only if $V(Q) > C_Q$.

In order to determine whether to ask a question prior to an action, we must analyze all possible succeeding sequences, and if there are several actions and questions, it is in general intractable. In the future, we will also have question options to interleave.

A workable approximation is the *myopic strategy*, where it is assumed at any stage of troubleshooting that we allow questions to be asked, but in the future we allow only repair actions. In that case, the task reduces to calculating expected costs given the various outcomes of the possible questions, and the approaches from the previous sections can be used.

The Myopic Repair–Observation Strategy

The following strategy is a workable approximation to the general troubleshooting task.

Algorithm 10.2 [Myopic repair–observation strategy] *To find a myopic repair-observation strategy, do:*

1. *Let $e :=$ "the device is not working properly".*
2. **While** *the device is not working properly* **do**
   a) *Calculate EGC (the expected cost of the greedy observationless repair sequence).*
   b) **For** *all O* **do**
      i. **For** *all states s of O* **do**
         A. *Calculate $P(O = s \,|\, e)$.*
         B. **For** *all a* **do**
            - *Calculate $p_a^s = P(a$ solves the problem $|\, O = s, e)$.*
         C. *Calculate EGC^s, the expected cost of the greedy observationless repair given $O = s$.*

ii. Calculate

$$EGC_O = c_O + \sum_s P(O = s \mid e)EGC^s.$$

c) Choose the observation or action with lowest expected greedy cost; update e according to the choice and result.

□

10.6 Solutions to Decision Problems with Unbounded Time Horizon

When solving a decision problem with an unbounded time horizon, we are looking for an optimal strategy for the decisions involved. However, as opposed to optimal strategies for bounded decision problems, an optimal strategy for an unbounded decision problem will specify the same optimal policy for all the decisions (see also Section 9.6.1). In what follows we will look at solution methods for unbounded decision problems. To keep things simple we will focus on the discounted reward model, and to simplify the exposition we shall assume that the reward received in a state is independent of the chosen action.

10.6.1 A Basic Solution

As described in Section 9.6.1 we look for a utility function U^* that specifies the value of any state s assuming that all subsequent actions maximize the expected discounted reward:

$$U^*(s) = \max_\Delta U^*(s, \Delta) = \max_\Delta \mathbb{E}\left[\sum_{i=0}^{\infty} \gamma^i R(s_i) \,\middle|\, \Delta, s \right].$$

Instead of calculating $U^*(s, \Delta)$ directly, it can be determined from its "step wise" specification: According to the principle of maximum expected utility we should always choose the action $\delta(s)$ that maximizes the expected utility of the subsequent states:

$$\delta(s) = \arg\max_a \sum_{s' \in \mathrm{sp}(S)} P(s' \mid s, a)U^*(s'). \qquad (10.10)$$

Hence, the value U^* of the current state s is the immediate reward collected at that state plus the maximum expected discounted reward of the subsequent states:

$$U^*(s) = R(s) + \gamma \max_a \sum_{s' \in \mathrm{sp}(S)} P(s' \mid s, a)U^*(s'). \qquad (10.11)$$

From equation (10.10) we see that if we can calculate the maximum expected utility U^* for each state, then we can also find the optimal policy. A way of calculating U^* is to consider the equations defined by equation (10.11) as a system of $|\mathrm{sp}(S)|$ nonlinear equations with $|\mathrm{sp}(S)|$ unknowns (corresponding to the utility of each state); the nonlinearity is due to the max operator. A solution to these equations then corresponds to the utility function U^*. Unfortunately, solving such a set of equations can be a very difficult task, and instead, iterative methods are usually applied. The two most commonly applied iterative methods are called *value iteration* and *policy iteration*.

10.6.2 Value Iteration

The idea of value iteration is to start out with an initial guess at the utility U^* for each state s, and then iteratively refine this guess. How this refinement could be done is suggested by equation (10.11): the utility of a state is determined by the immediate reward received at that state plus the maximum expected utility of all the neighboring states according to our current best guess at the utility function. To be more precise, if we let U^j denote our estimate of the utility function at step j, then we can define an updating function as

$$U^{j+1}(s) = R(s) + \max_a \sum_{s'} P(s'\,|\,a, s)U^j(s'). \tag{10.12}$$

The process of updating the utilities is continued for perhaps a fixed number of iterations or until the largest change is below a certain threshold value.

Example 10.3. In the robot navigation problem in Section 9.6.1, we may set the initial guess U^0 to 0. Then the first iteration sets the utilities U^1 equal to the rewards at the corresponding positions (see Figure 10.29(a)). During the next iteration we update, say position $(2, 1)$, as

$$U^2(2,1) = R(2,1) + \gamma \max \left\{ \sum_s P(s\,|\,north, (2,1))U^1(s)\,, \right.$$

$$\sum_s P(s\,|\,east, (2,1))U^1(s),$$

$$\sum_s P(s\,|\,south, (2,1))U^1(s),$$

$$\left. \sum_s P(s\,|\,west, (2,1))U^1(s) \right\}.$$

By setting the discount factor γ to 0.9 we get

$$U^2(2,1) = -0.1 + 0.9 \cdot \max\{0.7 \cdot -0.1 + 0.1 \cdot 10 + 0.1 \cdot -5 + 0.1 \cdot -0.1,$$
$$0.7 \cdot 10 + 0.1 \cdot -5 + 0.1 \cdot -0.1 + 0.1 \cdot -0.1,$$
$$0.7 \cdot -5 + 0.1 \cdot -0.1 + 0.1 \cdot -0.1 + 0.1 \cdot 10,$$
$$0.7 \cdot 0.1 + 0.1 \cdot -0.1 + 0.1 \cdot 10 + 0.1 \cdot -5\}$$
$$= -0.1 + 0.9 \cdot \max\{0.42, \mathbf{6.48}, -2.52, 0.56\}$$
$$= 5.73,$$

and the maximal value corresponds to going east. Similarly, for position $(2,2)$ we get

$$U^2(2,2) = -5 + 0.9 \cdot \max\{0.7 \cdot -0.1 + 0.1 \cdot -1 + 0.1 \cdot -0.1 + 0.1 \cdot -0.1,$$
$$0.7 \cdot -1 + 0.1 \cdot -0.1 + 0.1 \cdot -0.1 + 0.1 \cdot -0.1,$$
$$0.7 \cdot -0.1 + 0.1 \cdot -0.1 + 0.1 \cdot -0.1 + 0.1 \cdot -1,$$
$$0.1 \cdot 0.1 + 0.1 \cdot -1 + 0.1 \cdot -0.1 + 0.1 \cdot -0.1\}$$
$$= -5 + 0.9 \cdot \max\{\mathbf{-0.19}, -0.73, \mathbf{-0.19}, \mathbf{-0.19}\}$$
$$= -5.171,$$

and the optimal action is then either *north*, *south*, or *west* (ties are resolved according to the sequence *west*, *south*, *east*, and *north*).

By updating the remaining utilities in this fashion we get the utility function U^2 shown in Figure 10.29(b). Based on this utility function we can continue with the third iteration (the result is shown in Figure 10.29(c)) and so forth; the optimal strategies corresponding to U^2 and U^3 (according to equation (10.10)) are shown in Figures 10.29(d)–(e); the optimal policy for U^1 is completely random.

If we continue updating the utilities according to the procedure above, the method will eventually converge to the utility function and the strategy shown in Figure 10.30(a) and Figure 10.30(c), respectively. To see the effect of the discounting factor, Figures 10.30(b) and 10.30(d) show the utility function and the optimal strategy obtained for $\gamma = 0.1$. Observe that when the value of the discounting factor is reduced (the future becomes less significant) the robot cares less about the goal state and instead focuses on avoiding the immediate obstacles. Finally, Figure 10.31 shows the maximum $\log_2$-difference in the utilities after each iteration (using $\gamma = 0.9$), which indicates that the procedure converges exponentially fast.

The fact that value iteration converged to a solution for this particular problem is no coincidence. It can be shown that value iteration is guaranteed to converge, and the utility function that it converges to is the maximum expected discounted reward. Before we give an indication as to why value iteration exhibits these properties, we shall first state the algorithm in its general form.

Algorithm 10.3 [Value Iteration] *Let γ be the discounting factor, R the reward function, and P the transition function:*

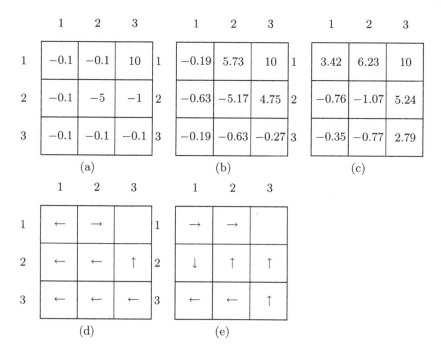

Fig. 10.29. Figures (a), (b), and (c) show the utility functions produced during the first three updates. Figures (d) and (e) show the corresponding optimal strategies; the arrows point in the direction of maximum expected discounted reward.

1. *Choose an $\epsilon > 0$ to regulate the stopping criterion.*
2. *Let U^0 be an initial estimate of the utility function (for example, initialized to zero for all states).*
3. *Set $i := 0$.*
4. **Repeat**
   a) *Let $i := i + 1$.*
   b) **For** *each $s \in \mathrm{sp}(S)$,*

$$U^i(s) := R(s) + \gamma \cdot \max_a \sum_{s' \in \mathrm{sp}(S)} P(s' \,|\, a, s) U^{i-1}(s').$$

5. **Until** $U^i(s) - U^{i-1}(s) < \epsilon$, *for all $s \in \mathrm{sp}(S)$.*

□

It can be shown that the updating step of the algorithm ensures that the difference between any two utility functions is guaranteed to get smaller after each update. To be more specific, if we measure the difference between two utility functions as the maximum distance between two components in the functions

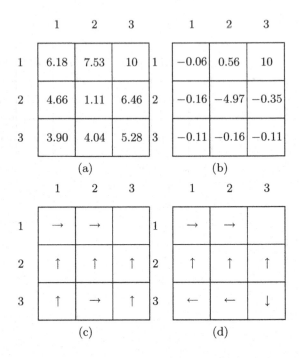

Fig. 10.30. Figures (a) and (c) show the utility function and the optimal strategy obtained upon convergence with discounting factor $\gamma = 0.9$. Convergence was achieved after 75 iterations. Figures (b) and (d) show the situation for $\gamma = 0.1$, where convergence was achieved after 18 iterations.

$$\text{dist}\,(U_1, U_2)_{\max} = \max_{s \in \text{sp}(S)} |U_1(s) - U_2(s)|,$$

then for two utility function U_1 and U_2 we have[1]

$$\text{dist}\,\left(U_1^{i+1}, U_2^{i+1}\right)_{\max} \leq \gamma \cdot \text{dist}\,\left(U_1^i, U_2^i\right)_{\max}.$$

In particular, if we set U_1 equal to the true utility function U^* (the solution to equation (10.11), which does not change during updates), we have

$$\text{dist}\,\left(U^*, U^{i+1}\right)_{\max} \leq \gamma \cdot \text{dist}\,\left(U^*, U^i\right)_{\max}.$$

This behavior allows us to derive two important properties of the updating function:

- There is only one true utility function (see Exercise 10.19).
- The value iteration algorithm is guaranteed to converge to the true utility function.

---

[1] The updating function is a contraction of a metric space with contraction constant γ.

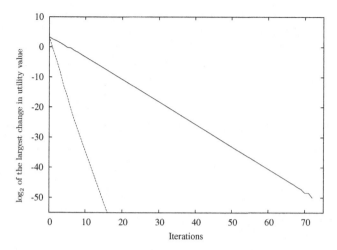

Fig. 10.31. The greatest ($\log_2$) difference in the utility values produced by a value iteration in the robot navigation problem. The solid line corresponds to the discounting factor $\gamma = 0.9$ and the dashed line corresponds to $\gamma = 0.1$.

In addition to these properties we can also find an upper bound on the number of iterations required for the distance between the true utility function and a candidate utility function to be less than ϵ. First of all, from equation (9.1) (page 329) we see that the utility of any state is bounded by $R_{\max}/(1-\gamma)$ ($R_{\max}$ is maximum *absolute* reward, $R_{\max} = \max_s |R(s)|$). Thus, for the initial iteration we have dist $(U^*, U^0)_{\max} \leq 2R_{\max}/(1-\gamma)$, and for the mth iteration we have dist $(U^*, U^m)_{\max} \leq \gamma^m \cdot 2R_{\max}/(1-\gamma)$. From the latter inequality we get

$$\text{dist}\,(U^*, U^m)_{\max} \leq \gamma^m \cdot 2R_{\max}/(1-\gamma) \leq \epsilon.$$

By taking the logarithm and isolating m we have $m = \log(\epsilon(1-\gamma)/2R_{\max})/\log(\gamma)$, which specifies an upper bound on the number of iterations required to achieve an error less than or equal to ϵ. In practice, however, this upper bound has a tendency to be overly conservative, and other methods have been devised to provide tighter bounds. Finally, from the equation above we can also see that the error fades away exponentially fast, but at the same time i will also quickly increase as γ approaches 1. These effects are demonstrated in Figures 10.30 and 10.31.

10.6.3 Policy Iteration

In value iteration you might say that we look for the true utility function as a means of finding an optimal policy. Another (more direct) approach, called *policy iteration*, is to perform an iterative refinement of the current best guess at an optimal policy. This method basically consists of two parts: calculate the utility function U_{Δ_i} corresponding to the current best guess Δ_i at an optimal policy [policy evaluation], and update Δ_i according to U_{Δ_i}, thereby producing an updated policy Δ_{i+1} [policy improvement]. See Figure 10.32.

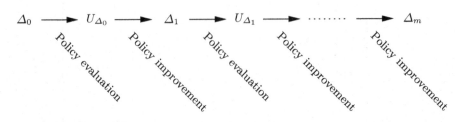

Fig. 10.32. Policy iteration alternates between two steps: policy evaluation and policy improvement.

The idea of policy improvement is to improve our current best guess at the optimal policy Δ_i by beginning in a single state s and finding the action that maximizes the expected utility for that state assuming that the current policy is optimal for all the other states:

$$\Delta_{i+1}(s) := \arg\max_a \sum_{s' \in \mathrm{sp}(S)} P(s' \mid a, s) U_{\Delta_i}(s').$$

That is, we can think of policy improvement as an updating procedure for Δ_i based on a one step look-a-head according to the utility function for Δ_i:

The utility function U_{Δ_i} used in policy improvement is found during policy evaluation, where the basic task is to calculate the expected discounted reward of following the strategy Δ_i for each state s:

$$U_{\Delta_i}(s) = R(s) + \gamma \sum_{s' \in \mathrm{sp}(S)} P(s' \mid \Delta_i(s), s) U_{\Delta_i}(s').$$

Since we are working with a fixed strategy, this equation does not involve a max-operator (as opposed to our initial specification of the problem in equation (10.11)) and the expression is therefore linear in the utilities. This also means that we can calculate the utility function for a specific strategy by treating it as a linear programming problem:

$$U_\Delta(s_1) = R(s_1) + \gamma \sum_{j=1}^{n} P(s_j \mid \Delta(s_1), s_1) U(s_j),$$

$$U_\Delta(s_2) = R(s_2) + \gamma \sum_{j=1}^{n} P(s_j \mid \Delta(s_2), s_2) U(s_j),$$

$$\vdots$$

$$U_\Delta(s_n) = R(s_n) + \gamma \sum_{j=1}^{n} P(s_j \mid \Delta(s_n), s_n) U(s_j),$$

consisting of n linear equations and n unknowns. For our robot navigation problem, n corresponds to the number of possible world positions. When the state space is small, this programming problem does not introduce any difficulties, but for larger state spaces it may be too time-consuming. Instead, we can go for an approximate solution to this problem using value iteration. In this case the time complexity can be controlled by specifying a suitable termination criterion (a value for ϵ) and then using the upper bound on the number of value iterations required to reach ϵ.

In general, the policy iteration method can be stated as follows:

Algorithm 10.4 [Policy iteration]

1. *Let Δ_0 be an initial randomly chosen policy.*
2. *Set $i := 0$.*
3. **Repeat**
   a) *Find the utility function U_{Δ_i} corresponding to the policy Δ_i* [policy evaluation].
   b) *Let $i := i + 1$.*
   c) **For** *each $s \in \mathrm{sp}(S)$*

   $$\Delta_i(s) := \arg\max_a \sum_{s' \in \mathrm{sp}(S)} P(s' \mid a, s) U_{\Delta_{i-1}}(s') \text{ [policy improvement]}.$$

4. **Until** $\Delta_i = \Delta_{i-1}$.

$\square$

The algorithm terminates when the current policy is not changed during an iteration. This also implies that the utility function U_{Δ_m} for the final policy Δ_m is the same as the utility function for the policy Δ_{m-1} found in the previous iteration, since they are both solutions to the same system of linear equations. Hence, U_{Δ_m} is a solution to equation (10.11):

$$U_{\Delta_m}(S) = R(S) + \gamma \sum_{S'} P(S' \mid \Delta_m(S), s_0) U_{\Delta_m}(S')$$

$$= R(S) + \gamma \sum_{S'} P(S' \mid \Delta_m(S), s_0) U_{\Delta_{m-1}}(S')$$

$$= R(S) + \gamma \max_a \sum_{S'} P(S' \mid a, s_0) U_{\Delta_{m-1}}(S')$$

$$= R(S) + \gamma \max_a \sum_{S'} P(S' \mid a, s_0) U_{\Delta_m}(S').$$

Since this solution is unique (see Section 10.6.2), we know that the policy returned by policy iteration is also an optimal policy.

10.6.4 Solving Partially Observable Markov Decision Processes*

As stated in Section 9.6.2, there is a fundamental difference between an optimal policy for an MDP and an optimal policy for a POMDP: an optimal policy for an MDP specifies an action for each possible state of the world, but an optimal policy for a POMDP specifies an action for each possible belief that we may have about the state of the world. A belief at step i corresponds to a probability distribution $P(S_i \mid d_1, o_1, \ldots, d_{i-1}, o_i)$, which summarizes the relevant information from the past (lowercase letters are used to denote specific observations and decisions). This means that $P(S_i \mid d_1, o_1, \ldots, d_{i-1}, o_i)$, our belief at step i, plays the same role as a state in an MDP, and this is also the reason why $P(S_i \mid d_1, o_1, \ldots, d_{i-1}, o_i)$ is called the *belief state* at time i (denoted by $b(S_i)$ or just b_i). Thus, if $\mathcal{B}_i$ denotes the set of all possible belief states (of which there are infinitely many), then an optimal policy for decision D_i is a function

$$\delta_{D_i} : \mathcal{B}_i \rightarrow \mathrm{sp}(D_i).$$

Since both value iteration and policy iteration for MDPs require a finite number of states, we cannot directly adopt these methods when working with POMDPs. Instead you might try to transform the POMDP into an "equivalent" MDP (see Figure 10.34), so that by solving the MDP we also obtain a solution to the original POMDP.

One possibility might be to simply construct a new finite belief space $\mathcal{B}'$ representing the original belief space $\mathcal{B}$. For example, in a POMDP with two world states, $\mathrm{sp}(S) = \{s_1, s_2\}$, we have a belief state for each probability of s_1; see Figure 10.33(a). This belief space can be partitioned into, for example, 10 equally wide intervals, $\mathcal{B}' = \{[0, 0.1), [0.1, 0.2), \ldots, [0.9, 1]\}$, which can be used as the world states in an MDP representation. To complete the specification you also need $P(\mathcal{B}'_i \mid \mathcal{B}'_{i-1}, D_{i-1})$ and $U(D_i, \mathcal{B}'_i)$, both of which can be derived from the original POMDP specification. An approximate solution to the POMDP can now be found by solving the MDP representation using either value iteration or policy iteration.

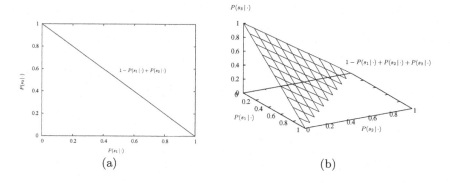

Fig. 10.33. The belief space for a POMDP with two and three world states, respectively.

Unfortunately, this partitioning/discretization procedure is infeasible for all but the smallest POMDPs, since the number of states in the MDP representation grows exponentially in the number of world states in the POMDP. Figure 10.33(b) shows a partitioning of the belief space for a POMDP with three states; with four states the belief space would be a hypercube in 4-dimensional space.

Rather than discretizing the belief space, a more common approach is to extend the MDP algorithms to infinite state spaces (see Figure 10.34). To give an idea of the procedure, let us first look at how a POMDP can be transformed into an MDP without dwelling on the potential complications of infinite state spaces.

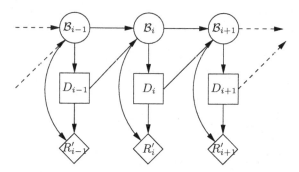

Fig. 10.34. The MDP representation of a POMDP. The state variable $\mathcal{B}_i$ contains one state for each possible belief state at step i (of which there are infinitely many).

First of all, let us start by establishing the fact that the belief state at step $i-1$ summarizes all the relevant information about the previous observations and decisions. This will also help us establish the conditional probabilities used in the MDP representation. Thus, we look for an independence relation formed by conditioning on a continuous variable. This type of conditioning is not an issue we have touched upon previously, but for the purpose of the subsequent derivations you may treat it as conditioning on a discrete variable. That is,

$$b_i = P(S_i \,|\, d_{i-1}, o_i, \text{past}(D_{i-1})) = P(S_i \,|\, d_{i-1}, o_i, b_{i-1}),$$

where $\text{past}(D_{i-1}) = (o_1, d_1, \ldots, d_{i-2}, o_{i-1})$ denotes all observations and decisions prior to decision D_{i-1}. By Bayes' rule we have that

$$
\begin{aligned}
&P(S_i \,|\, d_{i-1}, o_i, \text{past}(D_{i-1})) \\
&= \frac{P(o_i \,|\, S_i, d_{i-1}, \text{past}(D_{i-1})) P(S_i \,|\, d_{i-1}, \text{past}(D_{i-1}))}{P(o_i \,|\, d_{i-1}, \text{past}(D_{i-1}))},
\end{aligned}
\qquad (10.13)
$$

and since $P(o_i \,|\, d_{i-1}, \text{past}(D_{i-1}))$ is just a normalization constant, we get

$$
\begin{aligned}
&P(S_i \,|\, d_{i-1}, o_i, \text{past}(D_{i-1})) \\
&\quad \propto P(o_i \,|\, S_i, d_{i-1}, \text{past}(D_{i-1})) P(S_i \,|\, d_{i-1}, \text{past}(D_{i-1})).
\end{aligned}
$$

The third probability can also be expressed as

$$
\begin{aligned}
&P(S_i \,|\, d_{i-1}, \text{past}(D_{i-1})) \\
&\quad = \sum_{S_{i-1}} P(S_i \,|\, d_{i-1}, \text{past}(D_{i-1}), S_{i-1}) P(S_{i-1} \,|\, d_{i-1}, \text{past}(D_{i-1})).
\end{aligned}
$$

Since S_{i-1} is independent of d_{i-1}, and S_i is independent of $\text{past}(D_{i-1})$ given S_{i-1} (check the d-separation properties in the model) the above expression simplifies to

$$P(S_i \,|\, d_{i-1}, \text{past}(D_{i-1})) = \sum_{S_{i-1}} P(S_i \,|\, d_{i-1}, S_{i-1}) P(S_{i-1} \,|\, \text{past}(D_{i-1})).$$

By also exploiting that $P(o_i \,|\, S_i, d_{i-1}, \text{past}(D_{i-1})) = P(o_i \,|\, S_i, d_{i-1})$, equation (10.13) can now be expressed as

$$
\begin{aligned}
&P(S_i \,|\, d_{i-1}, o_i, \text{past}(D_{i-1})) \\
&\quad \propto P(o_i \,|\, S_i, d_{i-1}) \sum_{S_{i-1}} P(S_i \,|\, d_{i-1}, S_{i-1}) P(S_{i-1} \,|\, \text{past}(D_{i-1})).
\end{aligned}
$$

Since $P(S_{i-1} \,|\, \text{past}(D_{i-1}))$ is the belief state, b_{i-1}, for step $i-1$ we end up with

$$b_i = P(S_i \mid d_{i-1}, o_i, \text{past}(D_{i-1}))$$

$$\propto P(o_i \mid S_i, d_{i-1}) \sum_{S_{i-1}} P(S_i \mid d_{i-1}, S_{i-1}) b(S_{i-1}), \tag{10.14}$$

where the right-hand side of the expression does not depend on the past observations and decisions given the previous belief state $b(S_{i-1})$. We can therefore write

$$b(S_i) = P(S_i \mid d_{i-1}, o_i, \text{past}(D_{i-1})) = P(S_i \mid d_{i-1}, o_i, b(S_{i-1})).$$

It should also be noted that in equation (10.14) we have that $P(o_i \mid S_i, d_{i-1})$ is the observation function and $P(S_i \mid d_{i-1}, S_{i-1})$ is the transition function. Hence, equation (10.14) also provides a way to update our belief state based on the prior belief state, the decision d_{i-1}, and the observation o_i. This updated belief state corresponds to the observation of b_i.

Now, going back to our initial goal of describing the transformation of the POMDP model to the MDP model in Figure 10.34, we need to specify the transition function $P(b_i \mid b_{i-1}, d_{i-1})$ and the reward function $R'_i(b_i, D_i)$. The specification should ensure that the two models become equivalent, meaning that an optimal solution for one of the models is also an optimal solution for the other model.

The transition function $P(b_i \mid d_{i-1}, b_{i-1})$ can be expressed as

$$P(b_i \mid d_{i-1}, b_{i-1}) = \sum_{O_i} P(b_i \mid d_{i-1}, b_{i-1}, O_i) P(O_i \mid d_{i-1}, b_{i-1}), \tag{10.15}$$

where the probability $P(O_i \mid d_{i-1}, b_{i-1})$ corresponds to the normalization constant in equation (10.13) and can be calculated as

$$P(O_i \mid d_{i-1}, b_{i-1}) = \sum_{S_i} P(O_i \mid S_i, d_{i-1}) \sum_{S_{i-1}} P(S_i \mid d_{i-1}, S_{i-1}) b(S_{i-1}).$$

Again, the expression depends only on the observation function, the transition function, and the previous belief state. The function $P(b_i \mid d_{i-1}, b_{i-1}, o_i)$ is simply an indicator function defined as

$$P(b_i \mid d_{i-1}, b_{i-1}, o_i) = \begin{cases} 1 & \text{if } b(S_i) = P(S_i \mid d_{i-1}, o_i, b_{i-1}), \\ 0 & \text{otherwise.} \end{cases}$$

Next, we also have to specify the reward function $R'(b(S_i, d_i)$. Fortunately, this function can simply be calculated as

$$R'(b_i, D_i) = \sum_{S_i} R(S_i, d_i) b(S_i). \tag{10.16}$$

Thus, equation (10.15) together with equation (10.16) provides a complete specification of the transformed POMDP, and equation (10.14) describes how to find the observed belief state at each time step.

The final part is now to solve the MDP. However, we cannot immediately apply the algorithms described in the previous sections, since they work only on MDPs with finite state spaces. Instead these algorithms have to be modified to work with continuous MDPs. The overall approach is to partition the space of belief functions into regions, where each region is associated with a particular strategy and a corresponding linear utility function. A more thorough description of the algorithm is outside the scope of the present book.

10.7 Limited Memory Influence Diagrams

The major complexity problem for influence diagrams is that the relevant past for a policy may be intractably large. A way of addressing this problem is to restrict memory. This restriction can be introduced in the form of history variables or information blocking as described in Section 10.1. Another way is to pinpoint explicitly what is remembered when a decision is taken. That is, the no-forgetting assumption in interpreting an influence diagram is dropped, and instead memory is represented directly by information links.

Assume that for the fishing example in Figure 9.23 we add the restriction that we (the EU politicians) remember only last year's decision, but we can recall the T-observations up to two years back. This can be represented by the model in Figure 10.35.

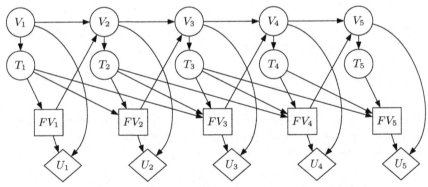

Fig. 10.35. Figure 9.23 modified to represent limited memory. Absent information arcs mean that the information is not remembered.

An influence diagram with direct representation of memory is called a *limited memory influence diagram* (LIMID). To stress the difference, influence diagrams can be called *perfect recall influence diagrams*. The advantage of LIMIDs is that they allow you to work with decision policies with small domains. If the domain of a policy does not include all the variables relevant for the associated decision, then the solution to the LIMID is an approximation to a solution for the corresponding perfect recall influence diagram.

The strong junction tree method automatically constructs cliques containing domains for perfect recall policies, and it is therefore not well suited for taking advantage of the space reduction offered by LIMIDs. Instead, a policy network can be used (see Section 10.2.3): substitute each decision variable D with a chance variable D^* having the same parents and children as D (we ignore that some informational parents may turn up nonrequired; see Section 11.2). A policy network representation of the LIMID in Figure 10.35 is shown in Figure 10.36.

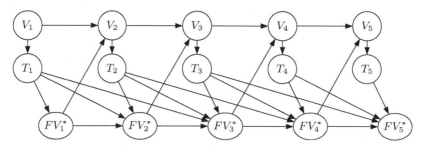

Fig. 10.36. The policy network for the LIMID in Figure 10.35.

We attach a set of initial conditional probability distributions $P_0(D^* \mid \text{pa}(D^*))$ to the D^* variables. These distributions represent our initial guess at the optimal policies of the decisions. The distributions need not be deterministic and could be chosen at random. Next, you change the policy network to a series of one-action networks and solve them as described in Section 9.1. It is natural to start with the last decision. The single-action network for the last decision in the fishing network is shown in Figure 10.37.

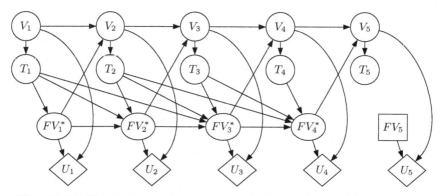

Fig. 10.37. The single-action network for the last decision in Figure 10.36.

To establish an optimal policy for FV_5 you need $P(V_5 \mid FV_4^*, T_5, T_4, T_3)$. To find this probability you can use any inference method for the underlying Bayesian network; there are no constraints on the elimination order.

Next, having found a new policy $\delta_{FV_5}(FV_4^*, T_5, T_4, T_3)$ for FV_5 you substitute the initially specified potential $P_0(FV_5^* \mid FV_4^*, T_5, T_4, T_3)$ with a chance variable representation of δ_{FV_5}:

$$P_1(FV_5^* = v \mid FV_4^*, T_5, T_4, T_3) = \begin{cases} 1 & \text{if } \delta_{FV_5}(FV_4^*, T_5, T_4, T_3) = v, \\ 0 & \text{otherwise,} \end{cases}$$

and construct the single-action network for FV_4. See Figure 10.38.

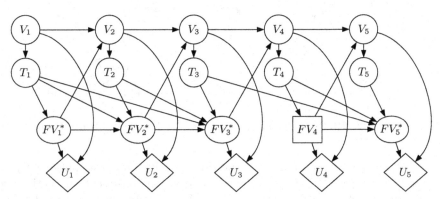

Fig. 10.38. A single-action network for FV_4.

To find a new policy for FV_4 we look for $\mathrm{EU}(FV_4 \mid FV_3, T_4, T_3, T_2)$, which is the sum of the expectations for U_4 and U_5. This requires the calculation of $P(FV_5^*, V_5 \mid FV_4, FV_3^*, T_4, T_3, T_2)$ and $P(V_4 \mid FV_4, FV_3^*, T_4, T_3, T_2)$, where the former joint probability can be found using, for example, variable propagation (see Section 5.2). Continue to FV_3 and down to FV_1.

Now, the initial policies for FV_1, FV_2, FV_3, and FV_4 were used in determining a new policy for FV_5. These initial policies also had an impact on $P(FV_5^*, V_5 \mid FV_4, FV_3^*, T_4, T_3, T_2)$ and $P(V_4 \mid FV_4, FV_3^*, T_4, T_3, T_2)$, and you need to repeat the process based on the new policies. That is, the procedure, called *single policy updating*, is iterative, and from the description above we see that it is closely related to policy iteration for MDPs. It can be shown that the procedure converges, and that it converges to an optimal strategy for the LIMID. However, this need not be an optimal strategy for the perfect recall influence diagram, and it is an issue of research to establish bounds on the distance between the LIMID optimal strategy and the perfect recall optimal strategy.

Algorithm 10.5 [Single policy updating] *Let I be a LIMID with decision variables $D_1, \ldots, D_n$, and let I' be a policy network for I, where the decision variables are represented by the chance variables $D_1^*, \ldots, D_n^*$.*

1. *Let $P_0(D_j^* \mid \mathrm{pa}(D_j^*))$ be a randomly chosen initial probability distribution for D_j^*, $1 \leq j \leq n$, in I'.*
2. *Let $i := 1$.*
3. *Repeat*
   a) *For $j := n$ to 1*
      i. *Let $\mathcal{U}_{D_j}$ be the utility descendants of D_j.[2]*
      ii. *Calculate a policy for D_j:*

$$\delta_{D_j}(\mathrm{pa}(D_j))^i = \arg\max_{D_j} \sum_{U \in \mathcal{U}_{D_j}}$$

$$\sum_{\mathrm{pa}(U)\backslash \mathrm{fa}(D_j)} P(\mathrm{pa}(U) \backslash \mathrm{fa}(D_j) \mid \mathrm{fa}(D_j))U(\mathrm{pa}(U)).$$

      iii. *Replace $P_{i-1}(D_j^* \mid \mathrm{pa}(D_j^*))$ in I' with*

$$P_i(D_j^* = d \mid \mathrm{pa}(D_j^*)) = \begin{cases} 1 & \text{if } \delta_{D_j}^i(\mathrm{pa}(D_j^*)) = d, \\ 0 & \text{otherwise.} \end{cases}$$

   b) *Set $i := i + 1$.*
4. *Until convergence.*

$\square$

The repeated construction of single-action networks and variable propagation can be performed in a unified framework saving a large number of repetitions of the same calculations. We shall not treat this further but refer the interested reader to the literature.

10.8 Summary

The Chain Rule for Influence Diagrams

Let ID be an influence diagram with universe $\mathcal{U} = \mathcal{U}_C \cup \mathcal{U}_D$. Then

$$P(\mathcal{U}_C \mid \mathcal{U}_D) = \prod_{X \in \mathcal{U}_C} P(X \mid pa(X)).$$

The Expected Utility and an Optimal Strategy

Let the temporal order of the variables in $\mathcal{U}$ be described as $\mathcal{I}_0 \prec D_1 \prec \mathcal{I}_1 \prec \cdots \prec D_n \prec \mathcal{I}_n$ and let $V = \sum_i V_i$. Then

(i) an optimal policy for D_i is

$$\delta_i(\mathcal{I}_0, D_1, \ldots, \mathcal{I}_{i-1}) = \arg\max_{D_i} \sum_{\mathcal{I}_i} \max_{D_{i+1}} \ldots \max_{D_n} \sum_{\mathcal{I}_n} P(\mathcal{U}_C \mid \mathcal{U}_D)V,$$

---

[2] No other utility nodes can influence the policy for D_j. See Section 11.2.

(ii) the expected utility from following the policy δ_i (and acting optimally in the future) is

$$\rho_i(\mathcal{I}_0, D_1, \ldots, \mathcal{I}_{i-1}) = \frac{1}{P(\mathcal{I}_0, \ldots, \mathcal{I}_{i-1} \mid D_1, \ldots, D_{i-1})}$$
$$\max_{D_i} \sum_{\mathcal{I}_i} \max_{D_{i+1}} \cdots \max_{D_n} \sum_{\mathcal{I}_n} P(\mathcal{U}_C \mid \mathcal{U}_D) V,$$

and the strategy for ID consisting of an optimal policy for each decision yields the maximum expected utility

$$\mathrm{MEU}(ID) = \sum_{\mathcal{I}_0} \max_{D_1} \sum_{\mathcal{I}_1} \max_{D_2} \cdots \max_{D_n} \sum_{\mathcal{I}_n} P(\mathcal{U}_C \mid \mathcal{U}_D) V.$$

Variable Elimination for Influence Diagrams

The influence diagram is solved by repeatedly eliminating the variables in reverse temporal order. When eliminating a variable, you work with two sets of potentials: Φ, the set of probability potentials; Ψ, the set of utility potentials. When a variable X is eliminated, the potential sets are modified in the following way:

1.

$$\Phi_X := \{\phi \in \Phi \mid X \in \mathrm{dom}\,(\phi)\};$$
$$\psi_X := \{\phi \in \Psi \mid X \in \mathrm{dom}\,(\phi)\}.$$

2. If X is a chance variable, then

$$\phi_X := \sum_X \prod \Phi_X;$$
$$\psi_X := \sum_X \prod \Phi_X \left(\sum \Psi_X\right).$$

If X is a decision variable, then[3]

$$\phi_X := \max_X \prod \Phi_X;$$
$$\psi_X := \max_X \prod \Phi_X \left(\sum \Psi_X\right).$$

3.

$$\Phi := (\Phi \setminus \Phi_X) \cup \{\phi_X\},$$
$$\Psi := (\Psi \setminus \Psi_X) \cup \left\{\frac{\psi_X}{\phi_X}\right\}.$$

These calculations can also be organized in a strong junction tree for the influence diagram. A strong junction tree is produced by eliminating the variables in reverse temporal order.

---

[3] When X is a decision variable, Φ_x is a constant function over X.

Policy Networks

Let D be a decision variable with policy $\delta_D(\text{req}(D))$. The chance-variable representation of D is the result of the following construction: Substitute D with a chance variable D^* with parents $\text{req}(D)$. The conditional probability potential $P(D^* \mid \text{req}(D))$ is

$$P(d|\bar{r}) = \begin{cases} 1 & \text{if } \delta_D(\bar{r}) = d, \\ 0 & \text{otherwise.} \end{cases}$$

Let ID be an influence diagram over $\mathcal{U} = \mathcal{U}_C \cup U_D$. A policy network for ID (denoted by ID^*) is a Bayesian network over $\mathcal{U} = \mathcal{U}_C \cup \mathcal{U}_D^*$ in which all decision variables D_i have been substituted with their chance-variable representations. The probability potentials from ID are kept (with D_js replaced by D_j^*).

Node Removal and Arc Reversal

The influence diagram is solved by iteratively removing nodes and reversing arcs according to the following rules:

Removal of barren nodes: A chance or decision node is *barren* if it has no children or all its children are barren. Since a barren node plays no role for any decision, it can safely be removed.

Removal of chance nodes: Let the only children of the chance node C be the utility nodes $U_1, \ldots, U_k$. Then C and the utility nodes can be removed by integrating them into one utility node with the utility potential

$$U^* = \sum_C P(C|\text{pa}(C)) \left[\sum_{i=1}^k U_i \right].$$

Removal of decision nodes. Let the only children of the decision node D be the utility nodes $U_1, \ldots, U_k$. Assume that all parents of $U_1, \ldots, U_k$ are known at the time of deciding on D. Then the optimal policy for D is

$$\delta_D = \arg\max_D \left(\sum_{i=1}^k U_i \right),$$

and D and $U_1, \ldots, U_k$ can be removed by substituting them with a new utility node having the potential

$$U^* = \max_D \left(\sum_{i=1}^k U_i \right).$$

If no nodes can be removed, then arc reversals can be performed to obtain another (EU-equivalent) influence diagram in which one of the rules above can be applied.

Arc reversal: Let A and B be chance nodes such that A is a parent of B and there are no other directed paths from A to B. Let $C, \ldots, D$ be the parents of A and let $A, E, \ldots, F$ be the parents of B. Then the arc from A to B can be reversed by assigning A and B the conditional probability distributions

$$P(B \mid C, \ldots, D, E, \ldots, F) = \sum_A P(B \mid A, E, \ldots, F)P(A \mid C, \ldots, D),$$

$$P(A \mid B, C, \ldots, D, E, \ldots, F) = \frac{P(B \mid A, E, \ldots, F)P(A \mid C, \ldots, D)}{P(B \mid C, \ldots, D, E, \ldots, F)},$$

respectively.

Unconstrained Influence Diagrams

An S-DAG can be constructed from a breadth-first procedure starting at the sink: add all the decisions that may come last, and after that you add the observables released by the decisions. By exploiting the following rule we need not construct the full S-DAG:

Let D be a decision node (or *Sink*) in an S-DAG, and let D_1 and D_2 be parents of D. If the set of observables released by D_1 is a subset of the set of observables released by D_2, then the path with D_2 as a parent of D can be removed without reducing the maximal expected utility.

A solution to the UID is found using variable elimination based on the S-DAG structure.

Troubleshooting

The expected cost of repair of a troubleshooting sequence $\mathbf{s} = \langle A_1, \ldots, A_n \rangle$ of repair actions is

$$\text{ECR}(\mathbf{s}) = \sum_i C_i(\mathbf{e}^{i-1})P(\mathbf{e}^{i-1}),$$

where $\mathbf{e}^j$ denotes the statement that the first j actions have failed.

For an optimal repair sequence, it holds that

$$C_i(\mathbf{e}^{i-1}) + C_{i+1}(\mathbf{e}^i)P(A_i = n \mid \mathbf{e}^{i-1})$$
$$\leq C_{i+1}(\mathbf{e}^{i-1}) + C_i(\mathbf{e}^{i-1}, A_{i+1} = n)P(A_{i+1} = n \mid \mathbf{e}^{i-1}).$$

The efficiency of a repair action is

$$\text{ef}(A \mid \mathbf{e}) = \frac{P(A = y \mid \mathbf{e})}{C_A(\mathbf{e})}.$$

If costs are independent of evidence, then for an optimal repair sequence it must hold that

$$\text{ef}(A_i \mid \mathbf{e}^{i-1}) \geq \text{ef}(A_{i+1} \mid \mathbf{e}^{i-1}),$$

and if for all $i < j$ it holds that

$$\text{ef}(A_j \mid \mathbf{e}) \leq \text{ef}(A_i \mid \mathbf{e})$$

for all simple evidence $\mathbf{e}$ (not involving A_i and A_j) of the type "actions $A, \ldots, B$ have failed," then the repair sequence $\langle A_1, \ldots, A_n \rangle$ is optimal (this does not necessarily hold when call service is an option).

Questions: The value of getting an answer of Q is

$$V(Q) = \text{ECR} - \sum_{s \in Q} P(Q = s \mid \mathbf{e}) \, \text{ECR}_s,$$

where ECR_s is the expected cost of repair for an optimal sequence given evidence $\mathbf{e}$ and "$Q = s$," and ECR is the expected cost of repair for an optimal sequence not starting with Q. Because neither ECR nor ECR_s is tractable, a myopic approach is often used.

Unbounded Decision Problems

Let γ be the discounting factor, R the reward function, and P the transition function.

Value iteration:

1. Choose an $\epsilon > 0$ to regulate the stopping criterion.
2. Let U^0 be an initial estimate of the utility function (for example, initialized to zero for all states).
3. Set $i := 0$.
4. **Repeat**
   a) Let $i := i + 1$.
   b) **For** each $s \in \text{sp}(S)$

$$U^i(s) := R(s) + \gamma \cdot \max_a \sum_{s' \, \text{sp}(S)} P(s' \mid a, s) U^{i-1}(s').$$

5. **Until** $U^i(s) - U^{i-1}(s) < \epsilon$, for all $s \in \text{sp}(S)$.

Policy iteration:

1. Let Δ_0 be some initial (randomly chosen) policy.
2. Set $i := 0$.
3. **Repeat**
   a) Find the utility function U_{Δ_i} corresponding to the policy Δ_i [policy evaluation].

b) Let $i := i + 1$.

c) **For each** $s \in \mathrm{sp}(S)$

$$\Delta_i(s) := \arg\max_a \sum_{s' \in \mathrm{sp}(S)} P(s' \mid a, s) U_{\Delta_{i-1}}(s') \text{ [policy improvement]}.$$

4. **Until** $\Delta_i = \Delta_{i-1}$.

Limited Memory Influence Diagrams (LIMIDs)

The no-forgetting assumption is dropped and instead, the informational arcs specify the variables observed before a particular decision (thereby controlling the size of the policy functions). A solution can be found using the single policy updating algorithm:

Single policy updating: Let I be a LIMID with decision variables $D_1, \ldots, D_n$, and let I' be a policy network for I in which the decision variables are represented by the chance variables $D_1^*, \ldots, D_n^*$.

1. Let $P_0(D_j^* \mid \mathrm{pa}(D_j^*))$ be an initial probability distribution (chosen at random) for D_j^*, $1 \le j \le n$, in I'.

2. Let $i := 1$.

3. Repeat
   a) For $j := n$ to 1
      i. Let $\mathcal{U}_{D_j}$ be the utility descendants of D_j.
      ii. Calculate a policy for D_j:

$$\delta_{D_j}(\mathrm{pa}(D_j))^i = \arg\max_{D_j} \sum_{U \in \mathcal{U}_{D_j}}$$

$$\sum_{\mathrm{pa}(U) \backslash \mathrm{fa}(D_j)} P(\mathrm{pa}(U) \setminus \mathrm{fa}(D_j) \mid \mathrm{fa}(D_j)) U(\mathrm{pa}(U)).$$

   iii. Replace $P_{i-1}(D_j^* \mid \mathrm{pa}(D_j^*))$ in I' with

$$P_i(D_j^* = d \mid \mathrm{pa}(D_j^*)) = \begin{cases} 1 & \text{if } \delta_{D_j}^i(\mathrm{pa}(D_j^*)) = d, \\ 0 & \text{otherwise.} \end{cases}$$

   b) Set $i := i + 1$.
4. Until convergence.

10.9 Bibliographical Notes

Various methods for solving influence diagrams have been constructed. Olmsted (1983) and Shachter (1986) introduced arc-reversal, and Shenoy (1992), Jensen *et al.* (1994), Cowell (1994), Ndilikilikesha (1994), and Madsen and

Jensen (1999a) used elimination and direct manipulation of potentials. Cooper (1988) presents a method that works well for scenarios with one decision variable. It substitutes the decision variable and the utility variables with chance variables and uses Bayesian network propagation. Zhang (1998) exploits Cooper's method to full influence diagrams.

The solution strategy for unconstrained influence diagrams was proposed in (Jensen and Vomlelova, 2002). A solution algorithm for sequential influence diagrams can be found in (Jensen *et al.*, 2006).

Troubleshooting based on decision theory was introduced by Kalagnanam and Henrion (1990), and it was further analyzed by Heckerman *et al.* (1995a). Section 10.5 is an extension of this work. Proofs that various versions of troubleshooting are NP-complete can be found in (Vomlelová, 2003).

The main ideas involved with solving Markov decision processes through value iteration originates with Shapley (1953). Policy iteration originates with Howard (1960).

LIMIDs were proposed in (Nilsson and Lauritzen, 2000).

10.10 Exercises

Exercise 10.1. Consider the influence diagram in Figure 9.22. Is L d-separated from E given I? Find a minimal set of nodes that d-separate A from D_3.

Exercise 10.2. Consider the influence diagram DI from Figure 10.1 but without the utility node V_1. Derive the formulas for an optimal strategy.

Exercise 10.3. Prove that during variable elimination, the potential $\prod \Phi_D$ is constant over D if all the variables following D in the partial ordering have already been eliminated.

Exercise 10.4. Construct a strong junction tree for the influence diagram in Figure 9.21 and determine the domains of the policies.

Exercise 10.5. Construct strong junction trees for the influence diagrams in Figures 9.23 and 9.24. Compare the clique sizes and the domains of the policies.

Exercise 10.6. Show that any strong triangulation of the influence diagram in Figure 10.10 will place E and B in the clique where D_1 is eliminated.

Exercise 10.7. Construct a strong junction tree for the influence diagram in Figure 10.39

(i) Is D_2 required for D_3?
(ii) Is B required for D_3?

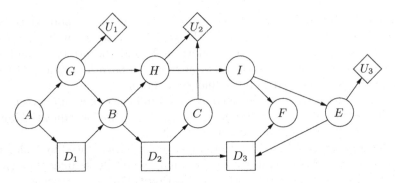

Fig. 10.39. Figure for Exercise 10.7.

(iii) Construct a join tree for the policy network and compare the size with the size of the strong junction tree.

Exercise 10.8. (i) Let $\{a_{ij}\}$ be an $n \times m$ matrix of reals. Prove that

$$\max_i \sum_j a_{ij} \leq \sum_j \max_i a_{ij}.$$

(ii) Use (i) to show that the MEU of an influence diagram will not increase by delaying an observation. (Hint: Look at the formulas for the two elimination orders.)

Exercise 10.9. Consider the arc-reversal solution method for influence diagrams, and a point where no node can be removed (because the only nodes with only utility nodes as children are decision nodes and these utility nodes have nonobservables as parents as well). To show that we can always find an arc to reverse, prove that there is at least one pair of chance nodes A and B such that A is a parent of B and there is no other directed path from A to B.

Exercise 10.10. Consider the simple influence diagram in Figure 10.40, where all variables are binary, and the probabilities for C_1 are given in Table 10.4, the probability of $C_2 = c_2$ is 0.8, and the utility functions U_1 and U_2 are given in Tables 10.5 and 10.6. Solve the influence diagram using node removal and arc reversal.

$D_1 \setminus C_2$	c_2	$\neg c_2$
d_1	0.2	0.7
$\neg d_1$	0.5	0.5

Table 10.4. $P(C_1 = c_1 \mid D_1, C_2)$.

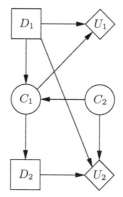

Fig. 10.40. A simple influence diagram.

$D_1 \setminus C_1$	c_1	$\neg c_1$
d_1	5	-2
$\neg d_1$	3	-10

Table 10.5. $U_1(D_1, C_1)$.

$D_1 \setminus C_2$	c_2	$\neg c_2$
d_1	$(0,0)$	$(8,-5)$
$\neg d_1$	$(5,-1)$	$(1,12)$

Table 10.6. $U_2(D_1, C_2, D_2)$. Entries should be interpreted as $(d_2, \neg d_2)$.

Exercise 10.11. Which steps would be carried out if the influence diagram in Figure 9.22 were solved using node removal and arc reversal? Assuming that each node has two states, what is the largest potential constructed during the solution process?

Exercise 10.12. Let I be an influence diagram, and I' be the influence diagram obtained by reversing an arc in I. Prove that if X and Y are variables d-separated by a set of variables $\mathcal{Z}$ in I', then X and Y are also d-separated given $\mathcal{Z}$ in I.

Exercise 10.13. Prove that when the node removal and arc reversal solution method is applied to an influence diagram, it eliminates the decision variables in an order consistent with the partial temporal ordering of the nodes in the diagram.

Exercise 10.14. Consider the UID in Figure 9.48. Construct the full S-DAG for the UID, and then reduce it as much as possible. Is the result the same when you do a roll-back construction of the S-DAG?

Exercise 10.15. Use the algorithm in Section 10.5.4 to solve the start problem in Example 9.2 (Page 293).

Exercise 10.16. ^{E} You are experiencing irregularities using your computer. There are several reasons why this can be: first, one of the programs you are running can be malfunctioning and interfering with your operating system; second, you can have attracted a virus; and third, you can have a hardware problem. Assuming that only one problem exists, the probabilities of the three problems are 0.8, 0.15, and 0.05, respectively. Your possible actions for fixing the problem are

1. Reboot the computer.
2. Run a virus removal tool.
3. Reformat your hard disk and reinstall your operating system.
4. Buy a new computer.

The costs of each option as an overall index of frustration, time usage, and money spent are 1, 2, 25, and 500, respectively. The probability of action 4 solving the problem is 1 no matter what the problem is and which other attempts to solve the problem have failed so far. Action 3 has a probability of 0.99 of fixing the problem if it is a nonhardware problem, and 0 if it is a hardware problem, no matter which other solutions that have failed previously. Action 2 solves the problem with probability 0.95 if it is a virus problem, and with probability 0 otherwise, again no matter what other solutions have unsuccessfully been tried. Finally, action 1 solves the problem with probability 1 if it is due to a malfunctioning program, and 0 otherwise, no matter what previous unsuccessful attempts at solving the problem were tried.

Formulate the above setting as a troubleshooting problem, and give an optimal sequence of repair actions. What is the expected cost of repair for the sequence?

Exercise 10.17. ^{E} Consider again the computer problem in Exercise 10.16, and assume further that you are given the option of buying a computer program that can scan the computer for hardware errors. The overall effort involved in doing this is 4. If there is a hardware error, the program has a 0.999 chance of discovering it, and there is no risk of false positives. Moreover, you are given the choice of having your computer scanned remotely on the Internet by some company for a price of 0.25. The scanning discovers a virus with a probability of 0.99 if there is one, but the scanner cannot remove it. For that you are given the option of downloading a special virus-removal program, which has a cost of 2 and which removes the identified virus with a probability of 1. Are the two offers individually worth the asking price? Are they worth the price in combination?

Exercise 10.18. Continue Example 10.3 and perform one more iteration of value iteration starting with the utility function shown in Figure 10.29(c).

Exercise 10.19. Show that there is only one true utility function representing the maximum expected discounted reward of a Markov decision process with an unbounded time horizon.

Exercise 10.20. [E] Consider the influence diagram in Example 10.10, but interpreted as a LIMID. Using the policies $D_1 = \neg d_1$ and $D_2 = d_2$, regardless of the state of C_1, run two iterations of policy updating.

Exercise 10.21. [E] Consider the LIMID in Figure 10.41, with its realization specified as in Example 10.10. Using the policies $D_1 = \neg d_1$ and $D_2 = d_2$, regardless of the states of C_1 and D_1, run two iterations of policy updating.

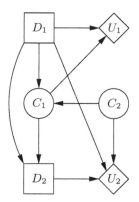

Fig. 10.41. A LIMID for Exercise 10.21.

Methods for Analyzing Decision Problems

The primary issue in dealing with a decision problem is to determine an optimal strategy, but other issues may be relevant. This chapter deals with value of information, the relevant past and future for a decision, and the sensitivity of decisions with respect to parameters.

11.1 Value of Information

As mentioned previously, there is a difference between action decisions and test decisions; action decisions may result in a state change for some of the variables, whereas test decisions are decisions to look for more evidence. A typical situation is that you may choose among some actions, but before deciding on the action you also have the option to perform some tests. The question is which test to perform, if any.

These types of decision problems can be characterized as asymmetric decision problems, since they contain at least two types of asymmetry: structural asymmetry (if you decide not to perform a test, the result is never observed), and order asymmetry (the sequence of tests may be unspecified). However, rather than looking at this as a general asymmetric decision problem we shall in this section deal directly with the problem by considering the actual value of information.

11.1.1 Test for Infected Milk?

Consider again the infected milk problem described in Example 9.1, where we assume that the farmer only has one test, which costs 6 cents and has a false positive/negative frequency of 0.01. The test situation corresponds to choosing between the two influence diagrams in Figure 11.1, where the leftmost influence diagram incurs an additional cost of 6 cents.

To establish the utilities, let us assume that the farmer has clean milk from the 49 other cows. If the farmer pours the milk into the container, he

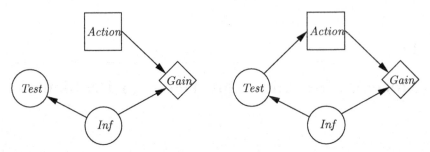

Fig. 11.1. The test scenario for infected milk corresponds to choosing between the influence diagrams, but by choosing the rightmost model you have to pay an additional 6 cents for the test.

will gain $100 if it is not infected, and he will gain nothing if it is infected. If he throws the milk away, he will gain $98 regardless of the state of the milk.

If the farmer does not perform a test, the probability of the milk being infected is 0.0007. The expected utility of pouring the milk into the container is

$$\text{EU}(pour) = P(Inf = no)U(Inf = no) + P(Inf = yes)U(Inf = yes)$$
$$= 0.9993 \cdot 100 + 0.0007 \cdot 0 = 99.93.$$

Because the expected utility of pouring the milk into the container is larger than 98, he will do this.

The reason for performing the test is that some outcome will make the farmer change the decision. To put it in another way, if the decision is the same regardless of the outcome of the test, then it is not worth the bother to perform it. Only a positive test result may change the current decision. An easy calculation yields $P(clean \mid pos) = 0.935$. The expected utility of pouring given a positive test result is

$$\text{EU}(pour \mid Test = pos) = P(Inf = no \mid Test = pos)U(Inf = no)$$
$$+ P(Inf = yes \mid Test = pos)U(Inf = yes)$$
$$= 0.935 \cdot 100 + 0.065 \cdot 0 = 93.5,$$

so if the test is positive, the farmer changes his decision. The next concern is whether the test is worth its price. There are two possibilities: the test is negative and the milk is poured, or the test is positive and the milk is thrown away. The probability of the first possibility can be calculated from the specified probabilities and is 0.9893, and the second possibility has the probability 0.0107. Hence, the expected benefit of performing the test is

$$\text{EU}(Test) = 0.9893 \cdot 100 + 0.0107 \cdot 98 = 99.98.$$

The farmer has an increase in expected utility only from 99.93 to 99.98 at the price of $0.06, so it is not worth while to perform the test.

11.1.2 Myopic Hypothesis-Driven Data Request

In the preceding example, we attached a value to the various information scenarios, namely the expected utility of the optimal action. The driving force for evaluating the information scenario was how the distribution of the variable *Infected?* was affected by the test. We call this kind of data request *hypothesis-driven*: the distribution of a hypothesis variable H is the target of the analysis. To formulate it in more general terms, there is a *value function* V attached to the distribution $P(H)$. Usually, the value function is a maximal utility for a decision variable D:

$$V(P(H)) = \max_{d \in D} \sum_{h \in H} U(d, h) P(h \mid d).$$

Note that here we use $V(P(H))$ rather than $EU(D)$ to emphasize that we are looking at the decision problem in a value-of-information context. If test T with cost C_T yields the outcome t, then the value of the new information scenario is

$$V(P(H \mid t)) = \max_{d \in D} \sum_{h \in H} U(d, h) P(h \mid t, d).$$

Since the outcome of T is not known, we can calculate only the *expected value*:

$$EV(T) = \sum_{t \in T} V(P(H \mid t)) \cdot P(t \mid d).$$

The *expected benefit* of performing test T is

$$EB(T) = EV(T) - V(P(H)).$$

The *expected profit* is

$$EP(T) = EB(T) - C_T.$$

The hard part in the calculations is the calculation of $P(H \mid T, D)$. This will usually require one propagation per state of T and D. Very often, the action has no impact on the hypothesis, and this reduces the work.

If there are several possible tests to perform, we are faced with a new problem. We may calculate the expected profit of each test, but we cannot be sure that the best choice is the one with the highest expected positive profit. A proper analysis of the data-request situation should consist in an analysis of all possible sequences of tests (including the empty sequence). To avoid such an intractable analysis, the so-called *myopic* approximation is often used: If you are allowed to perform at most one test, which one will you choose? The answer is the one with the highest expected profit if it is positive.

The myopic approach does not guarantee an optimal sequence (see also Section 10.5.4 in a troubleshooting context). Sometimes a single test does not yield anything by itself, whereas its outcome may be crucial for selecting a second very informative test.

Now, assume you have the tests $T_1, \ldots, T_m$, let H be the hypothesis variable, and assume that the action has no impact on H. To calculate the expected profit for all tests, you need $P(H \mid T_i)$ for each T_i. This can be achieved by propagating each possible outcome of each possible test. It can also be achieved in a simpler way. By propagating the states of H rather than the states of the tests, we get $P(T_i \mid H)$ for all T_i. Bayes' rule yields

$$P(H \mid T_i) = P(T_i \mid H)\frac{P(H)}{P(T_i)}.$$

Because $P(T_i)$ and $P(H)$ are available initially, we do not need more propagations than there are states in H.

The junction tree framework can also be used to perform some types of value of information analysis. For example, consider the influence diagram in Figure 11.2, where the variable C is observed prior to D_3.

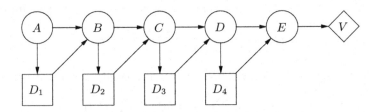

Fig. 11.2. An influence diagram.

The observation may improve the decision D_3 and yield a higher expected utility. The observation has a cost, though, but since it does not affect the strategy, it is not part of the model. Assume now that we wish to analyze how much the observation actually improves the expected utility. The situation in which C is not observed is reflected in the influence diagram in Figure 11.3. If the difference in MEU between the two influence diagrams is smaller than the cost of observing, then it does not pay to perform the test.

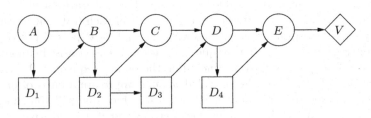

Fig. 11.3. An influence diagram for the scenario from Figure 11.2 but with C not observed.

If we assume that the cost of observing is not dependent on the timing, the MEU cannot get higher by delaying an observation that must eventually be performed. Therefore, the only option we have is either to observe as soon as possible or never to observe.

Using a method similar to propagation of variables as described in Section 5.2, the calculation of the various MEUs can be joined in one strong junction tree. Perform a strong triangulation for the influence diagram modeling that the observations have not been performed (that is, with the chance variables under analysis as members of $\mathcal{I}_n$) and construct the strong junction tree. When solving the influence diagram corresponding to an observation of the chance node C just before deciding on D_i, you use the same strong junction tree. However, you defer the elimination of C until D_i has been eliminated. Figure 11.4 shows the influence diagram from Figure 10.5, where an observation is optional for several variables as indicated by the dashed arrows. The reader may check that you can solve all influence diagrams corresponding to all combinations of possible observations through delayed elimination in the strong junction tree in Figure 10.8.

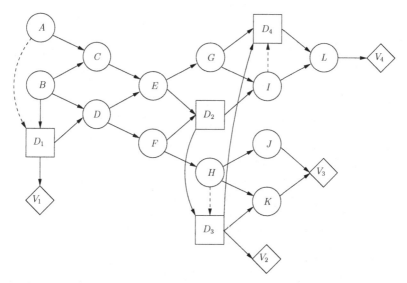

Fig. 11.4. An influence diagram with the option of not observing A, H, and I.

11.1.3 Non-Utility-Based Value Functions

If there is no proper model for actions and utilities, the reason for acquiring more information is to decrease the uncertainty of the hypothesis. This means that you will give high values to probabilities close to zero and one, while probabilities in the middle area should have low values. A classical function with this property is entropy (see Section 8.4).

The formula for the entropy of a distribution over H is in Section 8.4 defined as

$$\text{Ent}(P(H)) = -\sum_{h \in H} P(h) \log_2(P(h)),$$

where $p \log_2 p = 0$ if $p = 0$.

Because we want the value function to increase with preference, we let an entropy-based value function be

$$V(P(H)) = -\text{Ent}(P(H)) = \sum_{h \in H} P(h) \log_2(P(h)).$$

Variance

If the states of H are numeric, another classical measure can be used, namely the variance. Again, since small variances are preferred, the value function becomes

$$V(P(H)) = -\sum_{h \in H} (h - \mu)^2 P(h),$$

where $\mu = \sum_{h \in H} h P(h)$.

It is up to the modeler to specify the value function. If decisions with known utilities are attached to the hypothesis variable, then the utility value function should be preferred. If this is not the case, the user will mainly be interested in the precision of a diagnosis.

In the case of a Boolean hypothesis with states 0 and 1, the entropy function is $\log p^p(1-p)^{1-p}$, and the variance function is $-p(1-p)$. These two functions reflect that the value of p increases as it approaches its bounds 0 and 1. The entropy function is rather drastic in the way that the slope is infinite for 0 and 1. Therefore, small changes of p close to 0 and 1 will be highly valued. On the other hand, the variance is of polynomial degree 2, and the slope close to the bounds is 1 and -1, giving changes almost even value no matter how close they are to the bound.

Other Value Functions

In principle, any value function may be used. However, a particular class of functions called *convex functions* are best suited for the purpose.

Definition 11.1. *A function $f : R^n \to R$ is* convex *if for any two points P_1, P_2 on the graph of f, the line segment $P_1 P_2$ lies above the graph (see Figure 11.5). Mathematically, the property is expressed as follows:*

$$\forall t \in [0,1], \forall \mathbf{x}, \mathbf{y} \in R^n : t f(\mathbf{x}) + (1-t) f(\mathbf{y}) \geq f(t\mathbf{x} + (1-t)\mathbf{y}).$$

The reason why a convex function is well suited is due to the following theorem, which we will not prove.

Theorem 11.1. *If the value function is a convex function, then the expected benefit of performing a test is never negative.*

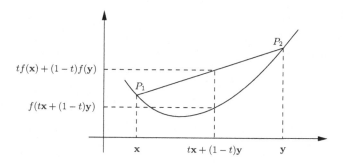

Fig. 11.5. A convex function. The line segment between two points of the graph lies above the graph.

Utility based functions are convex and so are entropy and variance.

11.2 Finding the Relevant Past and Future of a Decision Problem

When solving a decision problem we look for an optimal policy for each of the decisions. The optimal policy for a decision is in principle a function that for each possible configuration of the past, prescribes how to act in order to maximize the expected utility. Thus, for the poker domain modeled in Figure 11.6, a policy for the decision node D is a function over the entire past of D:

$$\delta_D : \mathrm{sp}(MH0, MFC, MH1, OFC, MSC, OSC) \to \mathrm{sp}(D).$$

In general, if we represent such a policy function as a table, then the size of the policy increases exponentially in the number of variables in the past, and the policy can therefore quickly become intractable to handle.

However, when analyzing the decision problem above, we find that not all variables can provide information influencing decision D. For example, if I know my current hand $MH2$, then knowledge about how many cards I discarded in the second round, MSC, will not affect my decision at D: at D I will try to maximize my profit represented by the utility function U. This utility function depends only on D and BH, and with knowledge of the state of $MH2$, the decision MSC becomes d-separated from BH. Hence MSC cannot tell me anything about BH (and therefore U), and it can therefore not affect my decision at D. By performing this type of analysis for the remainder of the variables in the past of D, we find that the only variables that can have

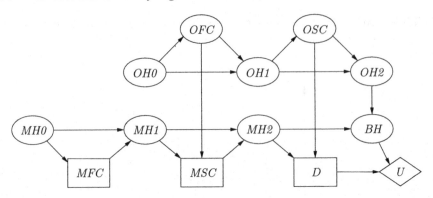

Fig. 11.6. An influence diagram representation of the poker domain described in Section 9.4.1. An optimal policy for decision D is a function over the past of D, namely *MH0*, *MFC*, *MH1*, *OFC*, *MSC*, and *OSC*.

an impact on D are *OFC*, *OSC*, and *MH2*. Hence, the optimal policy for D reduces to

$$\delta_D : \mathrm{sp}(\mathit{OFC}, \mathit{OSC}, \mathit{MH2}) \to \mathrm{sp}(D).$$

This policy contains only 96 configurations, as opposed to the full policy function containing 165888 configurations. By doing the same exercise for the two remaining decisions we find that only *MH1* and *OFC* are relevant for *MSC*, and *MH0* is relevant for *MFC*.

Definition 11.2 (Required variables). *Let I be an influence diagram and let D be a decision variable in I. The variable $X \in \mathrm{past}(D)$ is said to be required for D if there exist a realization $\mathcal{R}$ of I, a configuration $\bar{y}$ over $\mathrm{dom}\,(\delta_D) \setminus \{X\}$, and states x_1 and x_2 of X such that $\delta_D(x_1, \bar{y}) \neq \delta_D(x_2, \bar{y})$, where δ_D is an optimal policy for D with respect to $\mathcal{R}$. The set of variables required for D is denoted by $\mathrm{req}(D)$.*

To take another example, consider the influence diagram in Figure 11.7, which specifies the partial ordering

$$\{B\} \prec D_1 \prec \{E, F\} \prec D_2 \prec D_3 \prec \{G\} \prec D_4 \prec \mathcal{C}_4;$$

$\mathcal{C}_4$ denotes the variables not observed before the last decision.

When looking for an optimal policy for D_4 we should in principle consider all the variables in the past of D_4, i.e., B, D_1, E, F, D_2, D_3, and G. However, when analyzing the influence diagram, we see that deciding on D_4 has an impact only on V_4, and from the d-separation properties of the model we have that by conditioning on G and D_2, all the other variables in the past of D_4 become d-separated from V_4. Hence, only G and D_2 are required for D_4.

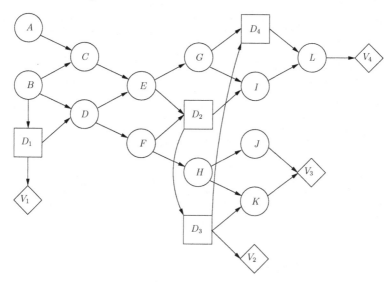

Fig. 11.7. The figure illustrates an influence diagram that specifies the partial order $\{B\} \prec D_1 \prec \{E, F\} \prec D_2 \prec D_3 \prec \{G\} \prec D_4 \prec \mathcal{C}_4$ ($\mathcal{C}_4$ denotes the chance variables observed after deciding on all the decisions.

11.2.1 Identifying the Required Past

In the examples above we informally characterized a variable as being required for D if it can provide information about the utility functions that we are trying to maximize when deciding on D. To test whether a variable X can provide information about these utility functions, we used the d-separation criterion. The question is then how to identify the utility functions that can influence D. To be on the safe side you might simply include all utility functions, but this may result in variables that are falsely identified as required for D. So we would like to identify the minimal set of utility functions to take into account when deciding on a particular decision.

Definition 11.3 (Relevant utility nodes). *The utility function U is relevant for decision D if there exists two realizations $\mathcal{R}_1$ and $\mathcal{R}_2$ of I that differ only on U such that the optimal policies for D are different in $\mathcal{R}_1$ and $\mathcal{R}_2$.*

Luckily, it turns out that this semantic definition also supports a simple syntactic characterization. For the last decision we have the following specification:

Proposition 11.1. *Let D_n be the last decision variable in the influence diagram I, and let U be a utility node in I. Then U is relevant for D_n if and only if there is a directed path from D_n to U.*

Proof. For the last decision D_n we know that the optimal policy is

$$\delta_{D_n}(\text{past}(D_n)) = \arg\max_{D_n} \sum_{\mathcal{C}_n} P(\mathcal{C}_n \mid \text{past}(D_n), D_n)\Big[U(\text{pa}(U))$$

$$+ \sum_{i=1}^{m} U_i(\text{pa}(U_i))\Big]$$

$$= \arg\max_{D_n} \Big[\sum_{\mathcal{C}_n} P(\mathcal{C}_n \mid \text{past}(D_n), D_n)U(\text{pa}(U))$$

$$+ \sum_{\mathcal{C}_n} P(\mathcal{C}_n \mid \text{past}(D_n), D_n) \sum_{i=1}^{m} U_i(\text{pa}(U_i))\Big].$$

Since

$$\sum_{\mathcal{C}_n} P(\mathcal{C}_n \mid \text{past}(D_n), D_n)U(\text{pa}(U))$$

$$= \sum_{\mathcal{C}_n \cap \text{pa}(U)} P(\mathcal{C}_n \cap \text{pa}(U) \mid \text{past}(D_n), D_n)U(\text{pa}(U)),$$

we have that U is relevant for D_n if and only if D_n is either a parent of U or D_n is d-connected to a variable in $\mathcal{C}_n \cap \text{pa}(U)$ given $\text{past}(D_n)$; otherwise, the above expression would be independent of D_n. In order for D_n to be d-connected to a variable $X \in \mathcal{C}_n \cap \text{pa}(U)$ given $\text{past}(D_n)$, there must be an active path between $\text{pa}(U)$ and D_n. Since a converging connection on such a path cannot be opened by evidence (a descendant of D_n cannot be observed), the path must be directed from D_n to a node in $\text{pa}(U)$. $\square$

Based on this proposition, we now have a full syntactic characterization of the variables required for the last decision.

Proposition 11.2. *Let D be the last decision variable in the influence diagram I and let X be a variable in $\text{past}(D)$. Then X is required for D if and only if X is d-connected to a utility node relevant for D given $\text{past}(D) \setminus \{X\}$.*

Proof. Follows the proof above. $\square$

For example, if we go back to the influence diagram shown in Figure 11.7, we see that V_4 is the only utility node to which there exists a directed path from D_4; hence V_4 is the only utility node relevant for D_4. Moreover, using Proposition 11.2 we find that only G and D_2 are required for D_4, $\text{req}(D_4) = \{G, D_2\}$.

Suppose now that we also want to identify the required variables for D_3. This can be done by substituting D_4 with its chance-variable representation (actually, we need not calculate the policy). This is done in Figure 11.8.

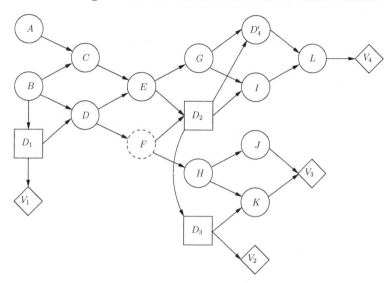

Fig. 11.8. The influence diagram obtained from the influence diagram in Figure 11.7 by substituting D_4 with its chance-variable representation. Since D_3 is the last decision, we see from Proposition 11.2 that F is the only variable required for D_3.

In this transformed influence diagram, D_3 appears as the last decision, and by applying the propositions we find that V_2 and V_3 are relevant for D_3 and that F is the only variable required for D_3.

By replacing D_3 in Figure 11.8 with its chance-variable representation we obtain the influence diagram in Figure 11.9, where D_2 is the last decision. From this model we find that V_4 is the only utility function relevant for D_2, and E is therefore the only variable required for D_2.

Finally, we can find the required variables for D_1 by substituting D_2 with its chance-variable representation . The resulting model is shown in Figure 11.10, where we see that all four utility functions are relevant for D_1, and since B is d-connected to V_2, V_3, V_4 we have that B is required for D_1.

More generally, we can specify an algorithm for finding the required variables for the decisions in an influence diagram as follows.

Algorithm 11.1 [Identify required variables] *Let I be an influence diagram and let $D_1, D_2, \ldots, D_n$ be the decision variables in I ordered by index. To determine* req(D_i), *the variables required for D_i ($\forall 1 \leq i \leq n$), do:*

1. *Set $i := n$.*
2. **For** *each decision variable D_i not considered ($i > 0$)*
    a) *Let $\mathcal{V}_i$ be the set of utility nodes to which there exists a directed path from D_i in I.*
    b) *Let* req(D_i) *be the set of nodes X such that $X \in$ past(D_i) and X is d-connected to a node in $\mathcal{V}_i$ given* past$(D_i) \setminus \{X\}$.

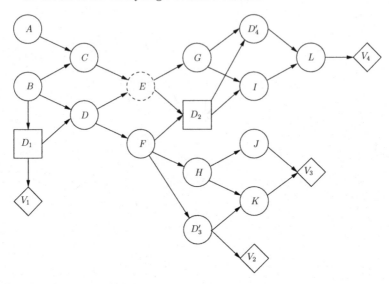

Fig. 11.9. The influence diagram obtained form the influence diagram in Figure 11.8 by substituting D_3 with its chance-variable representation.

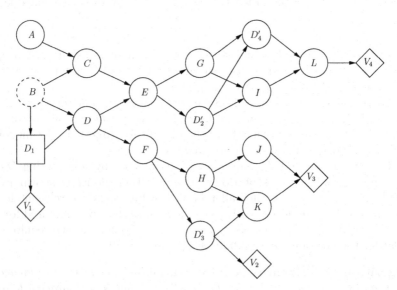

Fig. 11.10. The influence diagram obtained from Figure 11.9 by replacing D_2 with its chance variable policy.

c) *Replace D_i with a chance-variable representation of the policy for D_i, and let I be the resulting model.*

d) *Set $i := i - 1$.*   □

Identifying the Relevant Future

Analogously to the idea of identifying the required variables in the past of a decision, we can also identify the future variables that are relevant for that decision. By relevant variables we mean the variables whose probability distributions (or policies) should be taken into account when deciding on D. Having such a characterization will not reduce the complexity of the policies, but it may provide insight into the overall structure of the decision problem. For example, if some decision variable is of particular interest, then the relevant variables may pinpoint the part of the model that we should focus on when specifying the probabilities.

Definition 11.4. *Let I be an ID and let D be a decision variable in I. The future variable X is said to be* relevant *for D if either:*

- *X is a chance node and there exist two realizations $\mathcal{R}_1$ and $\mathcal{R}_2$ of I that differ only on the probability distribution associated with X such that the optimal policies for D are different in $\mathcal{R}_1$ and $\mathcal{R}_2$, or*
- *X is a decision variable and there exist a realization of I and two different policies δ_X^1 and δ_X^2 for X such that the optimal policies for D are different with respect to δ_X^1 and δ_X^2.*

Together with the required past, the relevant variables describe the part of a decision problem that is sufficient to take into account when one is focusing on a particular decision.

To complete the characterization, we need an algorithm for identifying the variables that are relevant for a decision D. The first thing to notice is that by using the chance-variable representation of a decision node, we again need to consider only the situation in which D is the last decision variable in the influence diagram. Hence we can identify the relevant future decisions as the decision variables whose chance-variable representations are relevant for D. This also means that in order to identify all the relevant variables we just need a method for identifying the relevant chance variables.

Theorem 11.2. *Let I be an ID and let D be the last decision variable in I. Then the future chance variable X is relevant for D if and only if*

- *X is not barren in the ID formed from I by removing all utility nodes that are not relevant for D, and[1]*
- *there exists a utility node U relevant for D such that X is d-connected to U in I given $\{D\} \cup \mathrm{past}(D)$.*

---

[1] If X is barren, then it does not affect any decisions and it can simply be removed.

By going back to the influence diagram in Figure 11.7, we see that I and L are the only future variables d-connected to the relevant utility function, V_4, for D_4. Hence, no other future chance variables are relevant, and the decision problem for D_4 can therefore be described by the utility node V_4, the required variables G and D_2, and the relevant chance variables I and L, see Figure 11.11(a). To determine the relevant variables for D_3 we substitute D_4 with a chance variable and apply the same procedure as above. That is, from Figure 11.8 we see that H, I, and K are relevant for D_3, and together with the relevant utility nodes and the required variables we can identify the part of the decision problem relevant for D_3. See Figure 11.11(b). By continuing to D_2, we use the influence digram in Figure 11.9. When performing the analysis, we identify the variable D_4' as relevant for D_2, which in turn means that the decision node D_4 is relevant for D_2 (the identification of the remaining variables is left as an exercise).

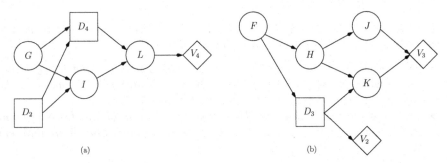

(a)   (b)

Fig. 11.11. The figures illustrate the parts of the influence diagram in Figure 11.7 relevant for D_4 and D_3, respectively.

11.3 Sensitivity Analysis

One of the main difficulties in modeling a decision problem is the elicitation of utilities and probabilities. This makes it desirable to be able to investigate how sensitive the solution is to variations in some utility or probability parameter, and how robust the solution is to joint variations over a set of parameters.

We distinguish between *value sensitivity* and *decision sensitivity*. Value sensitivity concerns variations in the maximum expected utility when a set of parameters is changed, and decision sensitivity refers to changes in the optimal strategy.

11.3.1 Example

Consider the following simplified binary version of the Oil Wildcatter Problem from Exercise 9.11. The influence diagram is shown in Figure 11.12. The hole can be *good* or *bad*. If the hole is good, the gain is $260,000, and if the hole is bad, the gain is $0. The test has no false negatives, and the probability of a false positive is 0.05. The prior probability for the hole being good is 0.2. The cost of drilling is $60,000, and the cost of the test is $5,000.

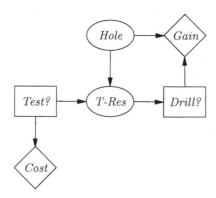

Fig. 11.12. An influence diagram for the Oil Wildcatter Problem.

The optimal strategy, Δ, is to test and then to drill if and only if the test is positive. However, although the oil wildcatter is quite certain of the specifics of the test, he is rather uncertain of the gain of a good hole as well as of the prior probability for this particular hole being good. If the gain and the prior for a good hole are large, he need not test, because he will drill regardless of the result of the test, and if the prior and the gain for a good hole are low, he will just leave the hole.

To be precise, the optimal strategy consists of two optimal policies, $\delta_{Test?} = y$ for *Test?*, and $\delta_{Drill?}(Test?, T')$ for *Drill?*, where $\delta_{Drill?}(y, pos) = y$, $\delta_{Drill?}(y, neg) = n$, $\delta_{Drill?}(n, no\text{-}test) = n$, and the values for other configurations are of no importance, since they will never be realized.

Let t denote $P(Hole = good)$ and let s denote $Gain(Hole = good) - 60000$. Then $\delta_{Drill?}$ is optimal for $(t, s) = (0.2, 200000)$, and the wildcatter would like to know which parameter values support this policy. To determine the support, we calculate the expected utilities of the various options. The relevant utilities are only the utilities on which *Drill?* has an impact, namely *Gain*; the descendant of *Drill?*. We now get

$$EU(Drill? \mid n, \textit{no-test}) = (P(good \mid \textit{no-test})s - P(bad \mid \textit{no-test})60000, 0)$$
$$= (ts - (1-t)6000, 0),$$
$$EU(Drill? \mid y, pos) = (P(good \mid pos)s - P(bad \mid pos)60000, 0)$$
$$= \left(\frac{ts - 0.05(1-t)60000}{0.95t + 0.05}, 0 \right),$$
$$EU(Drill? \mid y, neg) = (P(good \mid neg)s - P(bad \mid neg)60000, 0)$$
$$= (-60000, 0).$$

The policy $\delta_{Drill?}$ is optimal if

$$EU(Drill? = n \mid n, \textit{no-test}) \geq EU(Drill? = y \mid n, \textit{no-test}),$$
$$EU(Drill? = y \mid y, pos) \geq EU(Drill? = n \mid y, pos),$$
$$EU(Drill? = n \mid y, neg) \geq EU(Drill? = y \mid y, neg).$$

This gives the following inequalities:

$$0 \geq ts - (1-t)60000,$$
$$0 \leq ts - 0.05(1-t)60000,$$
$$0 \geq -6000.$$

That is,

$$ts + 3000t - 3000 \geq 0 \geq ts + 60000t - 60000. \tag{11.1}$$

For $s = 200000$ we get that $\delta_{Drill?}$ is optimal for $\frac{3}{203} \leq t \leq \frac{3}{13}$, and for $t = 0.2$ it is optimal for $12000 \leq s \leq 240000$. These intervals are called the *admissible domains* for the parameters in $\delta_{Drill?}$.

Next we analyze the first decision. The decision node *Drill?* is substituted with the chance node D (Figure 11.13), and $P(D \mid T')$ reflects the optimal policy (see Section 10.2.3).

Using the model in Figure 11.13 we calculate

$$EU(Test? = y) = -5000 + P(pos)(EU(Drill? = y \mid pos)$$
$$= -5000 + (0.95t + 0.05)\frac{ts - 0.05(1-t)60000}{0.95t + 0.05}$$
$$= -5000 + ts - 0.05(1-t)60000$$
$$= ts + 3000t - 8000,$$
$$EU(Test? = n) = 0.$$

This yields that testing is optimal if

$$ts + 3000t - 8000 \geq 0. \tag{11.2}$$

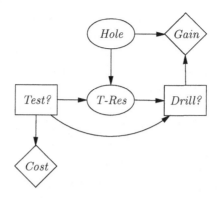

Fig. 11.13. The decision node *Drill?* is substituted by its chance-node representation.

For $s = 200000$ it holds for $t \geq \frac{8}{203}$ and for $t = 0.2$ it holds for $s \geq 37000$. The strategy is optimal in the intersection of the admissible domains of the two policies. That is, for $s = 200000$ the admissible domain for t is $[\frac{8}{203}, \frac{3}{13}]$. For $t = 0.2$, the admissible domain for s is $[37000, 240000]$.

11.3.2 One-Way Sensitivity Analysis in General

Let t be a parameter with initial value t_0 in an influence diagram, and let Δ be an optimal strategy for the value t_0. We wish to determine the admissible interval for t. The method starts determining the admissible interval for the policy $\delta_{Drill?}$ for the last decision D. Then D is substituted by its chance-variable representation, and the admissible interval for t is determined for the last decision in this influence diagram. The procedure is repeated until the first decision has been analyzed. The admissible interval for t in Δ is the intersection of the admissible intervals for all the policies. Since t_0 is a member of all intervals, we know that the intersection is nonempty.

In the example above it turned out that the expected utilities were simple expressions in the parameters. This holds in general.

Theorem 11.3. *Let s be a utility parameter in the influence diagram ID, let D be the last decision in ID, and let π be any configuration of the required past of D. Then for any d in D, the expected utility of d given π is a linear function in s.*

Let t be probability parameter in the influence diagram ID, let D be the last decision in ID, and let π be any configuration of the required past of D. Then for any d in D, the expected utility of d given π is a fraction of two linear functions in t.

Proof. [Sketch] The expected utility is calculated as

$$\sum_{Parents} P(\text{Parents} \mid \text{past}) U(\text{Parents}).$$

For utility parameters, this expression is linear. A probability parameter has an effect on $P(\text{Parents} \mid \text{past})$, and from Corollary 5.2, it can be expressed as a fraction of two linear functions. □

As for sensitivity analysis for Bayesian networks, this theorem can be exploited to establish a functional expression for the expected utilities. Assume that we analyze a utility parameter s with initial value s_0. We have a solution for ID with value s_0. That is, we have a value of the expected utility for the last decision D_n for each configuration of the required past. Next, substitute s_0 with s_1 and solve the influence digram. Again, we get the expected utility for each option and any configuration of the required past. Now, for each option and for each parent configuration we have two values of the expected utility, and the two coefficients in the linear expression can be determined.

The next step is to establish a new influence diagram, and do the same with D_{n-1} as the last decision. However, if the value s_1 lies in the admissible interval for the policy for D_n, the solution from before can be reused. The optimal policy for D_n is guaranteed, also for the value s_1, to be the same as the conditional probability for its chance-node representation. This holds for the next decisions too, so by careful choice of the new value, one extra solution of the influence diagram is sufficient for the calculation of all the expected utilities required for determining the admissible domain for the parameter. In the case of probability parameters, three extra solutions are sufficient.

We shall illustrate the method for the parameter s in the oil wildcatter example above.

Solving the influence diagram with $s = 200000$ we get the following expected utilities:

$$\text{EU}(Drill? \mid pos) = (156666, 0),$$
$$\text{EU}(Drill? \mid neg) = (-60000, 0),$$
$$\text{EU}(Drill? \mid no\text{-}test) = (-8000, 0),$$
$$\text{EU}(Test?) = (32600, 0).$$

Changing s to 150000 we get

$$\text{EU}(Drill? \mid pos) = (115000, 0),$$
$$\text{EU}(Drill? \mid neg) = (-60000, 0),$$
$$\text{EU}(Drill? \mid no\text{-}test) = (-18000, 0),$$
$$\text{EU}(Test?) = (22600, 0).$$

This yields the following expressions:

$$\text{EU}(Drill? = y \mid pos) = 0.833s + 10000,$$
$$\text{EU}(Drill? = y \mid neg) = -60000,$$
$$\text{EU}(Drill? = y \mid no\text{-}test) = 0.2s - 48000,$$
$$\text{EU}(Test? = y) = 0.2s - 7400,$$

which are the same as the result of the expressions in Section 11.3.1.

If you wish to find out how stable the strategy is to joint variations of several the parameters, one-way sensitivity analysis for each parameter may not provide the full picture and you may need to resort to n-way sensitivity analysis. However, the work becomes much harder. For example, in the case of a probability parameter t and a utility parameter s, the expected utilities have the form $\alpha s + \beta$, where α and β are fractions of linear expressions over t. This means that there are eight coefficients to determine. For illustration, the admissible area for (t, s) in the strategy from Section 11.3.1 is shown in Figure 11.14.

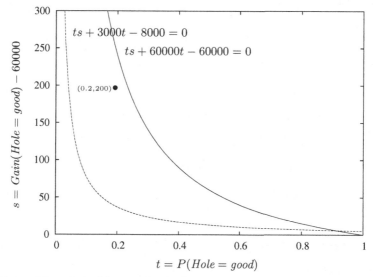

Fig. 11.14. The admissible area for (t, s) in the strategy for the the oil wildcatter. The y-axis is scaled by a factor of 1000.

If all parameters are utility parameters, $s_1, \ldots, s_n$, then the situation is much simpler. Since utilities are never multiplied, the expected utilities are linear expressions over $s_1, \ldots, s_n$. Therefore, there are only $n + 1$ coefficients to determine, and n extra solutions are sufficient.

11.4 Summary

Value of Information

Value function (one utility function U, one decision D):

$$V(P(H)) = \max_{d \in D} \sum_{h \in H} U(d, h) P(h \mid d).$$

Expected value of performing test T:

$$EV(T) = \sum_{t \in T} P(t) \max_{d \in D} \sum_{h \in H} U(d, h) P(h \mid t, d).$$

Expected profit:

$$EP(T) = EV(T) - V(P(H)) - C_T.$$

The value $EV(T)$ can be calculated for all tests T by entering the states of h as evidence and using Bayes' rule.

Myopic approach: Choose repeatedly a test with the highest positive expected profit, if any.

Nonutility value functions:

- Entropy: $V(P(H)) = \sum_{h \in H} P(h) \log_2(P(h))$;
- Variance: $V(P(H) = -\sum_{h \in H}(h - \mu)^2 P(h)$, where $\mu = \sum_{h \in H} h P(h)$.

The Required Past for a Decision

Let I be an influence diagram and let D be a decision variable in I. The variable $X \in \text{past}(D)$ is said to be *required* for D if there exist a realization of I, a configuration $\bar{y}$ over $\text{dom}(\delta_D) \setminus \{X\}$, and states x_1 and x_2 of X such that $\delta_D(x_1, \bar{y}) \neq \delta_D(x_2, \bar{y})$. The set of variables required for D is denoted by $\text{req}(D)$.

To determine $\text{req}(D_i)$ $(\forall 1 \leq i \leq n)$ do:

1. Set $i := n$.
2. **For** each decision variable not considered $(i > 0)$
   a) Let $\mathcal{V}_i$ be the set of value nodes to which there exists a directed path from D_i in I.
   b) Let $\text{req}(D_i)$ be the set of nodes X such that $X \in \text{past}(D_i)$ and X is d-connected to a node in $\mathcal{V}_i$ given $\text{past}(D_i) \setminus \{X\}$.
   c) Replace D_i with a chance-variable representation of the policy for D_i, and let I be the resulting model.
   d) Set $i := i - 1$.

Sensitivity Analysis

Value sensitivity: How much can the utility and probability parameters be varied without changing the optimal strategy? This question can be answered by performing an analysis of the expected utility as a function of these parameters.

Utility parameters: Let s be a utility parameter in the influence diagram ID, let D be the last decision in ID, and let π be any configuration of the required past of D. Then for any d in D, the expected utility of d given π is a linear function in s.

Probability parameters: Let t be probability parameter in the influence diagram ID, let D be the last decision in ID, and let π be any configuration of the required past of D. Then for any d in D, the expected utility of d given π is a fraction of two linear functions in t.

Calculating the coefficients: If there are only utility parameters to investigate, then all coefficients can be found by performing only one extra propagation for each parameter. This will also give all the information necessary for performing n-way sensitivity analysis (that is, sensitivity analysis in which you consider joint variations of the parameters).

11.5 Bibliographical Notes

Value of information is formally treated in (Howard, 1966) and (Lindley, 1971), where utilities are guiding the test selection. The myopic approximation was introduced by Gorry and Barnett (1968). In (Ben-Bassat, 1978), entropy and variance are used. Value of information for influence diagrams has been treated by Dittmer and Jensen (1997) and Shachter (1999). The required past of decisions in influence diagrams was introduced independently by Shachter (1999) and Nielsen and Jensen (1999). The relevant future of decisions was described in (Nielsen, 2002). Sensitivity analysis for multiple parameters in decision problems was investigated in (Felli and Hazen, 1999a). A method using value of information was given in (Felli and Hazen, 1999b). Sensitivity analysis for influence diagrams in particular was treated in (Nielsen and Jensen, 2003b).

The oil wildcatter's problem is due to Raiffa (1968). The used car buyer's problem is due to Howard (1962).

11.6 Exercises

Exercise 11.1. [E] Consider the insemination model from Exercise 3.8. Assume that you have the options to repeat the insemination or to wait another six-week period. The cost of repeating the insemination is 65 regardless of the pregnancy state of the cow. If the cow is pregnant and you wait, it will cost you nothing, but if the cow is not pregnant and you wait, it will cost you further 30 (that makes a total of 95 for waiting plus the eventual repeated insemination). The cost of BT is 1 and the cost of UT is 2. Perform a myopic value of information analysis.

Exercise 11.2. Solve the problem in Exercise 9.11 as a value of information problem.

Exercise 11.3. [E] Consider the influence diagram obtained by adding arcs from FC, SC, and MH to D in the network in Figure 9.3, using the probabilities found in Section 3.2.3 and the utilities found in Section 9.1.1. Assume that prior to the game, a shady-looking person at the table next to me offers to tell me the first hand of my opponent ($OH0$) for the price of $0.1. Ignoring ethical issues, should I take the offer?

Exercise 11.4. Consider the influence diagrams in Figures 9.23 and 9.24. What is the required past of decision FV_4 in the two diagrams?

Exercise 11.5. What are the relevant futures of decisions D_1 and D_2 in the influence diagram in Figure 11.7?

Exercise 11.6. Consider again the influence diagram in Example 10.10 and the strategy Δ, stating that one should always choose d_1 and d_2 if and only if C_1 is in state c_1. Denoting by t the utility parameter $U(d_1, c_2, \neg d_2)$, what is the support of Δ?

List of Notation

Variables have their names capitalized (X, A, *Fuel Meter*, ...). A state of a variable is written in lowercase. Sets are represented by caligraphic letters ($\mathcal{X}$, $\mathcal{A}$, ...) and boldface is used for vectors (**a**, **X**, **grad**). When we want to emphasize that we are working with a probability distribution we use, e.g., $P(A \mid B)$ or $P^{\#}(A \mid B)$; otherwise, we use $\phi(A, B)$ or $\psi(A, B)$ for probability potentials and utility potentials, respectively.

$\arg\max_D \rho$	A decision from D maximizing ρ.
BIC	Bayesian information criterion.
$\mathrm{ch}(A)$	The children set for variable A.
$\mathrm{CMI}(X, Y)$	The conditional mutual information for variables X and Y.
$\mathrm{conf}(e)$	Measure of conflict of evidence e.
$\mathcal{D}$	A database of cases.
Dir	Dirichlet distribution.
dist	Euclidean distance.
$\mathrm{dom}(\phi)$	The domain of potential ϕ.
e	Evidence e represented as a finding.
ECR	Expected cost of repair.
$\mathrm{ef}(A)$	Efficiency of action a.
$\mathrm{Ent}(X)$	The entropy of variable X.
EU	Expected utility.
$\mathbb{E}$	Expectation.
$\mathrm{fa}(A)$	The family set for variable A.
$\mathrm{hst}_G(N)$	The history of node N in the S-DAG G.
grad	Gradient vector.
$I(A, B, \mathcal{C})$	A and B are conditionally independent given $\mathcal{C}$.
KL	Kullbach-Leibler divergence.
$L(B \mid A)$	The likelihood of B given A, $P(A \mid B)$.
$LL(B \mid A)$	The log-likelihood of B given A.
$\max_A$	Max-marginal over variable A.
MEU	Maximal expected utility.
$\mathrm{MI}(X, Y)$	The mutual information between variables X and Y.

MPE	Most probable explanation.
$N(A = a, B = b)$	The number of cases containing $A = a$ and $B = b$.
nb(A)	The neighbor set for variable A.
O	Big-O notation.
$P(\mathcal{A})$	Probability of event $\mathcal{A}$.
$P(\mathcal{A} \mid \mathcal{B})$	Probability of event $\mathcal{A}$ given $\mathcal{B}$.
$P(A)$	Probability distribution for variable A.
$P(A \mid B)$	Probability distributions for variable A given the states of variable B.
$P(A, B)$	Joint probability distribution for variables A and B.
$P(A = a)$	The probability of variable A being in state a.
$P(a)$	The probability of outcome/state a (shorthand for $P(A = a)$).
$P(e)(t)$	$P(e)$ as a function of the parameter t.
$P^{\#}(X \mid Y)$	The frequency based conditional probability distribution for X given Y.
pa(A)	The parent set for variable A.
past(D)	The variable appearing in the past of variable D.
$\mathbb{R}$	The set of real numbers.
req(D)	The required past of variable D.
s^*	The effective sample size.
score	The score of a Bayesian network.
size(M)	The size of the Bayesian network M.
sp(X)	The state space of variable X.
$\mathcal{U}$	The set of all variables.
$\mathbf{1}$	The unit potential.
$\sum_A$	Summation over variable A.
$\phi^{\downarrow \mathcal{V}}$	Projection of potential ϕ down to domain $\mathcal{V}$.
$\Phi^{\downarrow \mathcal{V}}$	Projection of set of potentials Φ down to domain $\mathcal{V}$.
$\prod_i \psi_i$	The product of the potentials ϕ_i.
$\prod \Phi$	The product of all potentials in set Φ.
Φ^{-X}	The potentials resulting from elimination of variable X from the set of potentials Φ.
$\otimes$	Combination operator.
$\wedge$	Logical and.
$\vee$	Logical or.
$\neg$	Logical negation.
$|\mathcal{X}|$	The number of elements in the set $\mathcal{X}$.
μ	The mean value of a distribution.
ρ_D	The expected utility for decision variable D.
δ_D	A policy for decision variable D.
σ^2	The variance of a distribution.
$\hat{\theta}$	A maximum likelihood estimate of the parameter θ.

References

Andreassen, S. (1992). Knowledge representation by extended linear models. In E. Keravnou, editor, *Deep Models for Medical Knowledge Engineering*, pages 129–145. Elsevier Science Publishers B. V., Amsterdam.

Andreassen, S., Jensen, F. V., Andersen, S. K., Falck, B., Kjærulff, U., Woldbye, M., Sørensen, A. R., Rosenfalck, A., and Jensen, F. (1989). *MUNIN - an expert EMG assistant*, chapter 21, pages 255–277. Elsevier Science Publishers B. V. (North-Holland).

Andreassen, S., Falck, B., and Olesen, K. G. (1992). Diagnostic function of the microhuman prototype of the expert system munin. *Electroencephalography and Clinical Neurophysiology*, **85**, 143–157.

Åström, K. J. (1965). Optimal control of Markov decision processes with incomplete state estimation. *Journal of Mathematical Analysis and Applications*, **10**, 174–205.

Bangsø, O. and Wuillemin, P.-H. (2000). Top-down specification and compact representation of repetitive structures in Bayesian networks. In *Proceedings of the Thirteenth International Florida Artificial Intelligence Research Symposium Conference*.

Beeri, C., Fagin, R., Maier, D., and Yannakakis, M. (1983). On the desirability of acyclic database schemes. *Journal of the Association for Computing Machinery*, **30**(3), 479–513.

Ben-Bassat, M. (1978). Myopic policies in sequential classification. *IEEE Transactions of Computing*, **27**, 170–74.

Bertele, U. and Brioschi, F. (1972). *Nonserial Dynamic Programming*. Academic Press, London.

Boyen, X. and Koller, D. (1998). Tractable inference for complex stochastic processes. In *Proceedings of the 14th Annual Conference on Uncertainty in Artificial Intelligence (UAI-98)*, pages 33–42, San Francisco, CA. Morgan Kaufmann.

Buntine, W. L. (1996). A guide to the literature on learning probabilistic networks from data. *IEEE Transactions on Knowledge and Data Engineering*, **8**, 195–210.

Cano, A. and Moral, S. (1995). Heuristic algorithms for the triangulation of graphs. In *IPMU'94: Selected papers from the 5th International Conference on Processing and Management of Uncertainty in Knowledge-Based Systems, Advances in Intelligent Computing*, pages 98–107, London, UK. Springer-Verlag.

Castillo, E., Gutiérrez, J. M., and Hadi, A. S. (1996). A new method for efficient symbolic propagation in discrete Bayesian networks. *Networks*, **28**, 31–43.

Castillo, E., Gutiérrez, J. M., and Hadi, A. S. (1997). Sensitivity analysis in discrete Bayesian networks. *IEEE Transactions on Systems, Man and Cybernetics*, **27**(4), 412–423.

Cheng, J., Greiner, R., Kelly, J., Bell, D., and Liu, W. (2002). Learning Bayesian networks from data: An information-theory based approach. *Artificial Intelligence*, **137**, 43–90.

Chickering, D. M. (1995). A transformational characterization of Bayesian networks. In P. Besnard and S. Hanks, editors, *Proceedings of the Eleventh Conference on Uncertainty in Artificial Intelligence*, pages 87–98. Morgan Kaufmann Publishers.

Chickering, D. M. (2002). Optimal structure identification with greedy search. *Journal of Machine Learning Research*, **3**, 507–554.

Chickering, D. M. and Meek, C. (2002). Finding optimal Bayesian networks. In A. Darwiche and N. Friedman, editors, *Proceedings of the Eighteenth Conference on Uncertainty in Artificial Intelligence*, pages 94–102. Morgan Kaufmann Publishers.

Chickering, D. M., Heckerman, D., and Meek, C. (2004). Large-sample learning of Bayesian networks is NP-hard. *The Journal of Machine Learning Research*, **5**, 1287–1330.

Chow, C. and Liu, C. (1968). Approximating discrete probability distributions with dependence trees. *IEEE Transactions on Information Theory*, **14**(3), 462–467.

Cooper, G. F. (1987). Probabilistic inference using belief networks is NP-hard. *Artificial Intelligence*, **42**, 393–405.

Cooper, G. F. (1988). A method for using belief networks as influence diagrams. In G. F. Cooper and S. Moral, editors, *Proceedings of the Fourth Conference on Uncertainty in Artificial Intelligence*, pages 55–63.

Cooper, G. F. (March 1990). The computational complexity of probabilistic inference using Bayesian belief networks. *Artificial Intelligence*, **42**(2–3), 393–405.

Cooper, G. F. and Herskovits, E. (1991). A Bayesian method for constructing Bayesian belief networks from databases. In B. D. D'Ambrosio, P. Smets, and P. P. Bonissone, editors, *Proceedings of the Seventh Conference on Uncertainty in Artificial Intelligence*, pages 86–94. Morgan Kaufmann Publishers.

Cooper, G. F. and Herskovits, E. (1992). A Bayesian Method for Constructing Bayesian Belief Networks from Databases. *Machine Learning*, **9**, 309–347.

Coupé, V. M. H. and van der Gaag, L. C. (1998). Practicable sensitivity analysis of Bayesian belief networks. In M. Hušková, P. Lachout, and J. Víšek, editors, *Prague Stochastics '98 – Proceedings of the Joint Session of the 6th Prague Symposium of Asymptotic Statistics and the 13th Prague Conference on Information Theory, Statistical Decision Functions and Random Processes, Union of Czech Mathematicians and Physicists, Prague*, pages 81–86.

Covaliu, Z. and Oliver, R. M. (1995). Representation and solution of decision problems using sequential decision diagrams. *Management Science*, **41**(12), 1860–1881.

Cowell, R. G. (1994). Decision networks: A new formulation for multistage decision problems. Research Report 132, Department of Statistical Science, University College London, London.

Cowell, R. G. (2001). Conditions under which conditional independence and scoring methods lead to identical selection of Bayesian network models. In J. Breese and D. Koller, editors, *Proceedings of the Seventeenth International Conference on Uncertainty in Artificial Intelligence*, pages 91–97. Morgan Kaufmann.

Cowell, R. G., Dawid, A. P., Lauritzen, S. L., and Spiegelhalter, D. J. (1999). *Probabilistic Networks and Expert Systems*. Statistics for engineering and information science. Springer-Verlag New York, Inc. ISBN 0-387-98767-3.

D'Ambrosio, B. (1991). Local expression language for probabilistic dependence: a preliminary report. In B. D. D'Ambrosio, P. Smets, and P. P. Bonissone, editors, *Proceedings of the Seventh Conference on Uncertainty in Artificial Intelligence (UAI)*, pages 95–102. Morgan Kaufmann Publishers.

Darwiche, A. (2001). Recursive conditioning. *Artificial Intelligence*, **126**(1–2), 5–41.

Dawid, A. P. (1992). Applications of a general propagation algorithm for a probabilistic expert system. *Statistics and Computing*, **2**, 25–36.

de Dombal, F., Leaper, D., Staniland, J., McCann, A., and Harrocks, J. (1972). Computer-aided diagnosis of acute abdominal pain. *British Medical Journal*, **2**, 9–13.

Dechter, R. (1996). Bucket elimination: A unifying framework for probabilistic inference. In E. Horvitz and F. V. Jensen, editors, *Proceedings of the Twelfth Conference on Uncertainty in Artificial Intelligence*, pages 211–219. Morgan Kaufmann Publishers.

Dempster, A. P., Laird, N. M., and Rubin, D. B. (1977). Maximum likelihood from incomplete data via the EM algorithm. *Journal of the Royal Statistical Society, Series B*, **39**, 1–38.

Dittmer, S. L. and Jensen, F. V. (1997). Myopic value of information in influence diagrams. In D. Geiger and P. P. Shenoy, editors, *Proceedings of the Thirteenth Conference on Uncertainty in Artificial Intelligence*, pages 142–149. Morgan Kaufmann Publishers.

Domingos, P. and Pazzani, M. J. (1997). On the optimality of the simple Bayesian classifier under zero-one loss. *Machine Learning*, **29**(2–3), 103–130.

Drake, A. W. (1962). *Observation of a Markov process through a noisy channel*. Ph.D. thesis, Massachusetts Institute of Technology. Dept. of Electrical Engineering.

Druzdzel, M. and van der Gaag, L. (1995). Elicitation of probabilities for belief networks: Combining qualitative and quantitative information. In P. Besnard and S. Hanks, editors, *Proceedings of the Eleventh Conference on Uncertainty in Artificial Intelligence*, pages 141–148. Morgan Kaufmann Publishers.

Duda, R. O. and Hart, P. E. (1973). *Pattern Classification and Scene Analysis*. John Wiley & Sons, New York.

Edwards, D. and Havranek, T. (1985). A fast procedure for model search in multidimensional contingency tables. *Biometrika*, **72**(2), 339–351.

Felli, J. C. and Hazen, G. B. (1999a). Do sensitivity analysis really capture problem sensitivity? an empirical analysis based on information value. *Risk, Decision and Policy*, **4**(2), 79–98.

Felli, J. C. and Hazen, G. B. (1999b). Sensitivity analysis and the expected value of perfect information. *Medical Decision Making*, **18**, 95–109.

Friedman, N. (1998). The Bayesian Structural EM Algorithm. In G. F. Cooper and S. Moral, editors, *Proceedings of the Fourteenth Conference on Uncertainty in Artificial Intelligence*. Morgan Kaufmann Publishers.

Friedman, N. and Goldszmidt, M. (1998). Learning Bayesian networks with local structure. In M. Jordan, editor, *Learning in Graphical Models*, pages 421–459. Kluwer.

Friedman, N. and Koller, D. (2003). Being Bayesian about network structure. *Machine learning*, **50**(1–2), 95–125.

Friedman, N., Geiger, D., and Goldszmidt, M. (1997). Bayesian network classifiers. *Machine Learning*, **29**(2–3), 131–163.

Fung, R. M. and Chang, K.-C. (1990). Weighing and integrating evidence for stochastic simulation in Bayesian networks. In M. Henrion, R. Shachter, L. Kanal, and J. Lemmer, editors, *Proceedings of the Fifth Annual Conference on Uncertainty in Artificial Intelligence*, pages 209–220. North-Holland.

Geiger, D. and Pearl, J. (1988). On the logic of causal models. In *Proceedings of the 4th Annual Conference on Uncertainty in Artificial Intelligence (UAI-88)*, pages 3–14, New York, NY. Elsevier Science Publishing.

Geiger, D., Heckerman, D., and Meek, C. (1996). Asymptotic model selection for directed networks with hidden variables. In E. Horvitz and F. V. Jensen, editors, *Proceedings of the Twelfth Conference on Uncertainty in Artificial Intelligence*, pages 283–290. Morgan Kaufmann Publishers.

Geman, S. and Geman, D. (1984). Stochastic relaxation, Gibbs distributions, and the Bayesian restoration of images. *IEEE Transactions on Pattern Analysis and Machine Intelligence*, **6**(6), 721–741.

Gilks, W. R., Thomas, A., and Spiegelhalter, D. J. (1994). A language and a program for complex Bayesian modelling. *The Statistician*, **43**, 169–178.

Golumbic, M. C. (1980). *Algorithmic Graph Theory and Perfect Graphs*. Academic Press, London.

Gorry, G. A. and Barnett, G. O. (1968). Experience with a model of sequential diagnosis. *Computers and Biomedical Research*, **1**, 490–507.

Green, P. J. (1990). On use of the EM algorithm for penalized likelihood estimation. *Journal of the Royal Statistical Society, Series B*, **52**(3), 443–452.

Habbema, J. D. F. (1976). Models diagnosis and detection of diseases. In de Dombal *et al.*, editors, *Decision Making and Medical Care*, pages 399–411. Elsevier Science Publishers, Amsterdam.

Heckerman, D. (1990). Probabilistic similarity networks. *Networks*, **20**, 607–636.

Heckerman, D. (1998). A turorial on learning with Bayesian networks. In M. I. Jordan, editor, *Learning in Graphical Models*, pages 301–354. Kluwer Academic Publishers.

Heckerman, D., Horwitz, E., and Nathwani, B. (1992). Toward normative expert systems: Part i. the pathfinder project. *Methods of Information in Medicine*, **31**, 90–105.

Heckerman, D., Breese, J., and Rommelse, K. (1995a). Decision-theoretic troubleshooting. *Communications of the ACM*, **38**(3), 49–56.

Heckerman, D., Geiger, D., and Chickering, D. M. (1995b). Learning Bayesian networks: The combination of knowledge and statistical data. *Machine Learning*, **20**(3), 197–243.

Henrion, M. (1988). Propagating uncertainty in Bayesian networks by probabilistic logic sampling. In J. F. Lemmer and L. M. Kanal, editors, *Uncertainty in Artificial Intelligence 2*, pages 149–163. Elsevier Science Publishers, Amsterdam.

Howard, R. A. (1960). *Dynamic Programming and Markov Process*. MIT Press.

Howard, R. A. (1962). The used car buyer. In R. A. Howard and J. E. Matheson, editors, *The Principles and Applications of Decision Analysis*, volume 2, chapter 36, pages 691–718. Strategic Decision Group.

Howard, R. A. (1966). Information value theory. *IEEE Transactions on Systems Science and Cybernetics*, pages 22–26.

Howard, R. A. and Matheson, J. E. (1981). Influence diagrams. In R. A. Howard and J. E. Matheson, editors, *The Principles and Applications of Decision Analysis*, volume 2, chapter 37, pages 721–762. Strategic Decision Group.

Jaeger, M. (2003). Probabilistic classifiers and the concepts they recognize. In T. Fawcett and N. Mishra, editors, *Proceedings of the Twentieth International Conference on Machine Learning*, pages 266–273. AAAI Press.

Jensen, F., Jensen, F. V., and Dittmer, S. L. (1994). From influence diagrams to junction trees. In R. L. de Mantaras and D. Poole, editors, *Proceedings of the Tenth Conference on Uncertainty in Artificial Intelligence*, pages 367–373. Morgan Kaufmann Publishers.

Jensen, F. V. (1999). Gradient descent training of Bayesian networks. In A. Hunter and S. Parsons, editors, *Proceedings of the Fifth European Conference on Symbolic and Quantitative Approaches to Reasoning with Uncertainty*, Lecture Notes in Artificial Intelligence, pages 190–200. Springer-Verlag.

Jensen, F. V. and Vomlelova, M. (2002). Unconstrained influence diagrams. In A. Darwiche and N. Friedman, editors, *Proceedings of the Eighteenth Conference on Uncertainty in Artificial Intelligence*, pages 234–241. Morgan Kaufmann Publishers.

Jensen, F. V., Chamberlain, B., Nordahl, T., and Jensen, F. (1990a). Analysis in HUGIN of data conflict. In *Uncertainty in Artificial Intelligence 6*, pages 519–528. Elsevier Science Publishers, Amsterdam.

Jensen, F. V., Lauritzen, S. L., and Olesen, K. G. (1990b). Bayesian updating in causal probabilistic networks by local computations. *Computational Statistics Quarterly*, **4**, 269–282.

Jensen, F. V., Aldenryd, S. H., and Jensen, K. B. (1995). Sensitivity analysis in Bayesian networks. In C. Froidevaux and J. Kohlas, editors, *Proceedings of ECSQARU'95*, volume 946 of *Lecture Notes in Artificial Intelligence*, pages 243–250, Fribourg, Switzerland. Springer, Berlin.

Jensen, F. V., Nielsen, T. D., and Shenoy, P. P. (2006). Sequential influence diagrams: A unified asymmetry framework. *International Journal of Approximate Reasoning*, **42**(1–2), 101–118.

Jordan, M., editor (1998). *Learning in Graphical Models*. Kluwer.

Kalagnanam, J. and Henrion, M. (1990). A comparison of decision analysis and expert rules for sequential analysis. In P. Besnard and S. Hanks, editors, *Uncertainty in Artificial Intelligence 4*, pages 271–281. North-Holland, New York.

Kim, J. H. and Pearl, J. (1983). A computational model for causal and diagnostic reasoning in inference systems. In *Proceedings of the Eight International Joint Conference on Artificial Intelligence*, pages 190–193. William Kaufmann, Los Altos, CA.

Kim, Y.-G. and Valtorta, M. (1995). On the detection of conflicts in diagnostic Bayesian networks using abstraction. In P. Besnard and S. Hanks, editors, *Proceedings of the Eleventh Conference on Uncertainty in Artificial Intelligence*, pages 362–367. Morgan Kaufmann Publishers.

Kjærulff, U. (1990). Triangulation of graphs — algorithms giving small total space. Technical Report R 90-09, Department of Mathematics and Computer Science, Aalborg University.

Kjærulff, U. (1992). A computational scheme for reasoning in dynamic probabilistic networks. In D. Dubois, M. P. Wellman, B. D'Ambrosio, and P. Smets, editors, *Proceedings of the Eighth Conference on Uncertainty in Artificial Intelligence*, pages 121–129. Morgan Kaufmann Publishers.

Kjærulff, U. and van der Gaag, L. C. (2000). Making sensitivity analysis computationally efficient. In C. Boutilier and M. Goldszmidt, editors, *Proceedings of the Sixteenth Conference on Uncertainty in Artificial Intelligence*, pages 317–325. Morgan Kaufmann Publishers.

Koller, D. and Pfeffer, A. (1997). Object-oriented Bayesian networks. In *Proceedings of the Thirteenth Conference on Uncertainty in Artificial Intelligence (UAI-97)*, pages 302–313.

Lam, W. and Bacchus, F. (1994). Learning Bayesian belief networks. An approach based on the MDL principle. *Computational Intelligence*, **10**, 269–293.

Laskey, K. B. (1991). Conflict and surprise: Heuristics for model revision. In B. D. D'Ambrosio, P. Smets, and P. P. Bonissone, editors, *Proceedings of the Seventh Conference on Uncertainty in Artificial Intelligence*, pages 197–204. Morgan Kaufmann Publishers.

Laskey, K. B. (1995). Sensitivity analysis for probability assessments in Bayesian networks. *IEEE Transactions on Systems, Man and Cybernetics*, **25**, 901–909.

Lauritzen, S. L. (1995). The EM algorithm for graphical association models with missing data. *Computational Statistics and Data Analysis*, **19**, 191–201.

Lauritzen, S. L. (1996). *Graphical Models*. Oxford University Press. ISBN: 0-19-852219-3.

Lauritzen, S. L. and Jensen, F. V. (1997). Local computation with valuations from a commutative semigroup. *Annals of Mathematics and Artificial Intelligence*, **21**(1), 51–69.

Lauritzen, S. L. and Spiegelhalter, D. J. (1988). Local computations with probabilities on graphical structures and their application to expert systems. *Journal of the Royal Statistical Society, Series B*, **50**(2), 157–224.

Lauritzen, S. L., Dawid, A. P., Larsen, B. N., and Leimer, H.-G. (1990). Independence properties of directed Markov fields. *Networks*, **20**(5), 491–505.

Lindley, D. V. (1971). *Making Decisions*. John Wiley & Sons, New York.

Madsen, A. L. and Jensen, F. V. (1999a). Lazy evaluation of symmetric Bayesian decision problems. In K. B. Laskey and H. Prade, editors, *Proceedings of the Fifteenth Conference on Uncertainty in Artificial Intelligence*, pages 382–390. Morgan Kaufmann Publishers.

Madsen, A. L. and Jensen, F. V. (1999b). Lazy propagation: A junction tree inference algorithm based on lazy evaluation. *Artificial Intelligence*, **113**, 203–245.

Margaritis, D. and Thrun, S. (1999). Bayesian network induction via local neighborhoods. In *Advances in Neural Information Processing Systems 12*, pages 505–511. MIT Press.

Meek, C. (1995). Strong completeness and faithfulness in Bayesian networks. In *Proceedings of the 11th Annual Conference on Uncertainty in Artificial Intelligence (UAI-95)*, pages 411–41, San Francisco, CA. Morgan Kaufmann.

Michalewicz, Z. and Fogel, D. B. (2000). *How to Solve It: Modern Heuristics*. Springer Verlag.

Minsky, M. (1963). Steps toward artificial intelligence. In E. A. Feigenbaum and J. Feldman, editors, *Computers and Thoughts*, pages 406–450. McGraw-Hill.

Mitchell, T. M. (1997). *Machine Learning*. McGraw-Hill.

Ndilikilikesha, P. C. (1994). Potential influence diagrams. *International Journal of Approximate Reasoning*, **10**, 251–285.

Nielsen, T. D. (2002). Decomposition of influence diagrams. *Journal of Applied Non-Classical Logics – Symbolic and Quantitative Approaches to Reasoning with Uncertainty*, **12**(2), 135–150.

Nielsen, T. D. and Jensen, F. V. (1999). Welldefined decision scenarios. In *Proceedings of the 15th Annual Conference on Uncertainty in Artificial Intelligence (UAI-99)*, pages 502–551, San Francisco, CA. Morgan Kaufmann.

Nielsen, T. D. and Jensen, F. V. (2003a). Representing and solving asymmetric decision problems. *International Journal of Information Technology & Decision Making*, **2**(2), 217–263.

Nielsen, T. D. and Jensen, F. V. (2003b). Sensitivity analysis in influence diagrams. *IEEE Transactions on Systems, Man, and Cybernetics - Part A: Systems and Humans*, **33**(2), 223–234.

Nilsson, D. and Lauritzen, S. L. (2000). Evaluating influence diagrams using LIMIDs. In C. Boutilier and M. Goldszmidt, editors, *Proceedings of the Sixteenth Conference on Uncertainty in Artificial Intelligence*, pages 436–345. Morgan Kaufmann Publishers.

Olesen, K. G., Lauritzen, S. L., and Jensen, F. V. (1992). ahugin: A system creating adaptive causal probabilistic networks. In *Proceedings of the Eighth Conference on Uncertainty in Artificial Intelligence (UAI)*, pages 223–229.

Olmsted, S. M. (1983). *On representing and solving decision problems*. Ph.D. thesis, Department of Engineering–Economic Systems, Stanford University.

Pearl, J. (1982). Reverend Bayes on inference engines: A distributed hierarchical approach. In *Proceedings of the First National Conference on Artificial Intelligence*, pages 133–136. The AAAI Press.

Pearl, J. (1986). Fusion, propagation, and structuring in belief networks. *Artificial Intelligence*, **29**(3), 241–288.

Pearl, J. (1988). *Probabilistic Reasoning in Intelligent Systems*. Representation and Reasoning. Morgan Kaufmann Publishers, San Mateo California. ISBN 0-934613-73-7.

Pearl, J. (2000). *Causality: Models, Reasoning and Inference*. Cambridge University Press. ISBN 0-521-77362-8.

Puterman, M. L. (1994). *Markov Decision Processes: Discrete Stochastic Dynamic Programming*. John Wiley & Sons, Chichester, UK.

Quinlan, J. R. (1979). Discovering rules by induction from large collections of examples. In D. Michie, editor, *Expert Systems in the Micro Electronic Age*. Edinburgh University Press.

Quinlan, J. R. (1986). Induction of decision trees. *Machine Learning*, **1**, 81–106.

Raiffa, H. (1968). *Decision Analysis, Introductory Lectures on Choices under Uncertainty*. Addison-Wesley.

Raiffa, H. and Schlaifer, R. (1961). *Applied Statistical Decision Theory*. MIT press, Cambridge.

Rissanen, J. (1987). Stochastic complexity. *Journal of the Royal Statistical Society, Series B*, **49**(3), 223–239. With discussions.

Rubin, D. B. (1976). Inference and missing data. *Biometrika*, **63**(3), 581–592.

Russell, S. J., Binder, J., Koller, D., and Kanazawa, K. (1995). Local learning in probabilistic networks with hidden variables. In *Proceedings of the Fourteenth International Joint Conference on Artificial Intelligence*, pages 1146–1152.

Schwarz, G. (1978). Estimating the dimension of a model. *Annals of Statistics*, **6**, 461–464.

Shachter, R. D. (1986). Evaluating influence diagrams. *Operations Research*, **34**(6), 871–882.

Shachter, R. D. (1999). Efficient value of information computation. In K. B. Laskey and H. Prade, editors, *Proceedings of the Fifteenth Conference on Uncertainty in Artificial Intelligence*, pages 594–601. Morgan Kaufmann Publishers.

Shachter, R. D. and Peot, M. A. (1990). Simulation approaches to general probabilistic inference on belief networks. In M. Henrion, R. Shachter, L. Kanal, and J. Lemmer, editors, *Proceedings of the Fifth Annual Conference on Uncertainty in Artificial Intelligence*, pages 221–234. North-Holland.

Shafer, G. (1996). *Probabilistic Expert Systems*. Society for Industrial and Applied Mathematics, Philadelphia.

Shafer, G. R. and Shenoy, P. P. (1990). Probability Propagation. *Annals of Mathematics and Artificial Intelligence*, **2**, 327–352.

Shapley, L. S. (1953). Stochastic games. *Proceedings of the National Academy of Sciences*, **39**, 1095–1100.

Shenoy, P. P. (1992). Valuation-based systems for Bayesian decision analysis. *Operations Research*, **40**(3), 463–484.

Shenoy, P. P. (1996). Representing and solving asymmetric decision problems using valuation networks. In D. Fisher and H.-J. Lenz, editors, *Learning from Data: Artificial Intelligence and Statistics V*, volume 112 of *Lecture Notes in Statistics*, pages 99–108. Springer-Verlag.

Spiegelhalter, D. J. and Knill-Jones, R. P. (1984). Statistical and knowledge-based approaches to clinical decision-support systems. *Journal of the Royal Statistical Society, Series A*, **147**(1), 35–77.

Spiegelhalter, D. J. and Lauritzen, S. L. (1990). Sequential updating of conditional probabilities on directed graphical structures. *Networks*, **20**, 579–605.

Spirtes, P., Glymour, C., and Sheines, R. (1993). *Causation, Prediction and Search*. Lecture Notes in Statistics. Springer-Verlag.

Spirtes, P., Glymour, C., and Sheines, R. (2000). *Causation, Prediction and Search*. MIT Press, Cambridge, Massachusetts, second edition.

Spohn, W. (1980). Stochastic independence, causal independence, and shieldability. *Journal of Philosophical Logic*, **9**, 73–99.

Steck, H. (2001). *Constrained-based structural learning in Bayesian networks using finite data sets*. Ph.D. thesis, Institut für Informatik der Technischen Universität München.

Suermondt, H. J. (1992). *Explanation in Bayesian Belief Networks*. Ph.D. thesis, Knowledge Systems Laboratory, Medical Computer Science, Stanford University, California. Report No. STAN-CS-92-1417.

Tatman, J. A. and Shachter, R. D. (1990). Dynamic Programming and Influence Diagrams. *IEEE Transactions on Systems, Man and Cybernetics*, **20**(2), 365–379.

Titterington, D. M. (1976). Updating a diagnostic system using unconfirmed cases. *Applied Statistics*, **25**(3), 238–247.

Verma, T. (1987). Causal networks: Semantics and expressiveness. In *Proceedings of the Third Workshop on Uncertainty in Artificial Intelligence*, pages 352–359. Elsevier Science Publishers, New York.

Verma, T. and Pearl, J. (1991). Equivalence and synthesis of causal models. In *Uncertainty in Artificial Intelligence 6*, pages 255–268. Elsevier Science Publishers B.V.

Vomlelová, M. (2003). Complexity of decision-theoretic troubleshooting. *International Journal of Intelligent Systems*, **18**(2), 267–277.

von Neumann, J. and Morgenstern, O. (1944). *Theory of Games and Economic Behavior*. John Wiley & Sons, New York, first edition.

Wermuth, N. and Lauritzen, S. L. (1990). On substantive research hypotheses, conditional independence graphs and graphical chain models (with discussion). *Journal of the Royal Statistical Society*, **52**, 21–72.

Zhang, N. L. (1998). Probabilistic Inference in Influence Diagrams. In G. F. Cooper and S. Moral, editors, *Proceedings of the Fourteenth Conference on Uncertainty in Artificial Intelligence*, pages 514–522. Morgan Kaufmann Publishers.

Index

λ-message   153
π-message   153

A-saturated junction tree   *169, 170*
action sequence   373
acyclic directed graph   *33*
adaptation   83, *207*
  to structure   214
adjacent node   *119*
algebra of potentials   *13*
Allais' paradox   289
analysis
  data conflict   98
  relevant future   419
  required past   415
  SE   *99, 179*
  sensitivity   99, *184*, 420
  value of information   407
ancestral graph   *32*
associative law   *13*
asymmetric decision problems   *310*
  functional asymmetry   *315*
  order asymmetry   *315*
  structural asymmetry   *315*
attribute
  encapsulated   *86*
  input   *86*
  output   *86*

barren node   112, *130*
  rule   *130*
batch learning   *195*
Bayes' factor   *180*
Bayes' rule   *5*

for variables   *10*
Bayesian estimation   197
Bayesian information criterion   *243*
Bayesian network   *33*
  dynamic   *91*
  hybrid   *95*
  object-oriented   84, *85*
  parameters   *60*
  size   *240*
Bayesian score function   253
belief state   *388*
BIC   243
bucket elimination   *41*
BUGS   *156*
burn-in   *151*

call service   378
causal network   26, *26*
causality   *60*
chain graph   *74*
chain rule   *35*
  for Bayesian networks   36
  for influence diagrams   345
  general   36
chaining   24
chance node   305
chance variable   *33*
chance-variable representation   *360*
chord   *161*
Chow–Liu tree   *250*
  learning of   *250*
class variable   *265*
classification accuracy   *268*

classification tree   *272*
classifier
    accuracy   *268*
    classification tree   *272*
    confusion matrix   *268*
    evaluation of   268
    naive Bayes   266
    tree augmented naive Bayes   *270*
clique   *118*
collect evidence   *126*
commutative law   *13*
complete case   *195*
complete set of nodes   *118*
computation tree   *141*
conditional Gaussian distribution   *95*
conditional independence   *6*
    for variables   *10*
conditional probability   4
    for variables   8
conditioning   *164*
configuration of maximal probability
        171
conflict
    data   98, *174*
    local   *177*
    measure   99, *175*
    partial   *177*
confounding variable   *240*
confusion matrix   *268*
connected graph
    singly c. graph   *162*
connection
    converging   28
    diverging   27
    serial   26
constraint variable   *74*
constraint-based learning   230
continuous variable   *93*
converging connection   28
convex function   *412*, *see* value
        function
crucial evidence   *181*
crucial finding   *183*
cycle   *161*

d-connected   *30*
d-separation   26, *30*, 131
DAG   *33*
    neighborhood   245

data conflict   98, *174*
decision
    action   279
    node   305
    scenario   *290*
    test   279
    tree   *see* decision tree
decision tree   290
    coalesced   295
    no-forgetting   290
    strategy   296
decision variable   283
    chance -variable representation   *360*
decision/classification tree   *272*
default potential   *90*
density function   *16*
directed graph   *26*
distribute evidence   *126*
distributive law   *14, 174*
    for max   172
diverging connection   27
divorcing   *78*
domain
    finite-horizon   *92*
    infinite-horizon   *92*
    domain graph   *116*
    domain of variable   *13*
    domain set   *118*
dynamic Bayesian network   *91*
    time slice   *91*

ECR   *see* troubleshooting
effective sample size   *211*
elimination
    bucket   *41*
    of variable   *116*
    order   *110*
    variable   353
elimination order
    strong   *353*
elimination sequence
    perfect   *117*
EM algorithm   201, *206*
entropy   412
equivalence class search   248
Euclidean distance   *219*
evaluation of classifiers   268
event   *2*
    hypothesis   *51*

evidence *39*
  collect *126*
  crucial *181*
  distribute *126*
  hard 131
  important *181*
  likelihood *40*
  minimal sufficient *181*
  redundant *181, 190*
  sensitivity to *167*
  simple *374*
  sufficient *181, 190*
expectation step *201*
expected benefit *409*
expected profit *409*
expected utility 281, 346
  maximal *350, 396*
expected value *15, 409*
expert disagreements *81*
explaining away *28*
explanation *167*
  most-probable *98*

fading *211*
faithful sample *237*
false negative 18, *60*
false positive 18, *60*
fill-in *117*
finding *40*
  crucial *183*
finite-horizon domain *92*
fractional updating *210*
frequency function *16*
full junction tree *128*
fundamental rule *5*
  for variables *9*

Gaussian distribution 94
general chain rule 36
Gibbs sampling *150*
  burn-in *151*
global independence *195*
gradient descent *219*
graph
  acyclic directed *33*
  ancestral *32*
  chain *74*
  domain *116*
  moral *116*

nontriangulated *132*
  singly connected *162*
  triangulated *119*
  triangulation of *134*
graphical model 43
greedy approach 375
greedy equivalence search *248*

h-saturated junction tree 182
hard evidence 131
hidden Markov model *92*
hidden variable 200
history *319*
history variable *309*
horizon
  finite h. domain *92*
  infinite h. domain *92*
hybrid Bayesian networks *95*
hypothesis event *51*
hypothesis variable *51*

I-equivalence *48*
I-submap *48*
IEJ tree *168*
important evidence *181*
incremental updating *215*
independence 6
  conditional *see* conditional
    independence
  global *195*
  local *195*
  marginal 11
  structural *30*
infinite-horizon domain *92*
influence diagram 302, 305
  chain rule for 345
  limited memory *392*
  no-forgetting 306
  optimal policy 307
  optimal strategy 307
  partially observable Markov decision
    process 308
  policy 307, *307*
  policy network 360
  realized 305
  relevant future 419
  required past 358, 415
  solution *308*
  strategy 307, *308*

information
  blocking *309*
  hiding *86*
  link 305
  variable *52*
inheritance 88
inhibitor *77*
initial sample size *212*
instantiated potential *41*
instantiated variable *26*
instrumental rationality *287*
  axioms 287
intervention *96*

join tree *122*
joint probability *8*
joint probability table 8, 98
junction tree *124*
  A-saturated *169, 170*
  full *128*
  h-saturated 182
  strong *355*

Kalman filter *92*
Kullback-Leibler divergence *219*

latent variable 200
law
  associative 13
  commutative 13
  distributive 14, *174*
  distributive for max 172
lazy propagation *127*
LBP 152
likelihood 59, 196
  evidence *40*
  marginal *254*
  normalized *177*
  weighting *148*
LIMID *see* limited memory influence
    diagram
limited memory influence diagram
    392
  single policy updating *394*
link
  information 305
  moral *116*
  temporal *92*
local independence *195*

local conflict *177*
loopy belief propagation *152*
  λ-message 153
  π-message 153
lottery *287*
lower neighborhood *248*

mailbox *124*
MAR 200
marginal likelihood *254*
marginalization *9*
marginalize *115*
marginally independent *11*
Markov
  blanket *30*
  chain *92*
  hidden model *92*
  property *52*
Markov decision process 324, *326*
  average reward *330*
  discounting factor 329, *329*
  nonstationary strategy *328*
  policy iteration *see* policy iteration,
      387
  stationary strategy *328*
  terminal state *325*
  value iteration 381, *382*
max-marginal *172*
max-propagation *172*
maximal expected utility *350, 396*
maximization step *201*
maximum a posteriori parameters
    199, 201
maximum likelihood estimation 196
MCAR 200
MDP *see* see Markov decision process
mean value *15*
mediating variable *56*
message passing *127*
metric *219*
minimal sufficient evidence *181*
missing at random *200*
missing completely at random *200*
moral graph *32, 116*
moral link *116*
most-probable explanation *98*
MPE 98
multilinear polynomial function *186*
myopic repair strategy 379

myopic value of information   409

naive Bayes   *58*
  classifier   266
  tree augmented   *270*
NBC   267
necessary path condition   *237*
negative
  false   18, *60*
neighborhood   *245*, 248
  lower   248
  upper   248
network
  Bayesian   *33*
  causal   *26*
  fragment   *84*
network fragment   *84*
  instantiate   85
no-forgetting   *290*, 306
node
  adjacent   *119*
  barren   112, *130*, *362*
  barren n. rule   *130*
  chance   305
  decision   305
  misplaced   *368*
  simplicial   *119*
  utility   281, 305
node removal and arc reversal   362
  arc reversal   *365*
  removal of barren nodes   *362*
  removal of chance nodes   *362*
  removal of decision nodes   *362*
noisy functional dependence   *80*
noisy-and   *78*
noisy-or   *75*
nontriangulated graph   *132*
normal distribution   94
normalized likelihood   *177*, *180*
normative approach   vi

object-oriented Bayesian network   84,
    85
  attribute   *see* attribute
  interface   *87*
Ockham's razor   *240*
OOBN   *see* object-oriented Bayesian
    network
order

elimination   *110*
  perfect elimination   *117*
overfitting   *230*, 257

parameters   *60*
partial conflict   *177*
partially observable Markov decision
    process   *330*
  observation function   *331*
path   *161*
PC algorithm   *235*
perfect elimination sequence   *117*
policy   307, *307*
  decision   *320*
  optimal   307
  step   *320*
policy iteration   385, *387*
  policy evaluation   386
  policy improvement   386
policy network   360
POMDP   *see* partially observable
    Markov decision process
positive
  false   18, *60*
potential   *13*, 43
  default   *90*
  instantiated   *41*
pre-$\mathcal{J}$-tree   165
principle of maximum likelihood   *196*
probabilistic logic sampling   *146*
projection operator   174
propagation
  lazy   *127*
  loopy belief   *152*
  variable   *169*
proportional scaling   *185*

question   378

random variable   *15*
recursive conditioning   *140*
  cutset   *143*
red herring   *175*
redundant evidence   *181*, *190*
relevant future   419
relevant utility node   *415*
repetitive temporal model   *92*
required past   358, *415*
required variable   353, *414*

rule
　barren node　*130*
　chain　*35*

S-DAG　*319*
　decision policy　*320*
　dominating path　*369*
　history　*319*
　misplaced node　*368*
　optimal　*322*
　step policy　*320*
　step strategy　*320*
　strategy　*320*
sample size　*210*
　effective　*211*
　initial　*212*
satisfiability problem　*107*
scaling
　proportional　*185*
score equivalent　*248*
score function　242
　Bayesian　253
SE analysis　99, *179*
search
　equivalence　248
　greedy　246
　operator　245
second-order uncertainty　*207*
sensitivity analysis　99, *184*, 420
　decision sensitivity　*420*
　value sensitivity　*420*
sensitivity to evidence　*167*
separator　*123*
sequential influence diagram　322
　guard　*322*
　open link　*322*
　structural link　*322*
serial connection　*26*
SID　*see* sequential influence diagram
simplicial node　*119*
single fault assumption　*376*
singly connected graph　*162*
size
　Bayesian network　*240*
　effective sample　*211*
　initial sample　*212*
　sample　*210*
skeleton　*231*
solution　*308*, 334

stochastic simulation　*145*
strategy　*296*, 307, *308*, *320*, 334, 346
　myopic repair　*379*
　optimal　296, 307, *322*
　step　*320*
strictly repetitive model　*92*
strong elimination order　*353*
strong junction tree　*355*
strong root　*357*
strong triangulation　*355*
structural independence　*30*
subclass　*88*, *90*
subjective probabilities　*1*
sufficient evidence　*181*, *190*
sum-propagation　*172*
superclass　*89*
surprise index　*179*

TAN　270
temporal link　*92*
temporal model　*92*
time slice　*91*
time-stamped models　137
tree
　A-saturated junction　*170*
　augmented naive Bayes classifier
　　270
　Chow–Liu　*250*
　classification　*272*
　decision　*see* decision tree
　full junction　*128*
　IEJ　*168*
　join　*122*
　junction　*124*
triangulated graph　*119*
triangulation
　of graphs　*134*
　strong　*355*
triggered direction　*128*
troubleshooting　373
　call service　*378*
　expected cost of repair　*374*
　greedy approach　375
　observation step　*373*
　question　378
　repair step　*373*
　simple evidence　*374*
　single fault assumption　*376*

strategy   *see* troubleshooting
    strategy
troubleshooting strategy   *373*
    efficiency   375
    myopic   *379*
tuning   *218*

UID   *see* unconstrained influence
    diagram
uncertain region   *238*
unconstrained influence diagram   316,
    318
    admissible order   *319*
    free   *316*
    no-forgetting   318
    observable   *316*
    optimal strategy   *322*
    realized   *318*
    released   *316*
    S-DAG   *319*
    strategy   *see* S-DAG
undirected relations   73
unit potential   *13*
    property   *13*
upper neighborhood   *248*
utility   284
    expected   346
    maximal expected   *350, 396*
    node   305
    theory   *284*
utility node
    relevant   *415*

v-structure   231
valuation   *174*
valuation axiom   *174*
value

expected   *409*
value function   *409*
    convex   413
    entropy-based   412
    non-utility-based   411
    utility based   *409*
    variance-based   412
value iteration   381, *382*
value of information   407
    expected benefit   *409*
    expected profit   *409*
    hypothesis-driven   *409*
    myopic   409
    value function   *see* value function
variable   *7*
    chance   *33*
    class   *265*
    confounding   *240*
    constraint   *74*
    continuous   *93*
    decision   282, 283
    domain of   *13*
    elimination   42
    elimination of   *116*
    hidden   200
    history   309
    hypothesis   *51*
    information   *52*
    instantiated   *26*
    latent   200
    mediating   *56*
    random   *15*
    required   353, *414*
variable elimination   *42*, 353
variable propagation   *169*
variance   *15*, 412